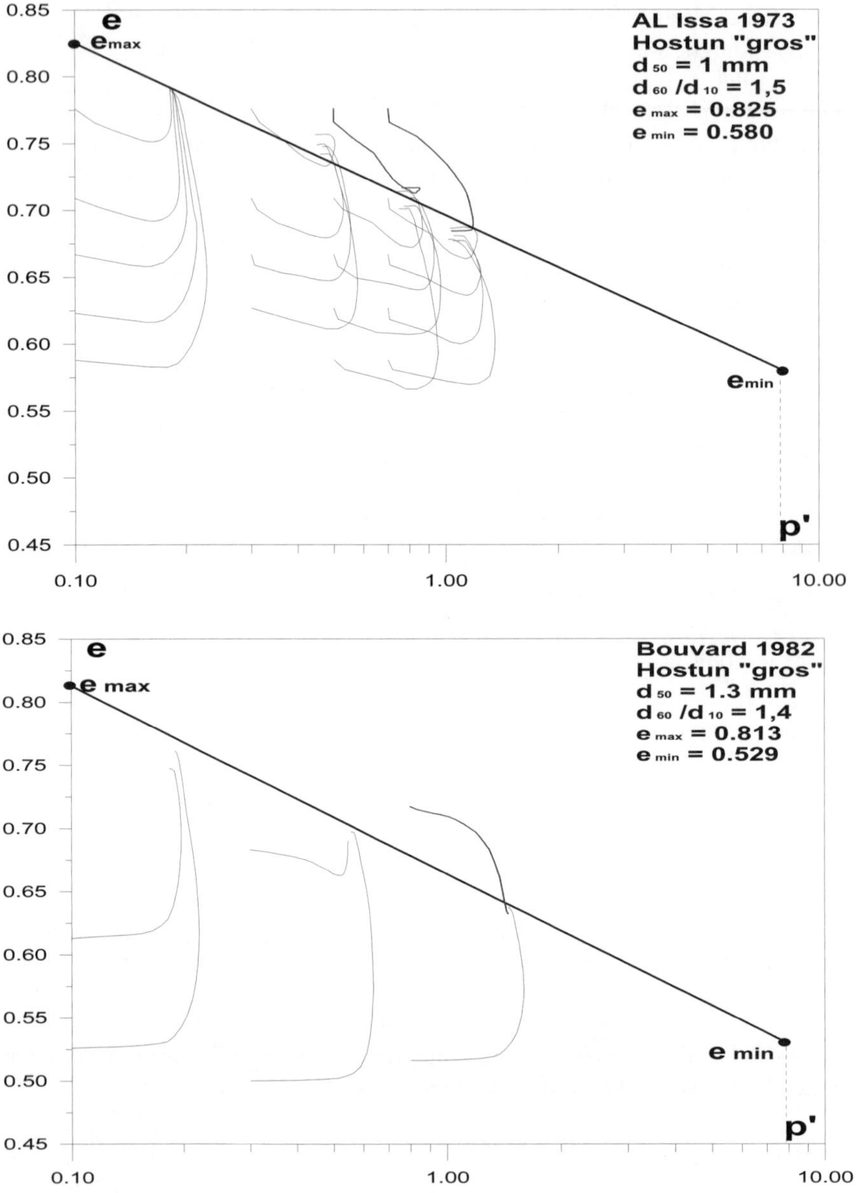

An examination of Figures 5.19a and 5.20a reveals a difference in the C_c (Biarez's approach) and C^*_c (Burland's approach) slopes, particularly between $\sigma'_v = 100$ and $\sigma'_v = 1{,}000$, when the liquid limit is high (w_L=100% to 160%). On the other hand, for the lower w_L, C_c and C^*_c are similar, and fit well the experimental data given by Burland for different values of w_L (see Figure 5.20a).

5.9.2. *Comparison of models and mixed model*

5.9.2.1. *Plotting Biarez's model in Burland's space (I_v – logσ'_v)*

According to correlation [5.7] we have:

$$I_L = \frac{e - e_P}{e_L - e_P} = 0.46\left(3 - \log\sigma'_v\right) \qquad [5.30]$$

Using [5.27] and [5.30] to express Biarez's model as a function of I_v, we obtain the NCRS equation [5.31] in Burland's space:

$$I_V = 1.38\frac{e_L - e_P}{C^*_c} - \frac{e^*_{100} - e_P}{C^*_c} - 0.46\frac{e_L - e_P}{C^*_c}\log\sigma'_v \qquad [5.31]$$

According to Biarez, hence [5.2] $e_P = e^*_{1000}$ and $e_L = e^*_{6,5}$, equation [5.31] becomes:

$$I_V = \frac{e^*_{6,5} - e^*_{1000}}{C^*_c}\left[1.38 - 0.46\ \log\sigma'_v\right] - 1. \qquad [5.32]$$

Considering a linear variation between Δe and $\Delta\log\sigma'_v$, we obtain:

$$\frac{e^*_{100} - e^*_{1000}}{\log 100 - \log 1{,}000} = \frac{e^*_{6.5} - e^*_{1000}}{\log 6.5 - \log 1{,}000}. \qquad [5.33]$$

Thus, Biarez's model given by relation [5.32] becomes:

$$I_V = 2 - \log\sigma'_v. \qquad [5.34]$$

Equation [5.34] is a straight line in the (I_v – logσ'_v) plane, passing through two specific points (I_v = 0, σ'_v = 100 kPa) and (I_v = -1, σ'_v = 1,000 kPa), represented by the NCRS line in Figure 5.21a. A representation of both the NCRS line and the ICL in Burland's plane shows that the two curves converge at a vertical stress range

materials (Schneebeli rolls or similar materials). Another possible approach was based on analytical calculations that could be carried out on simple arrays made of circular cylinders. These approaches often led to the development of qualitative analyses that provided a better understanding of the physical bases of a complex reality.

Of all the research topics developed by Jean Biarez we present the following:

– contact anisotropy;

– the evolution of the orientation of contact forces;

– the balance in local arrays of particles;

– the influence of particle breakage on the behavior of a granular material considered at the REV scale.

In addition to each of these points, some aspects of contemporary research results are presented. Today an approach of primary relevance for the scientific themes enumerated above is numerical discrete modeling, which has been developed into a tool called the discrete element method (DEM) based on the pioneering work by Cundal [CUN 79]. This very powerful tool has notably allowed us to surpass the pioneering work of Jean Biarez; particularly, to work on materials that are much closer to real materials, and above all to access the vast number of local variables far more easily. The main new results presented in this chapter are based on results obtained from this technique of numerical modeling.

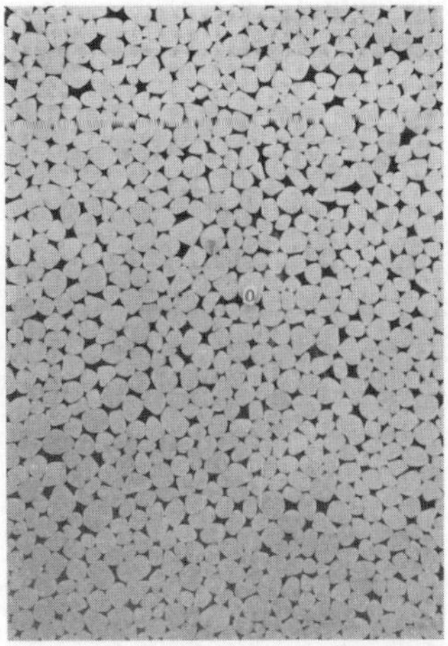

Figure 2.2. *Example of a loose sample of two-dimensional granular material made from non convex particles cut in a PVC plate 2 mm thick [WIE 64]*

2.3. Analysis of geometric anisotropy

Biarez and Wiendieck were the first to use a representation of the anisotropy of distribution of the orientation of tangent planes between particles in contact (Figure 2.3). These orientations can be measured in the two-dimensional projection of a model material composed, for example, of parallel cylinders with circular sections (Schneebeli rolls) or any other shape (as shown in [BIA 62]). Biarez and Wiendieck [BIA 63, WIE 64] suggested an approximation of the shape of the distribution obtained by an ellipse of major axis a and minor axis b. The ratio $A = \frac{a-b}{a+b}$ gives an overall indication of the value of the geometric anisotropy of the sample (or *texture*), which, combined with the value of the void ratio e, provides a better description of the material.

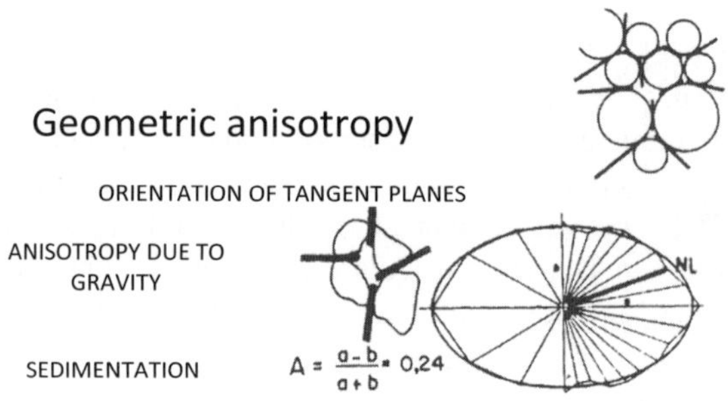

Figure 2.3. *Definition of geometric anisotropy from the direction of the tangent planes [BIA 63]*

This idea was subsequently taken up by Oda [ODA 72] and adapted to the distribution of the orientations of normal directions at contact points. This led, in the 1980s, to the definition of several fabric (or texture) tensors, among them those defined by Satake [SAT 82], Oda, Konishi and Nemat-Nasser [ODA 80], from the tensorial product of the normal n by itself. The elliptical approximation of the distribution has also been replaced by an approximation in Fourier series and the geometric anisotropy can be described by using a second order tensor, for example, defined by the relation:

$$H = \langle n \otimes n \rangle_C \qquad [2.1]$$

where C is the set of contacts on which the tensorial product is computed.

Biarez understood that this description of the geometric anisotropy of the granular assembly allowed not only the *initial anisotropy* to be characterized (due to the preparation of the sample or to the natural consolidation of the soil layer in the gravity field), but also the *induced anisotropy* (due to an anisotropic loading path) to be explained. Anisotropy in a granular material can be detected from an analysis of mechanical behavior or from geometric measurement, as defined above. Thus Figure 2.4 below shows, first, the mechanical demonstration of the anisotropic behavior of a sand sample (Figure 2.4a) when it was first subjected to triaxial compression and, second, the evolution of geometrical anisotropy in an analogical assembly composed of Schneebeli rolls as the sample was made to undergo deformations (Figure 2.4b).

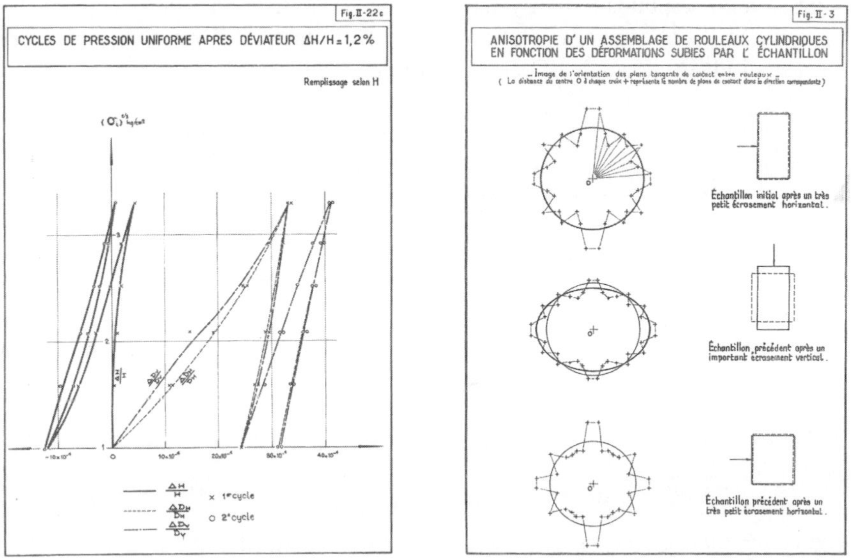

(a) *(b)*

Figure 2.4. *Effect of anisotropy for granular materials in Jean Biarez's thesis [BIA 62]: (a) Cycles with uniform pressure after a deviatoric strain ΔH/H=1.2%: macroscopic mechanical response; and (b) measurement of tangent plane orientations in cylindrical rods assemblies with respect to deformations*

The use of software based on the DEM has allowed us to analyze more accurately the relationship between changes in the geometric anisotropy and the mechanical behavior of numerical granular samples. For instance, the results of numerical simulations of biaxial tests, performed with software based on the *contact dynamics* method [JEA 95, MOR 94] on samples whose particles are polygonal [NOU 03, NOU 05] showed that:

– *critical anisotropy* is reached no matter what the initial density of the sample or the shape of the particles that compose it and no matter what the orientation of the stress with respect to the major principal direction of the initial anisotropy, with a coincidence of the major principal directions of anisotropy and loading (Figure 2.5a);

– initial anisotropy is great in the case of a deposit in the gravity field if the particles themselves are anisometric (i.e. elongated). For such materials, when the major principal directions of stress and initial anisotropy are orthogonal, the process of redrawing the geometric anisotropy is made through a slow and gradual rotation of the major principal direction of the tensor **H** (defined by equation [2.1]). This phenomenon corresponds to a physical rotation of elongated particles (Figure 2.5b), which causes a much slower evolution, in terms of plastic deformations, than it does in the case of materials made of particles with an isometric shape.

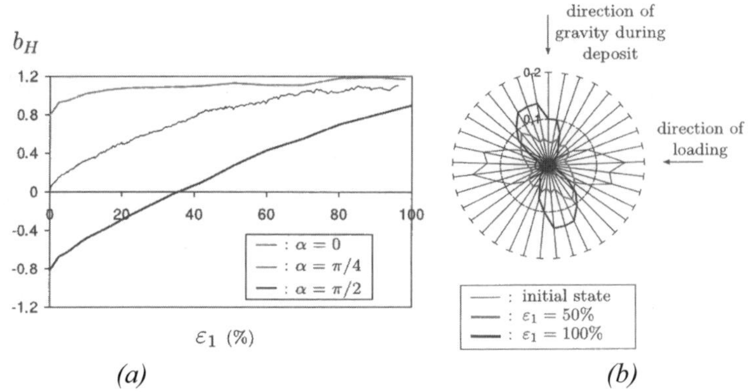

Figure 2.5. *(a) Evolution of the geometric anisotropy represented by the b_H parameter, during simulations of a biaxial test on a sample with elongated polygons (aspect ratio equal to 3) with different orientations of the initial anisotropy relative to the direction of loading (α angle); and (b) evolution of the orientation distribution of these polygons in the event where $\alpha = \pi/2$ (figures extracted from [NOU 05])*

Three-dimensional simulations of the triaxial test performed by Yunus [YUN 08, YUN 10] with the PFC software [ITA 02] on samples composed of spheres confirmed the existence of a critical anisotropy (Figure 2.6) and a high degree of similarity between the macroscopic response of the material ($q(\varepsilon_1)$ curve on the left) and the evolution of a measure of the geometric anisotropy induced by loading ($A_{1111}(\varepsilon_1)$ curve on the right). These curves clearly illustrate that the behavior of granular materials is strongly driven by the evolution of the geometric anisotropy, as foreseen by Biarez.

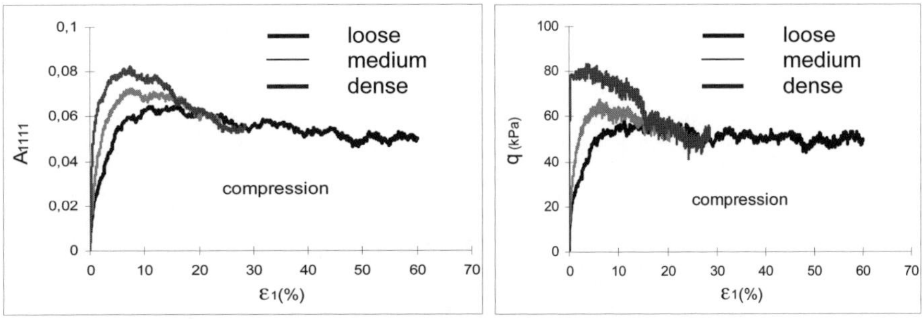

Figure 2.6. *Critical state obtained for three initial densities – numerical simulations of triaxial test on samples of spheres [YUN 08]*

2.4. Analysis of the distribution of contact forces in a granular material

The distribution of contact forces in granular materials is a topic widely discussed in scientific literature. The pioneering studies of Jean Biarez focusing on the orientation of the normal directions at contact provide a preliminary piece of information that can be described by the proposed anisotropy ratio (see section 2.3). Once the distribution of the orientation of the contact planes is identified, it is necessary to add two items of information: the distribution of the force intensity and the distribution of their orientations in the local axes defined at each contact. A great number of analyses, some of them experimental based on photoelastic measurement taken in the 1950s and, increasingly, numerical analyses since the 1980s, have shown a distribution of the contact force intensity linked to the applied stress tensor with maximal values in the direction of the major principal stress. Very early on, Biarez was interested in analyzing the orientation of contact forces with respect to the local axes defined at each contact. He showed in his thesis that this variable was linked to the loading history applied on a particular material and could thus be considered as a local variable able to explain the irreversible behavior of these kinds of materials. Figure 2.7, taken from Biarez's thesis, shows how he attempted to explain this point. Indeed, for a given stress state, let us consider one of the contact forces showing an orientation δ with respect to the normal direction at contact. An evolution of the deviatoric stress leads to an evolution of δ equal to δ'. If the considered evolution of the deviatoric stress corresponds to an increase, δ' will be positive and in the case where $(\delta + \delta')$ becomes equal to the local friction angle, contact and sliding will occur leading to irreversible (plastic) strains. If the evolution of the deviatoric stress shows a decrease, δ' is then negative and it would be far less probable that $(\delta + \delta')$ could be greater than the local friction angle at contact. Then, for unloading of a sample of granular material, very few irreversible (plastic) strains will occur.

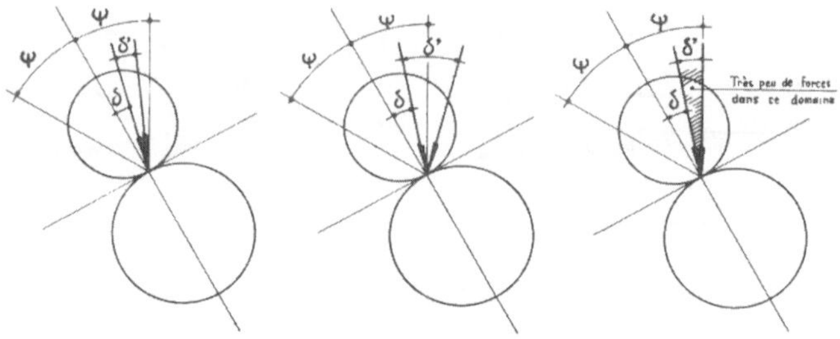

Figure 2.7. *Explanatory scheme on the evolution of the orientation of a contact force during a loading or unloading (figure taken from [BIA 62])*

With the numerical discrete modeling tool (DEM), it is possible to analyze this kind of local variable directly. It is then very convenient to construct the following three stress tensors:

– the stress tensor computed from the local contact forces using the equation of change of scale:

$$\sigma_{ij} = \frac{1}{V}\sum_k F_i^k l_j^k \qquad [2.2]$$

– the part of this tensor only linked to the normal components of the contact forces:

$$\sigma_{ij}^N = \frac{1}{V}\sum_k F_i^{Nk} l_j^k \qquad [2.3]$$

– the part of this tensor only linked to the tangential components of the contact forces:

$$\sigma_{ij}^T = \frac{1}{V}\sum_k F_i^{Tk} l_j^k \qquad [2.4]$$

where V is the volume in which the contact forces are considered, $\vec{F}^k$ is the contact force vector applied to contact k and $\vec{l}^k$ is the vector joining the centers of mass of the two particles in contact at contact-point k.

If component $\sigma_{11}{}^T$ is considered, it is clear that this component is positive when components $F_1{}^{Tk}$ and $l_1{}^k$ have essentially the same signs; whenever the global component is negative, the local components have essentially opposite signs.

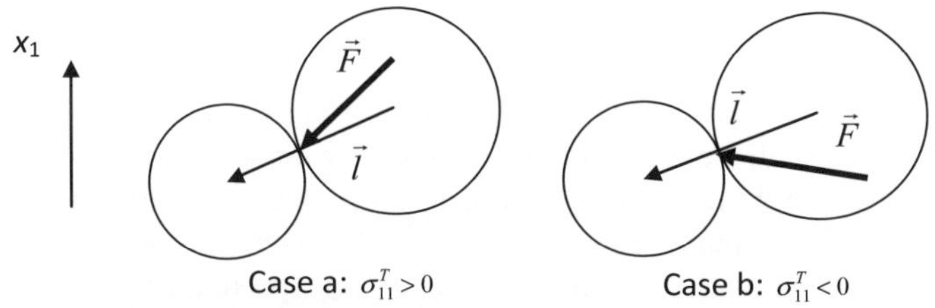

Figure 2.8. *Explanatory scheme of an orientation of a contact force leading to values of $\sigma_{11}{}^{T}$, positive and negative respectively*

The numerical simulation performed by Yunus [YUN 08, YUN 10] clearly showed that, for the same stress state, the orientation of contact forces depends strongly on the loading history. Indeed the numerical simulation of semi-alternated cyclic loadings with isotropic stress at the end of each cycle leads to values of $\frac{\sigma_{11}^{T}}{\sigma_{11}^{N}}$ which, for the maximum values of the deviatoric stress, are positive and increase significantly with the applied cycles and are negative when the stress state is equal to an isotropic state (except for the initial state, for which this value is equal to zero). Figure 2.9 clearly shows an evolution of the orientations of the contact forces throughout the cyclic loading in agreement with the local mechanisms described by Jean Biarez in his thesis [BIA 62].

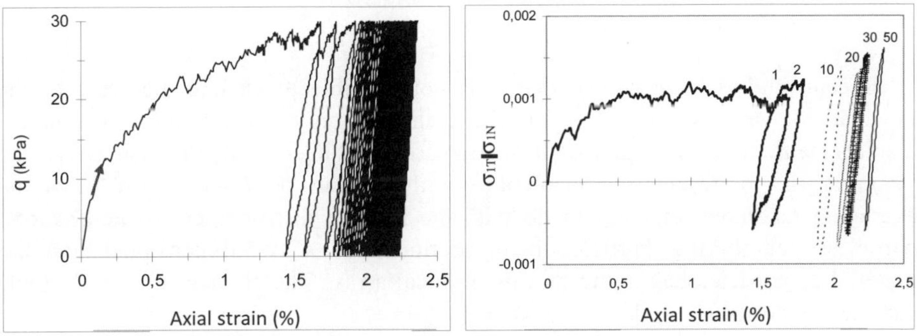

Figure 2.9. *Analysis of the evolution of the orientations of contact characterized by the ratio* $\frac{\sigma_{11}^{T}}{\sigma_{11}^{N}}$

2.5. Analysis of local arrays

In every study concerning the behavior of a granular material with a change of scale approach, one of the main questions is how to define the relevant local scale. Even to this day, we do not have a completely satisfactory answer but, in his thesis [BIA 62], Biarez had already posed this question. Indeed part of the analyses that he proposed was to be seen at the contact scale (particularly for defining the geometrical anisotropy of contact presented in section 2.3); he also understood that another interesting level of analysis could be the local arrays of particles. He analyzed, in particular, the balance of simple arrays consisting of four particles (Figure 2.10) and showed that, for a given stress state, the stability of these arrays depended on their geometry.

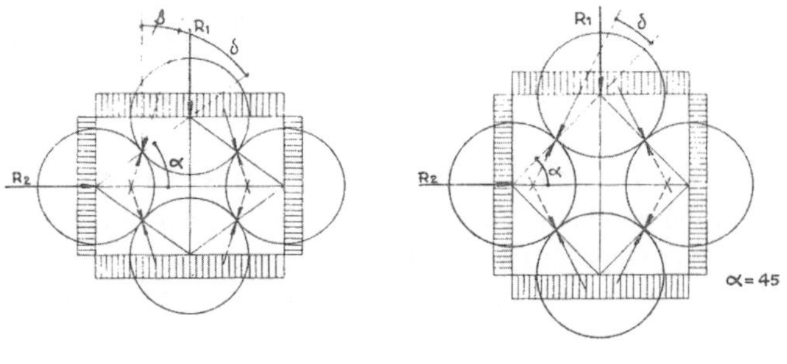

Figure 2.10. *Analysis of the equilibrium of regular arrays of discs (figure from [BIA 62])*

Recent studies have shown that it is not possible to change the scale of the kinematical variables in granular materials by only considering the relative displacements at contact points. It has nevertheless been well established that the strain at the representative elementary volume can be obtained by using an averaging technique that has to do with the relative displacement of neighboring particles; neighboring particles being defined for a two-dimensional granular material as particles belonging to a given local array. The change of scale formula can then be defined as follows [CAM 00]:

$$\delta\varepsilon_{ij} = \frac{1}{2N}\left\{\left(\sum_{k=1}^{N}\frac{\delta l_i'^k}{l'^k}n_n'^k\right)G'_{nj} + \left(\sum_{k=1}^{N}\frac{\delta l_j'^k}{l'^k}n_n'^k\right)G'_{ni}\right\}, \qquad [2.5]$$

where l'_k is the length of the branch vector defined by means of Delaunay discretization corresponding to neighboring particles; $\delta l_i'^k$ is the increment of the branch vector joining two neighboring particles $l_i'^k$; and G'_{nj} is the fabric tensor defined from the branch vectors joining neighboring particles from the following formula:

$$G'_{nj}H'_{jm} = \delta_{nm} \text{ with } H'_{jm} =< n_j'^k n_m'^k >. \qquad [2.6]$$

All local variables considered above are defined with "a prime" because they are defined for neighboring particles, which allows them to be differentiated from similar variables often used to define the *local mechanism* occurring at particle contact points. Neighboring particles are defined as particles belonging to a given array (forming a closed loop of particles in contact). These local arrays are very easy to conceive for two-dimensional materials, but much more difficult for three-dimensional materials. In Figure 2.11, the particles defined from their centers A, B, C and D constitute a closed loop. Two triangles can be defined in this quadrilateral: triangles ABD and BCD. The different vectors that combine these four points are what we call the branch vectors between neighboring particles defined in the quadrilateral ABCD. All the two-dimensional materials considered can be completely paved by polygons, such as ABCD, and it is then fairly easy to define all the variables appearing in equation [2.5].

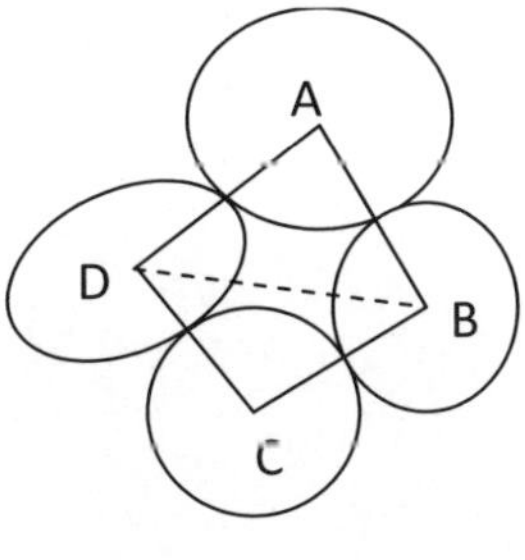

Figure 2.11. *Definition of the branch vectors between neighboring particles*

These results clearly show that the relevant local scale for analyzing strains corresponds to the local particle arrays, which agrees with the suggestions made by Biarez in his thesis. The studies by Nguyen [NGU 08, NGU 09] indicate the specific influence of local texture through a numerical modeling approach based on DEM. A sample consisting of circular particles is constructed numerically. It is then possible to define a complete discretization of the material considering all the branch vectors joining the particles in contact (Figure 2.12). Each local array corresponds to a closed loop of particles in contact with no other particle inside the loop. Each local array is characterized by a number of branch vectors forming a closed loop, these vectors being oriented in such a way that the loop is described anticlockwise. It is then possible to define a tensor characterizing the geometry of these arrays by the following equation:

$$A_{ij} = \frac{1}{n}\sum_{k=1}^{n} \frac{l_i^k l_j^k}{(l^m)^2}, \qquad [2.7]$$

$\vec{l}^{\,k}$ being a branch vector of a closed polygon having n sides and l^m being the mean value of the branch vectors in all the considered medium. The matrix of this tensor can then be diagonalized allowing two variables to be defined:

– the vector $\vec{l}_a$, which is the unit vector giving the major principal direction of this tensor and thus characterizes the general direction of the considered array (except in the particular case of a perfectly isotropic array);

– the ratio of the eigenvalues of the matrix allows what the author calls the "lengthening degree" to be defined.

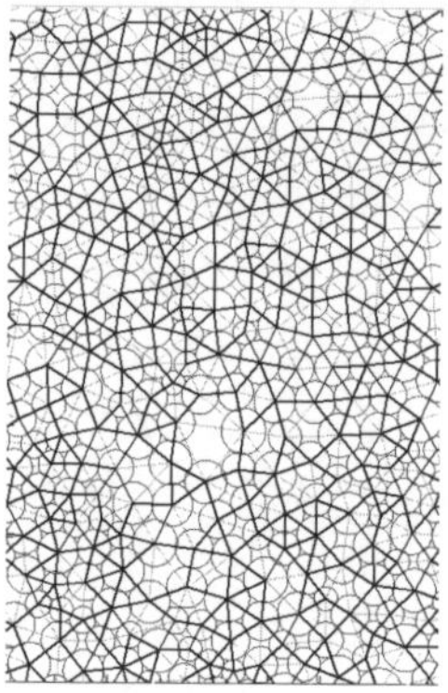

Figure 2.12. *Discretization of a granular two-dimensional material by a complete paving of local arrays (from [NGU 08])*

Figures 2.13a and 2.13b show the distributions of the volumetric strains computed in local arrays with respect to the orientation of these arrays $\vec{l}_a$ and their degree of elongation. Two increments of strain are analyzed:

– the first one is defined in the contractancy domain of a usual biaxial test;

– the second one in the dilatancy domain of this test.

It is clear from these figures that for the increment defined in the contractancy domain, the arrays orientated in the horizontal direction show negative strains (contractancy), this phenomenon being more marked for strongly elongated arrays. For the increment defined in the dilatancy domain, the arrays orientated in the vertical direction show positive volumetric strain (dilatancy), this phenomenon being more important for the strongly elongated arrays.

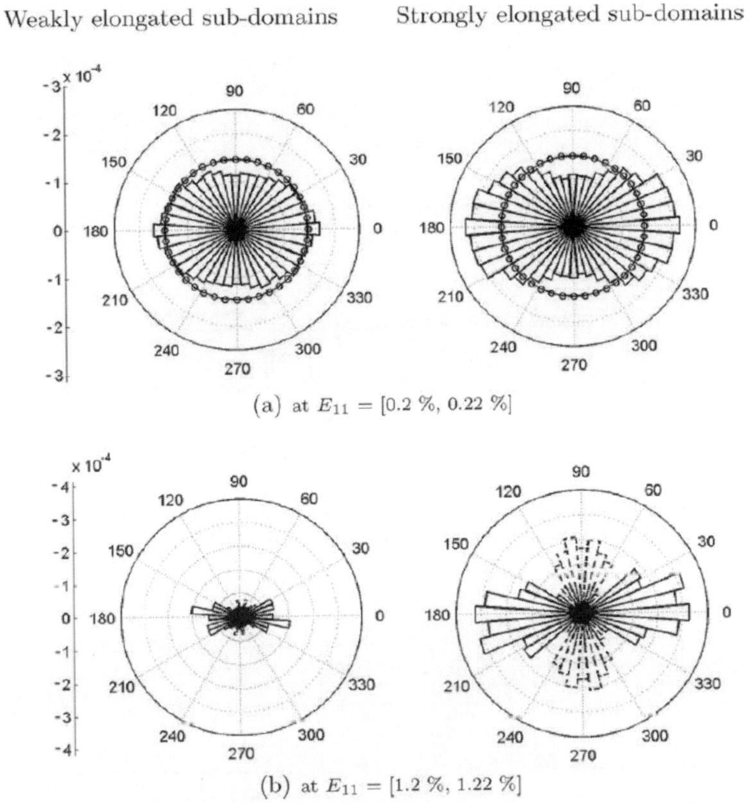

Figure 2.13. *Analysis of the volumetric strain of the local arrays with respect to their orientations and their degree of elongation (from [NGU 08])*

2.6. Particle breakage

Many rockfill dams were built in the 1950s and 1960s. Constructed by successive deposits of large-sized rocky blocks (whose diameter varied from around 10 cm to over 1 m), each layer was highly compacted before the next layer was deposited. For these construction projects, it was elemental to take into account the discrete nature of the material. Furthermore, an analysis of their deformation revealed delayed strains that were mainly related to breakage of the blocks. Biarez identified this phenomenon early on and studied the influence of grain breakage, especially in laboratory tests. Indeed, grain breakage caused the appearance of finer particles and the changes in the grading curve of the material significantly altered its behavior. This phenomenon usually resulted in an increase in compressibility (see Figure 2.14), a decrease in dilatancy and a decrease in the value of internal friction at peak.

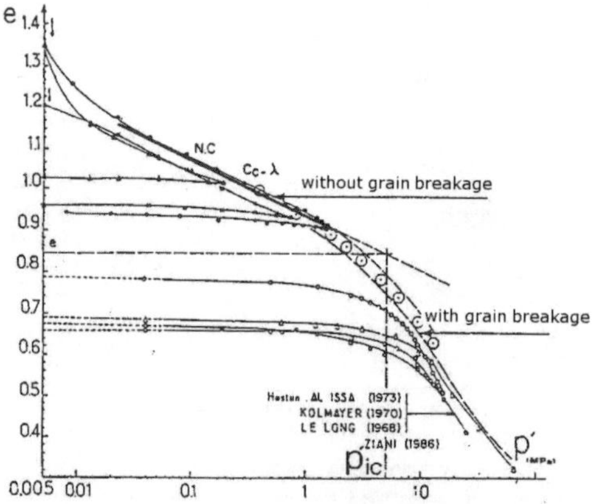

Figure 2.14. *Increase in compressibility for breakable grains during oedometer tests on narrowly graded silica sand (from [BIA 97])*

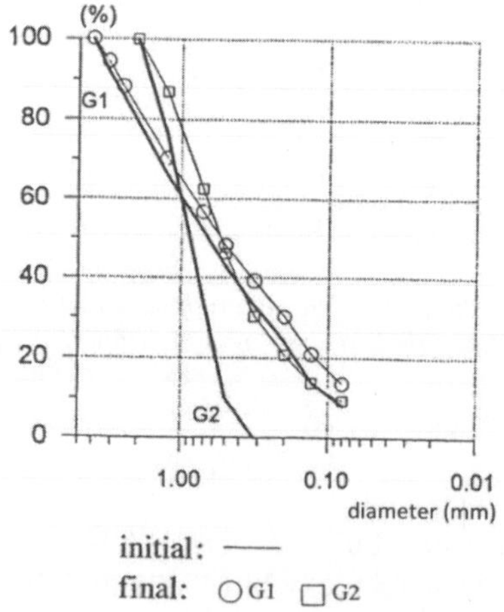

Figure 2.15. *Evolution of the grading curve during oedometer tests on samples of crushed granite with a broadly graded G1 ($d_{60}/d_{10} = 10$) or a narrowly graded G2 ($d_{60}/d_{10} = 2$) particle size distribution (from [BIA 97])*

In particular, experimental studies conducted by Biarez and his students [BIA 97, KIM 95] showed the influence of particle size distribution on the rate of breakages: subjected to the same loading, a sample of crushed granite with a broadly graded particle size distribution produces less breakage than a sample made from the same material but with a narrowly graded particle size distribution (see Figure 2.15).

These same studies also highlighted the role of the loading path: for the same particle size distribution and the same confining stress, the particles in the sample will break into a greater quantity if subjected to triaxial rather than isotropic compression or an oedometer test.

Due to the numerical simulation by DEM, breakable particles can also be modeled. For example, if PFC software is used [ITA 02], we can create an aggregate (called a *cluster*) consisting of several spheres (or disks in two-dimensions), according to a chosen geometry (elongated, with a surface more or less rough, with growths more or less pronounced, etc.) linked by bonds that break when the tensile force reaches a certain limit.

Attempting to simulate the behavior of dams that could take into account the breakage of blocks, Deluzarche used this model by considering breakable clusters [DEL 04, DEL 05]. In his study, he used a bond strength that was dependent on the size of the blocks, according to the results obtained by [MAR 73]. His numerical simulations of biaxial tests on samples with particle size distributions, either narrowly or broadly graded, led to the same result as the tests of [KIM 95], see Figure 2.16.

The explanation for this phenomenon is related to the transmission of forces through force chains and to the number of contacts per particle, easily measurable parameters in a discrete element simulation. Indeed, in a sample with a narrowly graded particle size distribution, the coordination number is generally low and all the particles have roughly the same number of contacts. The larger particles that are less resistant because they have more defects will thus break into greater quantities.

On the other hand, in a broadly graded sample, the coordination number is higher (the particles have more points of contact on average) because the largest particles may have a large number of contacts.

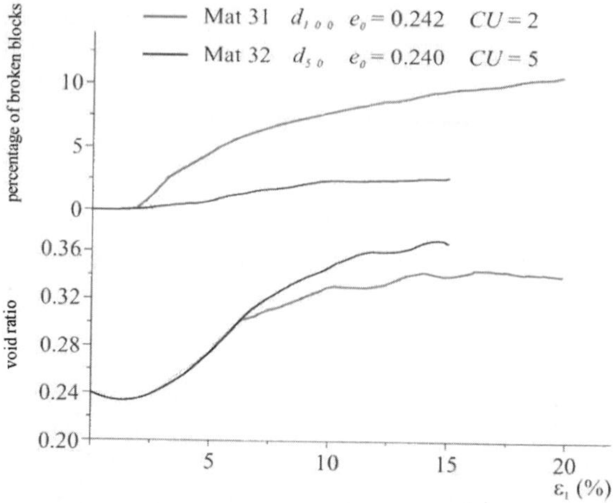

Figure 2.16. *Influence of grading on the percentage of broken blocks during the numerical simulation of a biaxial test, material 31 being narrowly graded and material 32 being broadly graded (from [DEL 04])*

The increase in compressibility during an oedometer test (Figure 2.17) and the decrease in dilatancy during a biaxial test linked to particle breakage were also found by means of these numerical simulations by discrete elements. Moreover, they showed that, during a biaxial test, the value ϕ of the peak friction angle decreases with the logarithm of the confining pressure when this pressure increases (Figure 2.18).

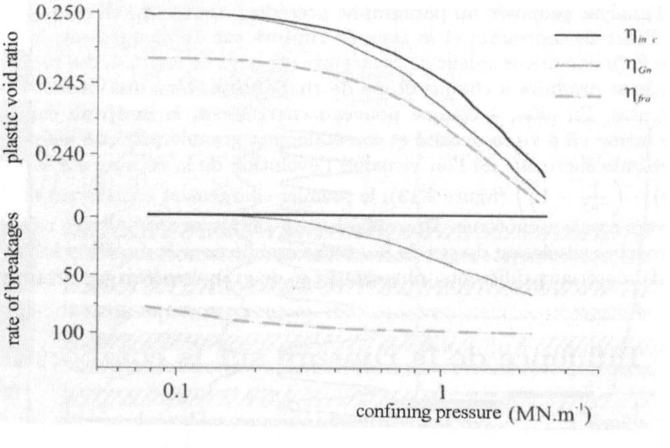

Figure 2.17. *Effect of the breakage of blocks on the slope of the compressibility curve during the numerical simulation of an oedometer test with materials of different strengths: η_{inc} (unbreakable material), η_{Gn} (moderately resistant material) and η_{fra} (weakly resistant material) (from [DEL 04])*

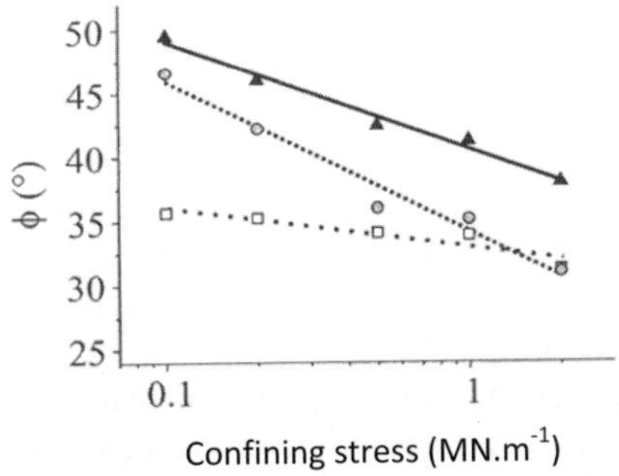

Figure 2.18. *Evolution of the peak friction angle ϕ with respect to the confining pressure for three materials with different strengths (triangles: unbreakable material; circles: moderately resistant material; amd squares: weakly resistant material) for which the relationship $\phi = A\log(B/\sigma_2) + 35°$ can be established, A and B being constants (from [DEL 04])*

Once the simulations at the REV scale validated the model, Deluzarche used it to simulate the behavior of a real dam, as Biarez would have done. The dam measured 20 m high, with an upstream impervious mask and a stone riprap (layer of stones arranged by hand to stabilize the rocky blocks on the surface) on the downstream side. Its two-dimensional modeling by breakable clusters required more than 20,000 blocks consisting of 10 disks each. The analysis of the results from numerical simulations shows that the dam deformations are due to the breakage of the blocks (see Figure 2.18) with, also, the stone riprap playing a stabilizing role.

Figure 2.19. *Displacements within the dam with breakable clusters during its building and the reservoir filling (from [DEL 04])*

2.7. Conclusion

It took little time for Biarez to understand that the complex behavior of granular materials derived from their discrete nature. Despite the very limited means at his disposal when he wrote his doctorate thesis, he did not hesitate to try to make the link between local phenomena and the behavior of the REV. He was therefore the first person to define geometric anisotropy from contact directions. He also analyzed the orientation of contact forces and the balance of local arrays, and drew certain information about plastic mechanisms in granular materials. He was also interested in the breakage of particles in a granular medium and showed that this phenomenon gave rise to a significant increase in the compressibility of these materials. To perform this analysis, very innovative for his day, Biarez used experimental approaches that were difficult to implement and performed analytical computations on very simplified structures that nevertheless accounted for the essential physics of the phenomenon.

In recent decades, the change of scale in granular media has become an area of research attracting much interest and activity. In part, this may be due to the emergence of a very powerful numerical tool: the DEM, which has enabled us to gain easy access to both local and global variables. This new tool has allowed researchers to explore the different paths chartered by Biarez more thoroughly and, more often than not, their results have demonstrated the relevance of his ideas.

2.8. Bibliography

[BIA 62] BIAREZ J., Contribution à l'étude des propriétés mécaniques des sols et des matériaux pulvérulents, Doctorate thesis, Faculté des Sciences, University of Grenoble, France, 1962.

[BIA 63] BIAREZ J., WIENDIECK K., "La comparaison qualitative entre l'anisotropie mécanique et l'anisotropie des milieux pulvérulents", *Compte Rendu de l'Académie des Sciences*, 256, pp. 1217-1220, 1963.

[BIA 97] BIAREZ J., HICHER P-Y., "Influence de la granulométrie et de son évolution par ruptures de grains sur le comportement mécanique des matériaux granulaires", *Revue Française de Génie Civil*, vol. 1, no. 4, pp. 607-631, December 1997.

[CAM 72] CAMBOU B., Compressibilité d'un milieu pulvérulent. Influence de la forme et de la dimension des particules sur les propriétés mécaniques d'un milieu pulvérulent, Doctorate thesis, Université scientifique et médicale de Grenoble, France, 1972.

[CAM 00] CAMBOU B., CHAZE M., DEDECKER F., "Change of scale in granular materials", *European Journal of Mechanics, A/Solids*, vol. 19, pp. 999-1014, 2000.

[CUN 79] CUNDALL P.A., STRACK O.D.L., "A discrete numerical model for granular assemblies", *Géotechnique*, vol. 29, no. 1, pp. 47-65, 1979.

[DEL 04] DELUZARCHE R., Modélisation discrète des enrochements – Application aux barrages, Doctorate thesis, Ecole Centrale de Lyon, France, 2004.

[DEL 05] DELUZARCHE R., CAMBOU B., "Discrete numerical modelling of rockfill materials. Application to rockfill dams", *International Journal for Numerical and Analytical Methods in Geomechanics*, vol. 1, pp. 1-23, 2005.

[JEA 95] JEAN M., "Frictional contact in collection of rigid or deformable bodies: numerical simulations of geomaterials", in: *Mechanics of Geomaterials Interfaces*, SALVADURAI A.P.S. and BOULON M.J. (eds), Elsevier, pp. 463-486, 1995.

[ITA 02] ITASCA CONSULTING GROUP INC., Mineapolis, PFC2D (Particle Flow Code in Two Dimensions), version 3.0, 2002.

[KIM 95] KIM M-S., Comportement mécanique des matériaux granulaires sous fortes contraintes, Doctorate thesis, Ecole Centrale de Paris, France, 1995.

[MAR 73] MARSAL R.J., "Mechanical properties of rockfill", in: *Embankment-dam Engineering – Casagrande Volume*, pp. 109-200, John Wiley & Sons, New York, USA, 1973.

[MOR 94] MOREAU J.J., "Some numerical methods in multibody dynamics: application to granular materials", *European Journal of Mechanics, A/Solids*, vol. 13, no. 4-suppl., pp. 93-114, 1994.

[NGU 08] NGUYEN N.S., MAGOARIEC H., DANESCU A., CAMBOU B., "On a physically relevant mesoscale for macroscopic modelling of granular assemblies", *Proceedings of the Inaugural International Conference of the Engineering Mechanics Institute*, Minneapolis, Minnesota, USA, May 19-21 2008.

[NGU 09] NGUYEN N.S., MAGOARIEC H., CAMBOU B., DANESCU A., "Analysis of structure and strain at the meso-scale in 2D granular materials", *International Journal of Solids and Structures*, vol. 46, no. 17, pp. 3257-3271, 2009.

[NOU 03] NOUGUIER-LEHON C., CAMBOU B., VINCENS E., "Influence of particle shape and angularity on the behaviour of granular materials: a numerical analysis", *International Journal for Numerical and Analytical Methods in Geomechanics*, vol. 27, pp. 1207-1226, 2003.

[NOU 05] NOUGUIER-LEHON C., VINCENS E., CAMBOU B., "Structural changes in granular materials: the case of irregular polygonal particles", *International Journal of Solids and Structures*, vol. 42, pp. 6356-6375, 2005.

[ODA 72] ODA M., "Initial fabrics and their relations to mechanical properties of granular materials", *Soils & Foundations*, vol. 12, no. 1, pp. 17-36, 1972.

[ODA 80] ODA M., KONISHI M., NEMAT NASSER S., "Some experimentally based fundamental results on the mechanical behavior of granular materials", *Géotechnique*, vol. 30, no. 4, pp. 479, 1980.

[SAT 82] SATAKE M., "Fabric tensor in granular materials", *IUTAM Conference on Deformation and Failure of Granular Material*, Delft, pp. 63, 1982.

[WIE 64] WIENDIECK K., L'anisotropie des milieux pulvérulents et son influence sur les tassements, Doctorate thesis, University of Grenoble, France, 1964.

[YUN 08] YUNUS Y., Modélisation discrète du comportement cyclique des matériaux granulaires, Doctorate thesis, Ecole Centrale de Lyon, France, 2008.

[YUN 10] YUNUS Y., VINCENS E., CAMBOU B., "Numerical local analysis of relevant internal variables for constitutive modelling of granular materials", *International Journal for Numerical and Analytical Methods in Geomechanics*, vol. 34, no. 11, pp. 1101-1123, 2010.

Chapter 3

Granular Materials in Civil Engineering: Recent Advances in the Physics of Their Mechanical Behavior and Applications to Engineering Works

Construction of large civil engineering infrastructures commonly involves large volumes of earthfill and rockfill, sometimes up to tens of millions of cubic meters, as in highways or railway platforms, marine infrastructures or large rockfill dams, see Figure 3.1. Examples of these include the Grand-Maison Dam in France (height 160 m, volume 1,400 m^3 materials) with a compacted central clay core, or the Campos Novos dam in Brazil (height 202 m, volume 1,300 m^3) with a watertight concrete slab on the upstream face, which will be discussed later on in the text.

For this last type of dam, which has become dominant in dam construction today, a major portion of the design methods has been based on the empirical extrapolation of the standard methods used (in the past) for lower dams. This empirical approach, based on North American expertise, has recently led to serious technical accidents during commissioning on very high dams. As a consequence, concern in the profession has arisen, leading to a return to more rational approaches in design, and particularly engineering approaches, through structural analysis and relevant material testing. This should be the case for any large civil engineering structure and highlights the need to improve our knowledge about the behavior of the coarse granular materials constituting these works, as well as the behavior of these large structures. This improvement can be done by integrating the local physical

---

Chapter written by Etienne FROSSARD.

phenomena within the materials up to the scale of the engineering structure, therefore considering a multiscale approach.

Figure 3.1. *Large earth and rockfill infrastructures in civil engineering*

This chapter presents the main results from personal research into the physics of granular materials, as well as results from a collaborative research in France investigating the mechanical behavior of large rockfill dams. This collaborative research program revisited previous experimental results on rockfills, applying present knowledge on the physics of materials, especially in the fracture mechanics of brittle materials [FRO 05], bringing to light a wide set of new results of practical use.

We will successively present the two physical phenomena at the basis of the mechanical behavior of these materials:

– first, the mobilization of the shear strength, which is the main aspect of structural stability due to energy dissipation by friction at the contact between grains;

– second, grain breakage leading to increasing compressibility of the granular assembly, additional settlements of the structure, and scale effects.

Practical applications will be presented throughout this chapter and the ones discussed in greatest detail will be the subject of a specific section (section 3.3).

3.1. Behavior resulting from energy dissipation by friction

3.1.1. *Introduction*

Since Coulomb's historical publication of 1773, many studies have been devoted to the role of physical friction in the mechanical behavior of granular materials. However, a direct link between the initial cause – friction at the grain contacts – and the elements of practical interest concerning behavior at macroscopic scale, such as the failure criterion or the 3D stress–strain relationship, has not been clearly established. Significant advances in this direction have been made, such as Rowe's stress-dilatancy theory (1962), enriched by Horne (1965-1969), and more recent statistical mechanics approaches. Their conditions of validity, however, which are limited to the axisymetric loading condition or 2D granular assemblies made of disks, etc., are too restrictive to be applied to the general case.

The approach presented has a larger scope and brings a solution to more general 3D static problems for granular media made from grains of irregular shape. It gives access to an explicit expression of numerous macroscopic properties, such as dilatancy law, failure criterion, strain localization with the internal structure of the shear bands, orientation and development of failure lines, etc. Only the main results will be presented here, a more detailed presentation being available in [FRO 01] and [FRO 04]. The approach is based on statistical physics ruling energy dissipation by friction, from elementary contact to macroscopic behavior, using an original mechanical concept of "internal action", materialized by a tensor formed by the product of internal forces and internal movements (see Figure 3.2). The first invariant of this tensor represents the mechanical work rate. This new concept allows us to:

– reformulate the friction laws in a form that is more functional in order to solve the mechanical problem;

– link the microscopic scale at grain contact to the macroscopic scale corresponding to the equivalent continuous medium, using an intermediate scale corresponding to the discontinuous granular mass.

This multiscale approach has been developed by using the concept of internal action combined with a rule of minimum dissipation based on the thermodynamics of dissipative processes. It leads to a wide set of results.

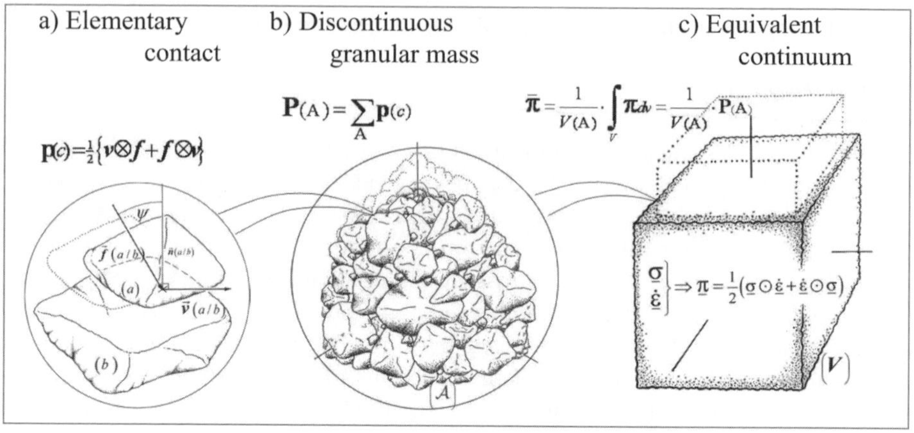

Figure 3.2. *The concept of Internal Actions*

3.1.2. *Fundamentals*

3.1.2.1. *Elementary intergranular contact*

Let us consider a simple contact, *c*, between two grains, *a* and *b*, sliding with a relative velocity, $\boldsymbol{v}(a/b)$, under a contact force, $\boldsymbol{f}(a/b)$, with an elementary friction angle at contact, ψ (see Figure 3.2a). The elementary law of friction gives the following relation between these two vectors, which corresponds to the energy dissipation during sliding:

$$\boldsymbol{f}(a/b) \bullet \boldsymbol{v}(a/b) = \sin\psi . \left\|\boldsymbol{f}(a/b)\right\| . \left\|\boldsymbol{v}(a/b)\right\| \qquad [3.1]$$

From the product of these two vectors, we can define a "tensor of elementary contact action", **p**(c), which is a symmetrical second-order tensor, its trace being the mechanical work rate produced by the contact force $\boldsymbol{f}(a/b)$ during sliding:

$$\begin{cases} \mathbf{p}(c) = \frac{1}{2}\left[\boldsymbol{f}(a/b) \otimes \boldsymbol{v}(a/b) + \boldsymbol{v}(a/b) \otimes \boldsymbol{f}(a/b)\right] \\ \quad \text{or in components } p_{ij} = \frac{1}{2}\left[f_i . v_j + v_i . f_j\right] \\ \quad \text{then } \quad Tr\{\mathbf{p}(c)\} = \boldsymbol{f}(a/b) \bullet \boldsymbol{v}(a/b) \end{cases} \qquad [3.2]$$

From the three eigenvalues of **p**(c)[1], we can now define the following symmetrical function, which is a tensorial norm of **p**(c):

$$N\{\mathbf{p}(c)\} = |\mathbf{p}_1(c)| + |\mathbf{p}_2(c)| + |\mathbf{p}_3(c)|.$$

This tensorial norm appears to be related to the Euclidian norm of the two vectors $\boldsymbol{v}(a/b)$ and $\boldsymbol{f}(a/b)$ by the following relation:

$$N\{\mathbf{p}(c)\} = \|\boldsymbol{f}(a/b)\| . \|\boldsymbol{v}(a/b)\| \quad [3.3]$$

Merging equations [3.1], [3.2] and [3.3], we can express the elementary law of friction using a relation between the eigenvalues of **p**(c) – piece-wise linear – which corresponds to the equation of the energy dissipation by friction at a contact point:

$$Tr\{\mathbf{p}(c)\} = \sin\psi \, . N\{\mathbf{p}(c)\} \quad [3.4]$$

3.1.2.2. *Discontinuous granular mass*

The granular mass in motion (see Figure 3.2b) can be considered a population of moving elementary contacts. Provided that the grains are rigid bodies, the work rate developed by internal forces in the assembly is the sum of the work rates produced by all contact forces. This leads us to consider a tensor of "internal actions in the granular mass". This tensor is a symmetrical second-order tensor resulting from the sum of all the tensors of elementary contact actions within the granular assembly:

$$\mathbf{P}(A) = \sum_{A} \mathbf{p}(c) \quad [3.5]$$

The trace of tensor **P**(A) is, therefore, the work rate corresponding to all the contact forces, and the eigenvalues of **P**(A) can be considered fluxes of mechanical energy exchanged with the outside, flux coming from the outside for positive eigenvalues, and flux going to the outside for negatives values.

If we assume that there is no other dissipative process within the granular assembly in motion, we can demonstrate that the eigenvalues of **P**(A) are linked by a relation of energy dissipation induced by the friction law at contact, including a population effect here:

---

1 **p**(c) can be easily diagonalized in its natural base formed by the two bisecting lines of the directions of $\boldsymbol{v}(a/b)$ and $\boldsymbol{f}(a/b)$, and their common normal.

$$Tr\left\{\mathbf{P}(A)\right\} = \frac{\sin\psi}{1 - R(A).(1-\sin\psi)}.N\left\{\mathbf{P}(A)\right\} \qquad [3.6]$$

where function *R(A)* corresponds to this population effect and is related to an energy exchange between neighboring moving contacts. We can call this function the "internal feedback". Comprised between 0 and 1, the internal feedback function is related to the degree of disorder of the statistical distribution of the moving contact orientations. Its mathematical representation verifies the following conditions:

$$\begin{cases} a) \quad R(A).(1-\sin\psi) = \left[1 - \dfrac{N\left\{\sum\limits_{A} \mathbf{p}(c)\right\}}{\sum\limits_{A} N\left\{\mathbf{p}(c)\right\}} \right] \\ b) \quad 0 \le R(A) \le 1 \end{cases} \qquad [3.7]$$

By comparing the dissipation rate, $Tr\{\mathbf{P}(A)\}$, with the input power, which is the sum of the positive eigenvalues of tensor **P**(*A*), we can show that the lower and upper bounds of inequality [3.7b] define theoretical solutions to the dissipation equation [3.6] in the cases of minimum and maximum relative energy dissipation.

For *R(A)*=0, i.e. for solutions corresponding to minimal dissipation, the granular assembly follows the same dissipation equation as the elementary contact (see equation [3.3]).

Minimum dissipation solutions correspond to ordered patterns in the distribution of moving contact orientations within the granular assembly, consisting in two 3D modes with signatures (+,-,-) and (+,+,-), separated by a plane strain border mode (+,0,-). In plane strain condition, this pattern coincides with Rankine's slip lines (1857). These theoretical solutions imply a complete polarization of the sliding contact orientation distribution at a given time.

These theoretical results, combined with various experimental observations, suggest that the discontinuous granular assembly verifies a "minimum dissipation rule", which can be expressed as follows: *under regular, monotonic, quasi-equilibrium boundary conditions, the moving medium tends toward a regime of minimum energy dissipation compatible with the imposed boundary conditions; this regime is independent of the initial particular conditions*. For granular media in slow motion, such a rule has been shown [FRO 04] to be a corollary of the minimum entropy production theorem based on the thermodynamics of dissipative systems near equilibrium [PRI 77].

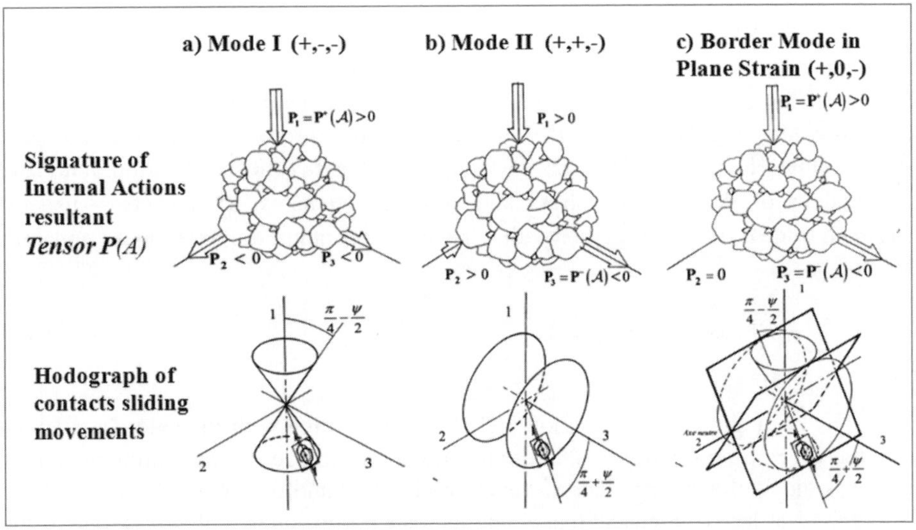

Figure 3.3. *Theoretical minimum energy dissipation solutions*

These results concern a region close to the theoretical minimum of the dissipation, with a complete polarization of the relative movements at contact. However, in the movements within usual granular media, a certain amount of disorder is necessary to make kinematic compatibility possible. This leads us to envisage larger domains of minimum dissipation having small but non-null values of *R(A)*, as well as an incomplete material polarization. We have been able to analyze heterogeneous movements in this enlarged framework [FRO 04].

Finally, let us denote "apparent friction", ψ^*, the scalar defined by:

$$\sin\psi^* = \frac{\sin\psi}{1 - R(A).(1 - \sin\psi)} \qquad [3.8]$$

The dissipation equation [3.6] for the granular assembly becomes:

$$Tr\{\mathbf{P}(A)\} = \sin\psi^* .N\{\mathbf{P}(A)\} \qquad [3.9]$$

This relation is formally identical to the one corresponding to the elementary contact (equation [3.3]), in which the physical contact friction, ψ, is replaced by the apparent friction, ψ^*, which includes the population effect, *R(A)*.

3.1.2.3. *Equivalent continuum*

Experimental results have shown that the macroscopic behavior of granular media verifies an energy-dissipation relationship, provided that boundary conditions are sufficiently regular, such as the ones applied during triaxial or plane strain tests, for example [FRO 79, FRO 83, FRO 86]. This experimental relation links the eigenvalues of the Eulerian stress and strain rate tensors, $\boldsymbol{\sigma}$ and $\dot{\boldsymbol{\varepsilon}}$ – usually assumed to be coaxial – together with a material constant, interpreted as an apparent friction, ψ^*:

$$\sum_i \sigma_i \dot{\varepsilon}_i = \sin\psi^* . \sum_i \left|\sigma_i \dot{\varepsilon}_i\right| \qquad [3.10]$$

It has been shown that equation [3.10] can be written with the sole eigenvalues of a tensor, $\boldsymbol{\pi}$, representing the "internal actions for the equivalent continuum". This tensor is the contracted product of the internal forces and the internal movements; it is a second-order symmetrical tensor and its trace corresponds to the work rate of the internal forces in the continuum (see Figure 3.2c):

$$\begin{cases} \boldsymbol{\pi} = \frac{1}{2}\left\{\boldsymbol{\sigma} \otimes \dot{\boldsymbol{\varepsilon}} + \dot{\boldsymbol{\varepsilon}} \otimes \boldsymbol{\sigma}\right\} \ \textit{contracted} \\ \quad \textit{then} \ \ Tr\left\{\boldsymbol{\pi}\right\} = \sum_{i,j} \sigma_{ij} \dot{\varepsilon}_{ij} \end{cases} \qquad [3.11]$$

With this definition, the experimental dissipation relation (equation [3.10]) can be written as follows:

$$Tr\left\{\boldsymbol{\pi}\right\} = \sin\psi^* . N\left\{\boldsymbol{\pi}\right\} \qquad [3.12]$$

Let us note that there is a formal identity between equations [3.12] and [3.9] that corresponds to the expression of the energy dissipation within the discontinuous granular mass.

3.1.2.4. *Correspondence between equivalent continuum and discontinuous granular mass: the equivalence of internal actions*

Comparing discontinuous granular mass A with the equivalent *continuum V(A)*, it appears that the work rate produced by the internal forces within the continuum must be equal to the work rate developed within the discontinuous medium:

$$\int_V Tr\left\{\boldsymbol{\pi}\right\} dv = Tr\left\{\mathbf{P}(A)\right\} \qquad [3.13]$$

To be more precise in this comparison of internal actions, we have to be sure that the internal movements that have been accounted for within the discontinuous medium correspond to all the macroscopic internal movements observable when considering the equivalent continuum. This requires the two following conditions to be met:

– the macroscopic deformations in the continuum are due to intergranular sliding within the discontinuous medium;

– the component of the macroscopic movements due to eventual rolling or spinning movements at the grain scale can be neglected (no significant "roller bearing" motion of the grains).

Kinematic analyses of numerical simulations by the discrete element method [NOU 05] show that these assumptions are realistic as long as the grain shapes are sufficiently irregular, which is the case for granular media used in civil engineering. Therefore, we obtain complete equality between the mean values of the internal actions within the discontinuous granular mass, and the corresponding equivalent continuum:

$$\overline{\pi} = \frac{1}{V(A)} \cdot \int_V \pi dv = \frac{1}{V(A)} . \mathbf{P}(A) \qquad [3.14]$$

As a result, the inner actions within the equivalent continuous medium follow the same dissipation relation as the inner actions within the discontinuous medium. Therefore, the phenomenological relationship found for the equivalent continuum (equations [3.10] and [3.12]) can be seen as a direct explicit consequence at the macroscale of energy dissipation due to friction within the discontinuous granular mass.

3.1.3. *Main practical consequences*

3.1.3.1. *Compatibility with heterogeneity*

Under regular boundary conditions, when a granular medium evolution is sufficiently close to the minimum dissipation, some specific properties are derived from this proximity to a minimum. One of them is the compatibility of the internal actions with the heterogeneity of the medium: dissipation relation [3.12] is verified by the local variables as well as by the mean macroscopic values of the same variables. With stress and strain rate heterogeneity, a different scale of local fluctuations of the stresses and strain rates was shown [FRO 04], making the dissipation equation [3.10] compatible with the macroscopic heterogeneity: this equation obtained for the local variables also holds for the mean values. Furthermore, it has been demonstrated that heterogeneities in internal movements and internal forces compatible with the different forms of the dissipation equation represent the eventual patterns of strain localization or stress concentration [FRO 04].

This compatibility of the energy dissipation relation with the heterogeneity justifies consideration of the mean values of the stress and strain rates.

3.1.3.2. *Dilatancy rule: the effects of two-way cyclic loading*

Introducing the auxiliary variables $\alpha_i = \dot{\varepsilon}_i / |\dot{\varepsilon}_i|$ for $\dot{\varepsilon}_i \neq 0$, equation [3.10] becomes:

$$\sum_i (1-\alpha_i \sin\psi^*).\sigma_i\dot{\varepsilon}_i = 0 \qquad [3.15]$$

Under axisymetric loading, assuming that the radial strains have the same sign (so $\alpha_2 = \alpha_3$), we obtain:

$$\frac{\sigma_1}{\sigma_3} = \left(1-\frac{\dot{\varepsilon}_v}{\dot{\varepsilon}_1}\right).\left(\frac{1-\alpha_3 \sin\psi^*}{1-\alpha_1 \sin\psi^*}\right) \qquad [3.16]$$

where index 1 corresponds to the axial direction.

For the usual triaxial compression test, $\alpha_1 = +1$ and $\alpha_3 = -1$; equation [3.16] becomes:

$$\frac{\sigma_1}{\sigma_3} = \left(1-\frac{\dot{\varepsilon}_v}{\dot{\varepsilon}_1}\right).\tan^2\left(\frac{\pi}{4}+\frac{\psi^*}{2}\right) \qquad [3.17]$$

Equation [3.17], directly derived from the dissipation equation [3.10], appears to be Rowe's dilatancy rule in axisymetric compression. We can see that equation [3.17] gives the expression of the mobilized shear as a product of two factors: the dilatancy rate and the apparent intergranular friction coefficient.

If, at a given point of the loading in compression, we reverse the strain direction in order to impose loading in extension, the change in the sign of $\dot{\varepsilon}_i$ in equation [3.10] also produces a sign-change in the auxiliary variables α_i in equation

[3.16]. We therefore obtain a stress–strain relationship similar to equation [3.17], but with a different coefficient: $\tan^2\left(\frac{\pi}{4}-\frac{\psi^*}{2}\right)$ instead of $\tan^2\left(\frac{\pi}{4}+\frac{\psi^*}{2}\right)$. If the direction of loading is changed again, returning to compression, equation [3.17] will hold again. Therefore, the energy dissipation equation [3.10] predicts that, under two-way loading, the stress–strain relationship depends on the dilatancy rule, which oscillates between two conjugate relations dependant on the sign of the loading direction. This behavior, which is well established experimentally, is the consequence of the irreversibility of the deformation process resulting from energy dissipation by friction. The response during unloading is not the exact inverse of the response during loading, and the difference between the two responses is due to the energy dissipation by friction.

An important practical consequence is the fact that, during a two-way loading close to the isotropic stress state, the volume change due to a cycle is always contractive and the cyclic loading creates a densification of the granular medium [FRO 83]. This is the basic principle of compaction by two-way cyclic loading.

Finally, under plane strain condition, for the same reason, equation [3.10] leads to similar forms of the dilatancy rule [FRO 01]. Under two-way cyclic loading, the alternative response of the medium between two different dilatancy rules in loading and unloading has been demonstrated by numerical discrete simulations [NOU 05] for grains with sufficiently irregular shapes.

3.1.3.3. *Experimental validation from triaxial test results*

Figure 3.4a shows an experimental validation of the dilatancy rule (equation [3.17]) for a triaxial compression test on crushed limestone with very angular particles of irregular shape at three different initial densities and under the same confining pressure. The stress–strain curves in the lower left-hand diagram and the volume change in the upper left-hand diagram show the influence of the initial density on the material behavior, especially at peak strength. The dilatancy diagram on the right-hand side shows a linear relation between the stress ratio, σ_1/σ_3, and the dilatancy rate, which is in agreement with equation [3.17]. The slope of this straight line gives a value of the apparent friction $\psi^* = 42.5°$. This straight line is independent of the initial density, in agreement with the least dissipation rule.

We observe that the three samples exhibit the same contractancy rate at the beginning of the loading. This is also in agreement with equation [3.17] in the vicinity of an isotropic stress state.

Other experimental data plotted in the same diagram have shown the same linear relation with a slope – function of apparent friction – essentially being dependent on the mineralogy and shape of the grains [FRO 79].

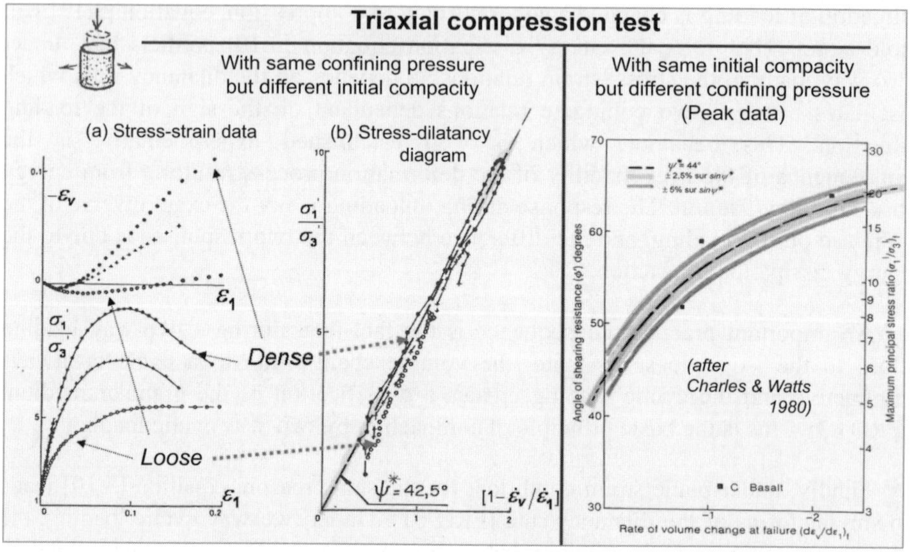

Figure 3.4. *Experimental validation of energy dissipation relation based on friction*

In Figure 3.4b, we show the results of triaxial compression tests on crushed basaltic reduced rockfills published by Charles and Watts [CHA 80]. The granular material is well graded, with grain sizes varying from sand to gravel with maximum grain size equal to 38 mm. The samples, of diameter equal to 230 mm, were tested at the same initial density and confining pressures, varying from 27 to 700 kPa. The diagram shows the evolution of the friction angle at peak with the dilatancy rate. From the experimental points, we can fit a curve corresponding to equation [3.17], which gives a value of the apparent friction equal to 44° with a dispersion equal to plus or minus 5%.

In these tests, the effect of grain breakage reduces the dilatancy rate at peak when the confining stress increases. Despite the phenomenon of grain breakage, the experimental values remain close to the ones predicted by the equation of energy dissipation by friction, even if the results obtained at peak are more dispersed than those deduced from a fitting over the whole test data.

If we plot these results in the Mohr plane (see Figure 3.5) we can observe the effect of dilatancy on shear strength by comparing two circles at the same confinement. The decrease in dilatancy at elevated stresses due to grain breakage reduces the apparent friction. As a consequence, the maximum strength envelope is no longer a straight line but has a curvature with a reduced slope when the stresses increase, which can be fitted by a power law with a coefficient equal to 0.77.

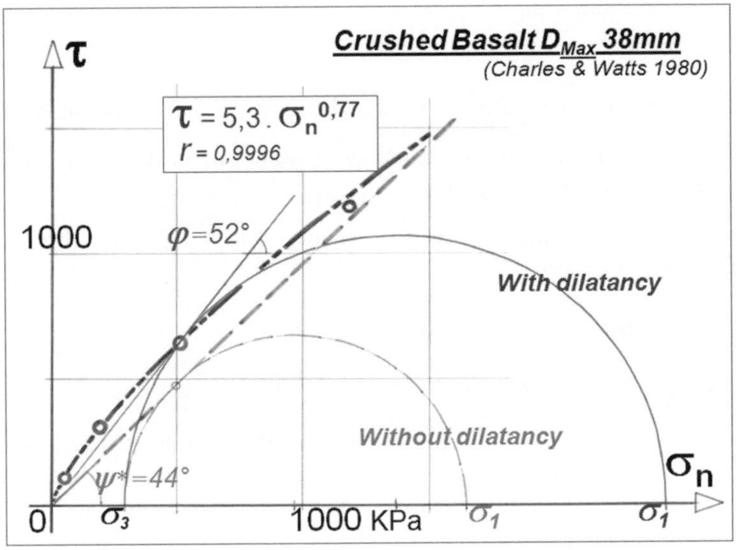

Figure 3.5. *Shear strength envelope for "reduced" basalt rockfill*

3.1.3.4. *Other experimental validations*

Other validations of the dissipation equation [3.10] were made for different experimental conditions, including the plane strain condition with rotation of the loading directions, and large-amplitude cyclic loading in a true triaxial apparatus [FRO 86].

These experimental results, in particular, provide access to the internal feedback rate, R, which takes the following values:

– between 0.05 and 0.20 according to grain angularity, roughness and elongation in triaxial compression tests;

– between 0.15 and 0.30 according to grain angularity, roughness and elongation in plane strain condition.

3.1.3.5. *Failure criterion at critical state*

At critical state, the failure criterion corresponding to the dissipation equation is obtained by introducing the condition of constant volume into equation [3.10].

Under axisymmetrical stress or plane strain conditions, the dilatancy equation established above leads to the following expression of the failure criterion:

$$\frac{Sup(\sigma_1,\sigma_2,\sigma_3)}{Inf(\sigma_1,\sigma_2,\sigma_3)} = \tan^2(\frac{\pi}{4}+\frac{\psi^*}{2}) \qquad [3.18]$$

For more general 3D conditions, we can demonstrate that at critical state:

– equation [3.18] gives the minimum in strength and in dissipation under usual boundary conditions at the same time;

– the corresponding strain mode is the plane strain mode in the direction of the intermediate principal stress.

These conditions define the well-known Coulomb's criterion with precision.

3.1.3.6. *Link with fluid mechanics*

Another interesting result from the use of this approach to energy dissipation by friction can be stressed: if the friction between grains tends toward zero, the dissipation vanishes within the medium and, as a consequence, the macroscopic apparent friction also vanishes. The expression of the failure criterion shows that, under this condition, the stress states converge toward isotropy, even at large deformations; the dissipation relation means that, in the vicinity of an isotropic state of stress, the specific volume becomes constant. A remarkable solution therefore arises for this particular behavior, which includes no dissipation during the movement, isotropic states of stress and a constant specific volume for any given state of deformation: this is the behavior of a "perfectly incompressible fluid", which is the basis of hydraulics. This result appears to be quite understandable, because the granular medium in motion is in this case an assembly of grains sliding without friction against each other while staying in contact: this is the classical microscopic description of a liquid.

3.1.3.7. *Strain localization – shear band internal structure*

A strong compatibility with the heterogeneity being inscribed in the energy dissipation relation, the strain localization in shear bands can be studied using the internal actions without particular caution. The macroscopic shear bands found with this approach tend to converge toward a plane strain kinematic, with an angle close to $(\pi/4 - \psi/2)$ with the direction of the major principal stress [FRO 04] (see Figure 3.6).

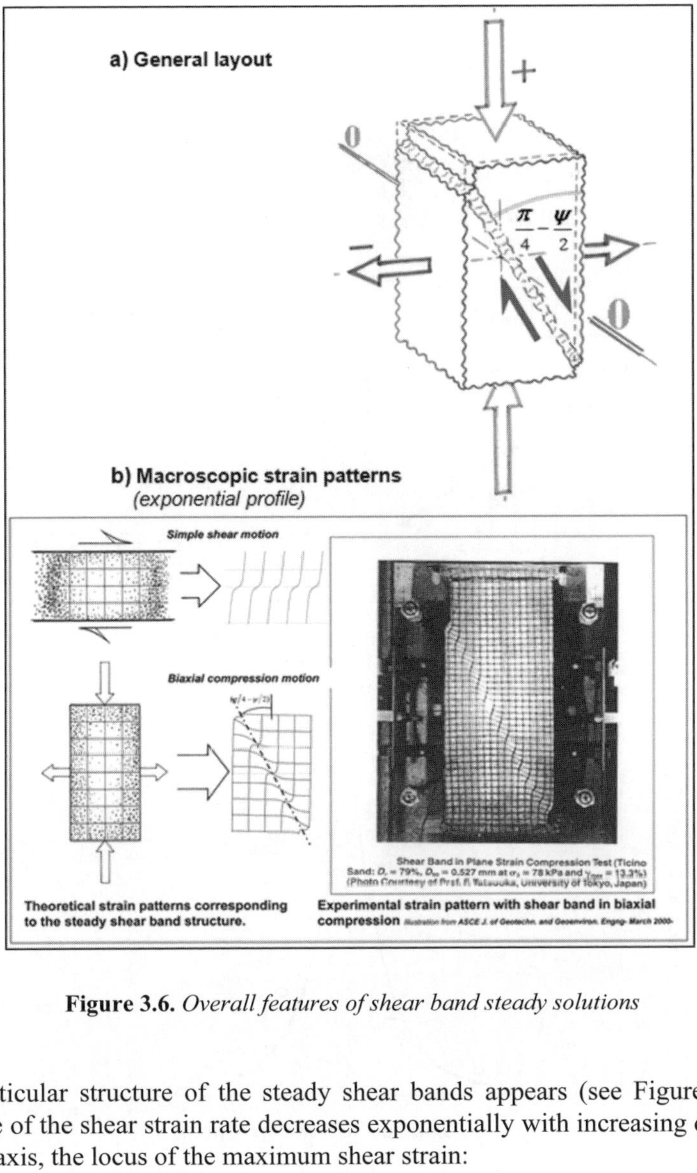

Figure 3.6. *Overall features of shear band steady solutions*

A particular structure of the steady shear bands appears (see Figure 3.7): the amplitude of the shear strain rate decreases exponentially with increasing distance to the band axis, the locus of the maximum shear strain:

$$\dot{\gamma}(x) \approx \dot{\gamma}(0).\exp(-k|\mathbf{x}|) \; ; \; with \; k \approx \frac{1}{5\bar{d}}.\text{Ln}2\left\{\frac{1+\sin\psi}{R(1-\sin\psi)}-1\right\} \qquad [3.19]$$

This relation has been well verified by the experimental results obtained by Nemat-Nasser and Okada [NEM 01], see Figure 3.7b.

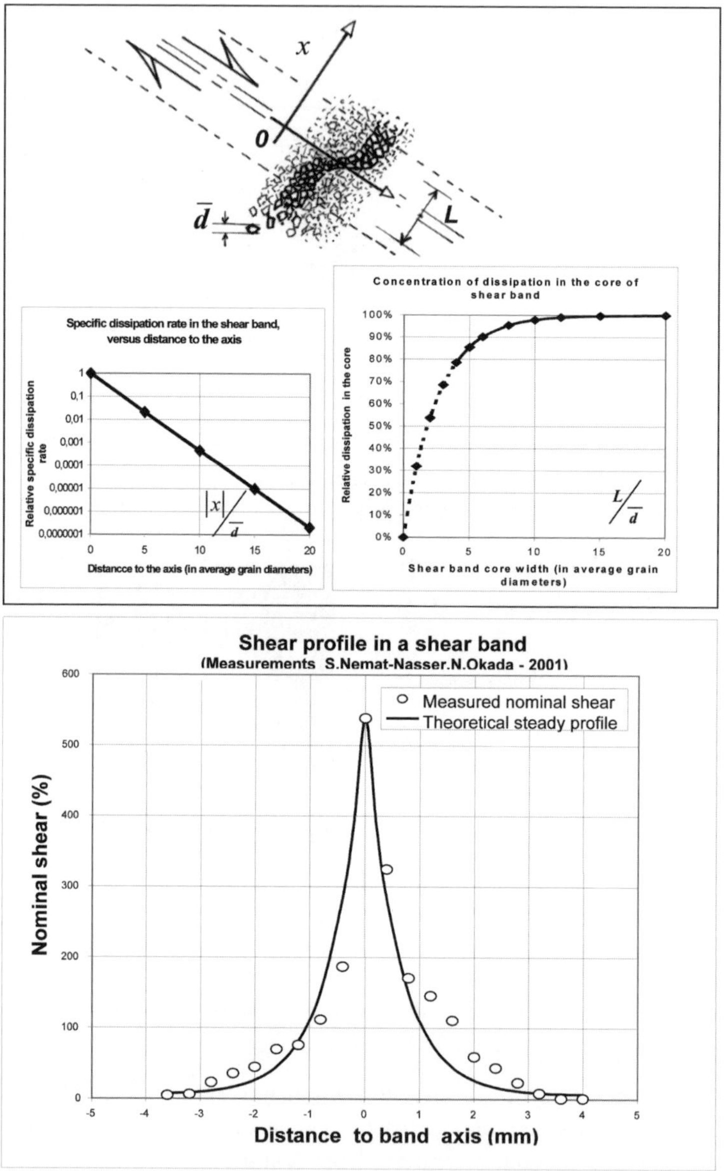

Figure 3.7. *Detailed internal structure of shear bands steady solution (exponential profile): a) dissipative structure; b) comparison between theoretical and experimental shear profiles*

The energy dissipation structure induced shows that, for badly graded granular media, 99% of the energy dissipation within the shear band takes place in the center of the band having a width equal to 12 times the mean diameter of the grains (see Figure 3.7a). This is in agreement with the experimental results presented by Oka and Kazama [OKA 98].

Finally, let us notice that the form of equation [3.19] is compatible with a homogeneous far-field movement, which also indicates that, if a mean shear strain rate $\bar{\dot{\gamma}}$ is imposed at the boundary of a domain of width L, the maximum shear rate possible within a shear band concentrating all the movements in the domain is limited by the following expression:

$$Max\left\{\left(\frac{\dot{\gamma}_{Max}}{\bar{\dot{\gamma}}}\right)(t)\right\} \approx \frac{kL}{2} \quad [3.20]$$

3.1.3.8. *Localization criterion – shear band evolution*

We now consider the localization process as a particular mode of heterogeneity growth in the motion, induced by the minimum dissipation rule. A localization criterion can then be established, based on the non-convexity of the specific energy dissipation rate depending on the correlated fluctuations in strain and density [FRO 04]. Under usual regular monotonic boundary conditions, the specific energy dissipation is almost proportional to the strength of the medium expressed in terms of stress, whose evolution during deformation usually takes one of the forms presented in the diagram in Figure 3.8a.

As a result, the localization criterion is met early during the movement. This criterion therefore predicts that the initialization of the strain localization starts well before the peak of the stress–strain curve, which is experimentally observed [DES 96].

Once created, their evolution during movement is driven by a criterion that can also be obtained by the minimum energy dissipation rule. For steady structures in a permanent regime, the analysis of the heterogeneity by the minimum energy dissipation rule shows that the shear band evolution tends to maximize the variance of the deformation amplitudes, i.e. the variance of the shear strain rates [FRO 04].

The comparison of the variances obtained for various admissible deformation schemes shows that:

– a family of parallel bands tends to converge into a single band that concentrates the whole movement;

– this single band will narrow until the maximum amplification ratio is reached (see section 3.1.3.7), with the disappearance of all the movements outside the band.

In other words, the predicted evolution converges towards the progressive formation of what is usually called the "failure line" (see Figure 3.8b).

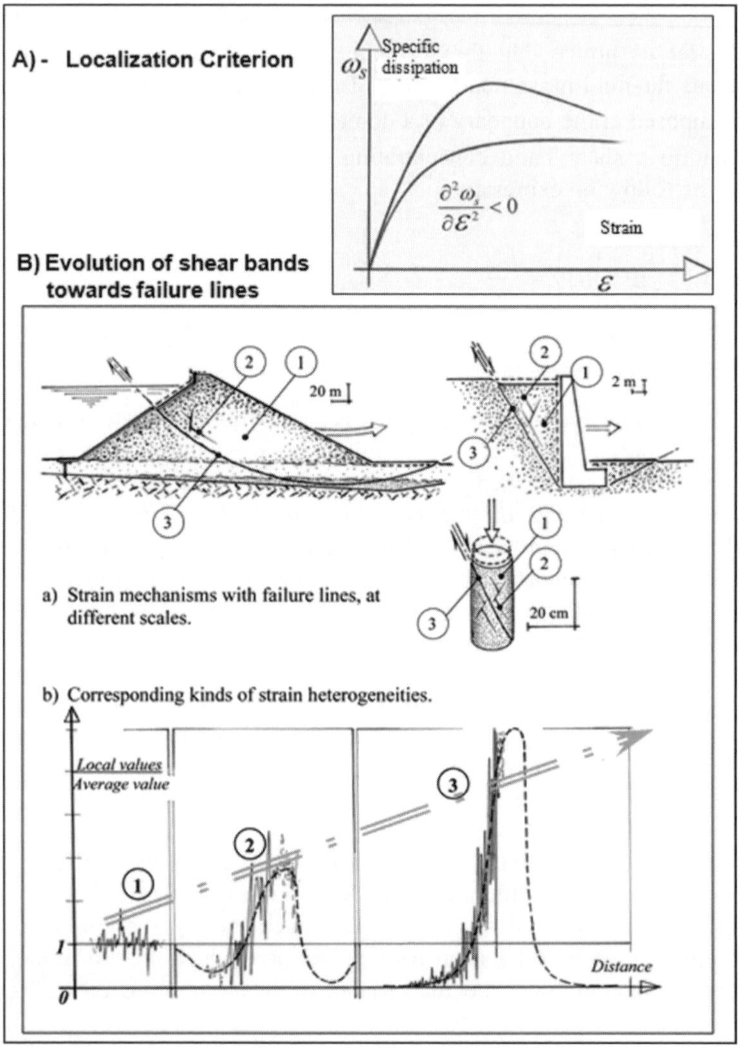

Figure 3.8. *Localization process*

3.1.4. *Conclusions*

This representation of granular media as a set of simple contact points by means of the concept of internal actions allows us to predict a large number of properties specific to these materials. We have to underline that these results are obtained by using a simplified view of the internal actions at contact points:

– the kinematic screws of the local relative contact displacements are reduced to simple sliding, neglecting rolling and spinning motions;

– the static screws of the local contact forces are reduced to resultant contact forces, neglecting rolling and spinning torques.

This simplified representation of the contact actions, associated with the concept of internal actions, appear therefore to be very efficient, at least when usual civil engineering granular materials are concerned.

As for practical consequences, we can connect the results obtained in the analysis of strain localization and shear bandings, the failure criterion corresponding to Coulomb's criterion, and the tendency of the movement to converge towards the kinematics of plane strain. All these characteristics are the basis of classical studies of stability of slopes and bearing capacity of foundations by 2D limit analysis. This represents an essential part of the possible applications of this theoretical approach. Some other practical properties of granular media will be discussed later on.

3.2. Influence of grain breakage on the behavior of granular materials

Grain breakage in granular materials during loading was underlined long ago by studies of grain size distributions before and after testing, see for example Marsal's work [MAR 73]. However, the analysis of the influence of this phenomenon on granular materials' mechanical behavior is more recent, as the results presented by Bolton and Mc Dowell [BOL 98] show, giving the impact on the non-reversible compressibility. The original approach presented here consists in analyzing the effect of local breakage on the maximum shear strength of the granular assembly and the induced scale effect. The latter can be explicitly expressed by a "scale effect relation" acting on the expression of the intrinsic failure criterion. This result leads to practical applications in the stability analysis and the design of engineering works [FRO 09], as shown later on in sections 3.3.2 and 3.3.3.

3.2.1. *Introduction to the grain breakage phenomenon*

3.2.1.1. *Elementary grain breakage*

The basic description of elementary grain failures when internal forces evolve in granular media may be summarized as follows (see Figure 3.9a):

– as contact forces increase, maximum strength is reached in some of the grains that fail;

– these local failures result from sudden propagation of pre-existing microcracks in the grains, controlled by the theoretical laws developed in fracture mechanics;

– these failures are mainly indirect tensile failures, similar to those that develop in the "Brazilian test". Failure mode is mainly Mode I, due to tensile stresses induced in the grains by the loading of the granular assembly;

– in a grain with cracks of width *c*, the failure stress in Mode I is given by the following expression:

$$\sigma_{ft} \approx K_{Ic} \Big/ \sqrt{\pi c} \qquad [3.21]$$

where k_{Ic} is the propagation threshold.

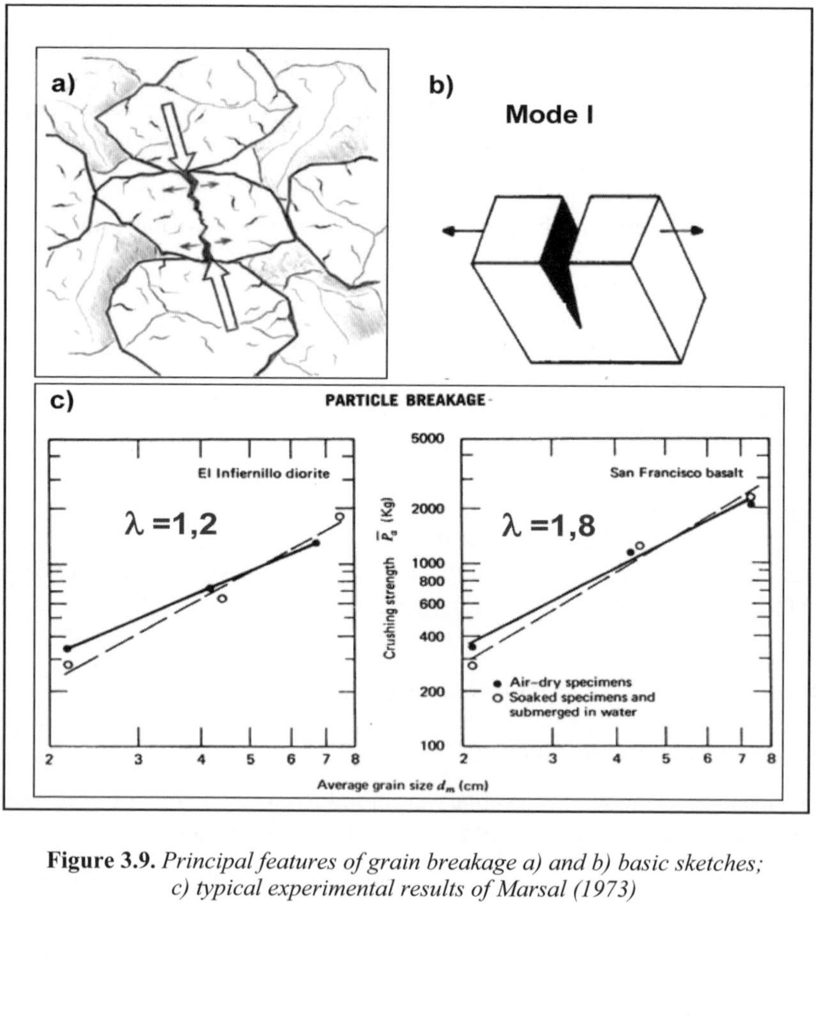

Figure 3.9. *Principal features of grain breakage a) and b) basic sketches; c) typical experimental results of Marsal (1973)*

This basic description of grain breakage raises several questions:

– What is the relation between the microcrack size and the grain size?

– What is the relation between local stresses within the grains and the macroscopic stresses applied on the granular assembly?

– Are there relations with the grain size distribution?

The answers to these questions, often found through simplifying assumptions, have led to the description of the macroscopic effects of grain breakage.

3.2.1.2. *Statistical representations*

The influence of grain size was investigated long ago [MAR 72], the results showing that the average crushing force, F_{cr}, of gravels or rock fragments is a power function of the average grain diameter (see Figure 3.9c):

$$\overline{F}_{cr} = \eta . \overline{d}^{\lambda} \qquad [3.22]$$

This expression can be connected to Weibull's theory, which gives the probability of survival within a population of brittle objects subjected to stress condition near failure:

$$P_s(V) = \exp\left[-\frac{V}{V_0}\left(\frac{\sigma}{\sigma_0}\right)^m\right] \qquad [3.23]$$

As the volume, V, of a grain is proportional to the cube of its diameter, Weibull's approach for a given value of the probability of survival, P_s, also leads to an average crushing force proportional to a power function of the grain diameter:

$$\sigma_{cr} \propto \overline{\boldsymbol{d}}^{-3/m} \qquad [3.24]$$

The average crushing force being proportional to the crushing stress multiplied by the average grain section, the comparison of the two approaches gives a simple means to fit a Weibull distribution for a given material from a set of crushing tests on grains of different sizes:

$$\lambda = 2 - \frac{3}{m} \quad \text{or} \quad m = \frac{3}{2-\lambda} \qquad [3.25]$$

3.2.1.3. *Central trend in the statistics of failure*

In the wide range of materials investigated by Marsal, values of exponent λ (lambda) between 1.2 and 1.8 were found, which corresponds to values of Weibull's parameter, m, between 4 and 15, with a mean value of λ equal to 1.5 and a corresponding mean value of m equal to 6. The subject of the suitability of Weibull's theory to natural grains has recently been updated by Lobo-Guerrero Vallejo [LOB 06].

This central value of $\lambda = 1.5$ has physical meaning. It corresponds to the configuration of a microcrack distribution within grains, which verifies the geometrical similarity with their diameter. For this specific distribution, the size of the significant cracks is statistically proportional to the grain diameter. In this case, the failure stress given by fracture mechanics in equation [3.21] becomes inversely proportional to the square root of the grain diameter. If we also assume that the significant stress in a grain is proportional to the average macroscopic stress within the granular medium, then the maximum macroscopic stress endurable by the granular medium before significant grain crushing develops is inversely proportional to the square root of a characteristic grain diameter.

These central assumptions are the core of the "clastic process" theory developed by Bolton and McDowell [BOL 98], which provides a clear relation between the compressibility of granular materials and grain breakage, depending on well defined material parameters.

3.2.2. *Scale effect in shear strength*

3.2.2.1. *Shear strength of rockfill*

Numerous experimental results on the shear strength of rockfill have been gathered and published by different authors, such as Leps [LEP 70], Charles and Watts [CHA 80] (see Figure 3.10), Duncan [DUN 04] and others. They show a wide dispersion of the values and a pronounced curvature of the shear strength envelope: the friction angle measured in triaxial compression tests significantly decreases when the confining stress increases. This reduction of the friction angle has been attributed to grain breakage that induces a decrease in dilative behavior (see sections 3.1.3.2 and 3.1.3.3). The amplitude of grain breakage was measured by comparing the grain size distribution before and after testing.

In Figure 3.10, the results presented by Charles and Watts – curves A, B1, and C – appear shifted towards the upper right-hand side of the diagram when compared to the results obtained by Marsal or Marachi [MAR 69]. This remark also holds for material B2, which corresponds to very weak grain strength and was chosen by the authors for this peculiar property; therefore, curve B2 should be located below the range of usual rockfill properties.

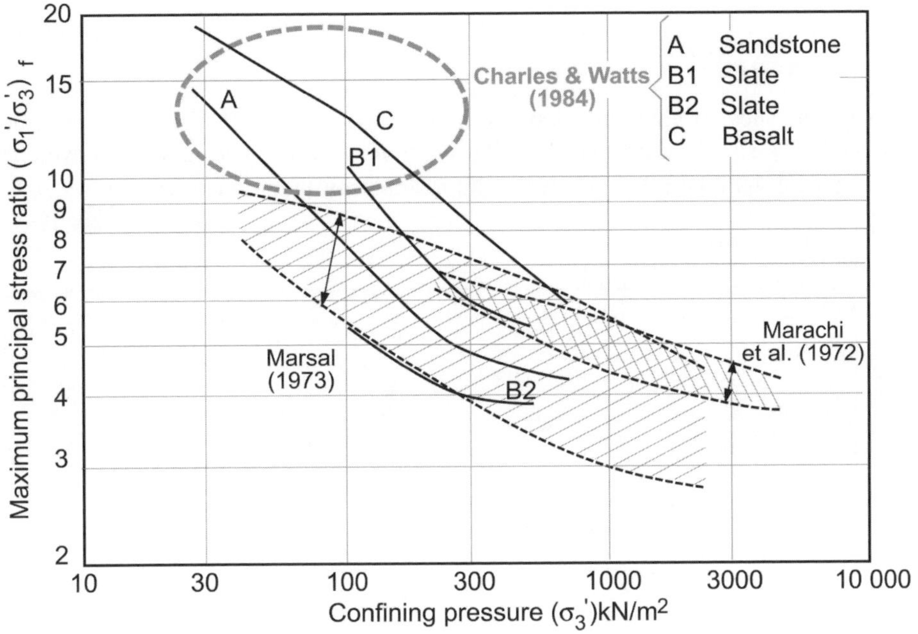

Figure 3.10. *Charles & Watts compilation of rockfill shear strength envelopes (1980)*

3.2.2.2. *Evidence of scale effect*

In another study, Barton [BAR 81] displayed a very peculiar set of large triaxial test results. The tests have been performed on granular materials issued from the same mineral stock, with homothetic grain size distributions, and tested at the same density. In Figure 3.11a, the results obtained on two different groups show a clear size effect: the shear strength is lower when the grain size is coarser.

These two groups having parallel grain size distribution curves, a comparison of their properties may be examined in light of the grain strength resulting from the probabilistic approach with the central trend distribution (see section 3.2.1.3):

– for homothetic grain size distributions, the grain resistance is in principle inversely proportional to the square root of a given characteristic diameter, for example equal to D_{Max};

– therefore, to obtain in a granular material *1* with characteristic grain size D_1, the same amount of grain breakage and, then, the same shear strength as in material

2 with characteristic grain size D_2 tested under a confining stress, $\sigma_{(2)}$, we have to apply a confining stress on material 1 equal to:

$$\sigma_{(1)} = \sigma_{(2)} \cdot \sqrt{D_2/D_1} \qquad [3.26]$$

– so, the shear strength envelope of the rockfill material (D_1 = 150 mm) can be obtained by a simple geometrical similarity applied on the shear strength envelope of the gravel material (D_2 = 12 mm) with a similarity factor equal to $\sqrt{12/150} \approx 1/3.5$ applied on the confining stress.

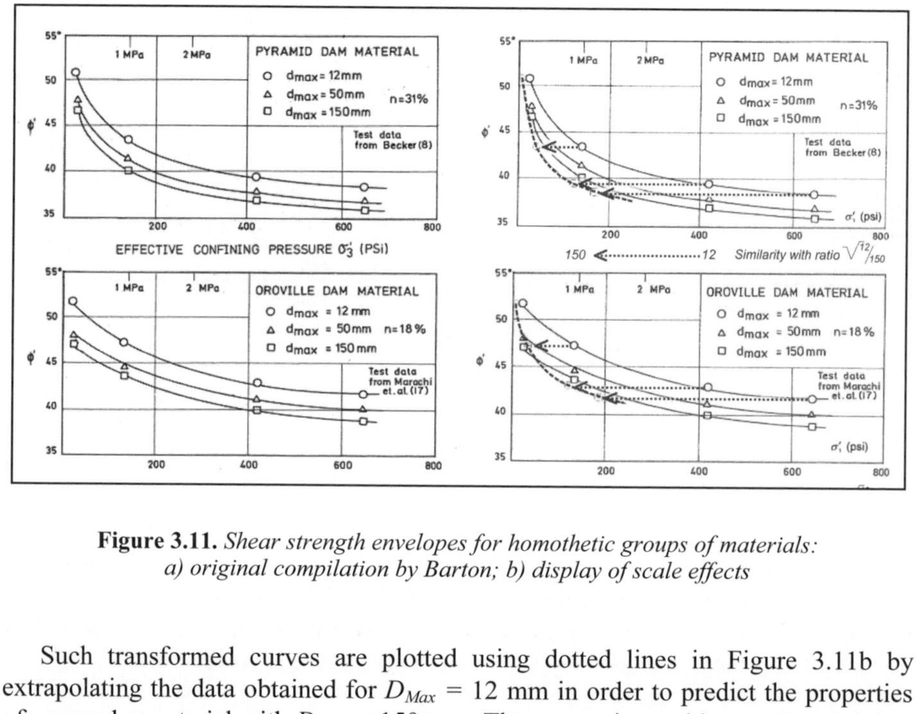

Figure 3.11. *Shear strength envelopes for homothetic groups of materials: a) original compilation by Barton; b) display of scale effects*

Such transformed curves are plotted using dotted lines in Figure 3.11b by extrapolating the data obtained for D_{Max} = 12 mm in order to predict the properties of a granular material with D_{Max} = 150 mm. The comparison with experimental data shows a quite satisfactory agreement. It appears, therefore, to be possible to determine the shear strength of very coarse granular materials (here 150 mm maximum) by using the results obtained on finer materials (here 12 mm maximum), provided that they have the same mineral origin, parallel grain size distribution, and the same density.

The original data reported by Barton come from an exceptional experimental study by Marachi *et al.* [MAR 69] performed on three groups of materials:

– a rockfill made of fine grained argillite produced by quarry blasting, having very angular particles of comparatively low strength, used for the construction of the Pyramid Dam in California;

– a rockfill made of crushed basalt extracted from a quarry, with angular sound particles;

– a coarse alluvium, predominantly made of sound unweathered rounded gravels and cobbles of fine-grained amphibolite, used for the construction of the Oroville Dam in California.

These three groups of materials, although strongly different, display the same qualitative trend concerning grain size influence, suggesting the existence of a fairly general scale effect rule.

An independent validation of this approach can be made by using the data gathered by Charles and Watts (see Figure 3.10). Their data was obtained on materials with maximum grain sizes of D_{Max}= 38 mm, whereas the results previously obtained by Marsal and Marachi concerned coarser materials (D_{Max} = 150 to 200 mm). A correction of Charles and Watts' curves by the method presented above on the confining stress corresponds here to a simple shift of their results towards the left of the diagram, as confining stresses in abscissae are displayed in logarithmic scale. Figure 3.12 presents the curves *A*, *C* and *B2* from Charles and Watts' [CHA 80] corrected by an horizontal similarity factor equal to $\sqrt{^{38\ \text{mm}}/_{150\ \text{mm}}}$. The transformed curves *A'* and *C'* in the diagram are now in far better agreement with the results obtained by Marsal and Marachi than the original curves *A* and *C*. For the slate, B2, with low-strength particles, the transformed curve is now located below the usual range for rockfill materials, which is more in agreement with the poor quality of this material, which was why it was selected.

Thus, the scale effect in Barton's data and the apparent distortion in Charles and Watts' diagram appear to be due to this physical similarity rule resulting from fracture mechanics laws governing grain breakage. The adjusted data compilation in Figure 3.12 also suggests a typical shear strength envelope for coarse granular materials corresponding to the dotted line plotted in the central part of the experimental data range. This central line, corresponding to the shear strength envelope range of granular materials with grain size D_{Max} = 150 mm, can be fitted by a power law:

$$\tau \approx 4 \cdot \sigma_n^{0,77} \quad ; \tau \text{ and } \sigma_n \text{ in } kPa \qquad [3.27]$$

We notice that we find the same value of the exponent as the one already obtained in section 3.1.3.3.

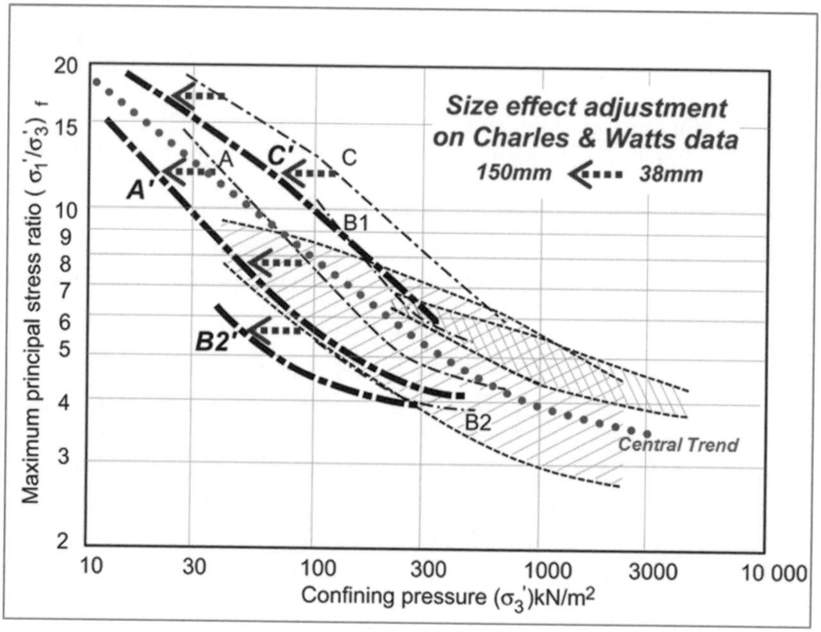

Figure 3.12. *Evidence of scale effects in Charles and Watts compilation*

3.2.2.3. *The scale effect rule on the shear strength envelope*

3.2.2.3.1. General case

In the above analysis, reasoning is based on the "central trend distribution" where grain crushing resistances follow Weibull's theory with a parameter m = 6. Therefore, the failure stresses vary with the grain diameter, according to a power law with an exponent $3/m$ = -1/2 (see section 3.2.1.3).

All the data reported by [MAR 72] and others on grain breakage show a scattering in the values of parameter m for the different materials. This scattering can be attributed to differences in the defect distribution within the grain material, according to the nature and origin of these materials. A direct use of parameter m determined on each material appears to improve the representation of particle breakage statistics. So, the analysis presented above can be reconsidered with the use of $-3/m$ instead of 1/2 in the effect of the characteristic grain diameter, which leads to the following result that may be called the "scale effect rule on shear strength envelope".

Let us consider two granular materials coming from the same homogeneous mineral stock, compacted at the same density, with homothetic grain size distributions, G_0 and G_1, and characteristic diameters, D_0 and D_1 (for example D_{Max}). The two materials are, then, geometrically similar in a ratio D_1/D_0:

– In order to mobilize the same internal friction within the two materials, the maximum dilatancy rate has to be the same during shearing, so the amount of grain breakage, or the probability of survival, also has to be the same, which means that the stresses applied to the grains must verify the following relation of similitude:

$$\{ \sigma_{grains} (G_1) \}= \{\sigma_{grains}(G_0) \} \times (D_1/D_0)^{-3/m} \quad [3.28]$$

The link between macroscopic stresses and stresses applied on the grains being enforced by the geometrical similitude of the two materials, the macroscopic stress states, necessary for mobilizing the same internal friction, must verify the second relation of similitude, which is identical to the first one:

$$\{\sigma(G_1) \}= \{\sigma(G_0) \} \times (D_1/D_0)^{-3/m} \quad [3.29]$$

Equation [3.29] represents the scale effect rule which generalizes equation [3.26][2].

In this expression, the scale effect rule resulting from grain breakage in Mode 1 is not fundamentally linked to a particular expression of the shear strength envelope. Thus, it can be applied for different criteria, such as De Mello's or Hoek and Brown's criteria.

3.2.2.3.2. Shear strength envelope $\tau = \mathbf{f}(\sigma_n, D)$: De Mello's criterion

If the expression of the shear strength envelope of material G_0 is given by

$$\tau_{G0} = f(\sigma_n, D_{Max}) \quad [3.30]$$

then, the shear strength envelope of material G_1 from the same mineral stock, having a homothetic grain size distribution and the same density, is given by

$$\tau_{G1} = \left(\frac{D_1}{D_0}\right)^{\frac{-3}{m}} \cdot f\left\{\sigma_n \cdot \left(\frac{D_1}{D_0}\right)^{\frac{3}{m}}, D_0\right\} \quad [3.31]$$

2 The result found here from a reasoning at the macroscale can also be demonstrated by a reasoning at the microscale between two homothetic granular media by using the relations that give the macroscopic stress function of the intergranular forces and the geometry of granular arrangement (Weber's relationship).

In the particular case of a power law $\tau(\sigma_n) = A \,.\, \sigma_n{}^{b}$ with $b < 1$ (De Mello's criterion [DEM 77]), the scale effect rule leads to the following expression of the shear strength envelope:

$$\tau_{G_1}(\sigma_n) = \left(\frac{D_1}{D_0}\right)^{\frac{-3(1-b)}{m}} \cdot \tau_{G_0}(\sigma_n) = A_{G_0} \cdot \left(\frac{D_1}{D_0}\right)^{\frac{-3(1-b)}{m}} \cdot \sigma_n{}^{b_{G_0}} \qquad [3.32]$$

It shall be outlined that, in this particular condition of a power law, the scale effect incidence is only on coefficient A, but not exponent b.

3.2.2.3.3. Shear strength envelope $\sigma_1 = \mathbf{h}(\sigma_3, D)$ – Hoek-Brown's criterion

If the expression of the shear strength envelope of material G_0 is given by

$$\sigma_{1\,(G_0)} = \boldsymbol{h}(\sigma_3, D_{Max}) \qquad [3.33]$$

then, for the same characteristics of material G_1 as the ones given above, combining equations [3.29] and [3.33] gives the expression of the shear strength envelope of material G_1:

$$\sigma_{1\,(G_1)} = \left(\frac{D_1}{D_0}\right)^{\frac{-3}{m}} \cdot \boldsymbol{h}\left\{\sigma_3 \cdot \left(\frac{D_1}{D_0}\right)^{\frac{3}{m}}, D_0\right\} \qquad [3.34]$$

In the case of Hoek-Brown's criterion [HOE 80]:

$$\sigma_1 = \sigma_3 + \sigma_c \cdot \left[m_{HB} \cdot \left(\frac{\sigma_3}{\sigma_c}\right) + s\right]^{\boldsymbol{a}} \text{ with } \boldsymbol{a} < 1,$$

The scale effect rule leads to the following expression of the shear strength envelope:

$$\sigma_1(G_1) = \sigma_3 + \sigma_c(G_0) \cdot \left(\frac{D_1}{D_0}\right)^{\frac{-3}{m}} \cdot \left[m_{HB}(G_0) \cdot \left(\frac{\sigma_3}{\sigma_c(G_0)}\right) \cdot \left(\frac{D_1}{D_0}\right)^{\frac{3}{m}} + s(G_0)\right]^{\boldsymbol{a}(G_0)} \qquad [3.35]$$

The influence of the scale effect, in this case, affects only parameter σ_c, but not the other parameters m_{HB}, S, and $\boldsymbol{a}$.

3.2.2.3.4. Link between grain breakage and curvature of the shear strength envelope: linear envelope

These results raise the question of the influence of the scale effect on a linear shear strength envelope, such as the Mohr-Coulomb criterion. An interesting answer can be obtained by making the values of b and $\boldsymbol{a}$ converge towards 1 in equations [3.32] and [3.35], where these two exponents control the curvature of the envelope. In both cases, the envelope becomes linear; the scale effect disappears from the friction angle and only affects the value of the pseudo-cohesion[3].

The integration of the scale effect in the expression of the shear strength envelope allows us to analyze its incidence on the stability of engineering construction, in a second step, and to take it into account in the design and optimization of embankment works.

3.3. Practical applications to construction design

3.3.1. *A new method for rational assessment of rockfill shear strength envelope*

Rockfill materials are seldom tested to determine their shear strength (see for instance ICOLD (International Commision on Large Dams) [ICO 02]) because of the size of the apparatus required for this purpose and the corresponding costs and duration of the tests (for a 0–250 mm rockfill, a representative sample for a triaxial test should measure 1.5 m in diameter and 3 m in height, and weigh approximately 3 tons!) The construction trend towards higher rockfill dams with steeper slopes requires better knowledge of their shear strength for stability assessment. In this situation, the above results on scale effects give access to a new method by which to assess the shear strength of a given rockfill. The procedure for applying this method can be divided into four steps, including (see Figure 3.13) classical laboratory testing:

– preparation of a “reduced” material that is geometrically similar to the rockfill, coming from the same mineral stock, but with smaller grains limited to gravel size;

– realization of a series of normalized crushing tests (Protodiakonov or Franklin test) on grains of different sizes in order to obtain a Weibull’s distribution of the statistics of grain crushing strength;

3 This result can also be found directly by applying equation [3.31] to a linear expression of the shear strength envelope.

– realization of a series of shear tests (triaxial or others) on the homothetically-reduced samples with the same density in order to obtain their shear strength envelope;

– application of the scale effect rule with Weibull parameter m, measured from the crushing tests to assess the rockfill shear strength envelope from the envelope measured on the "reduced" gravel material.

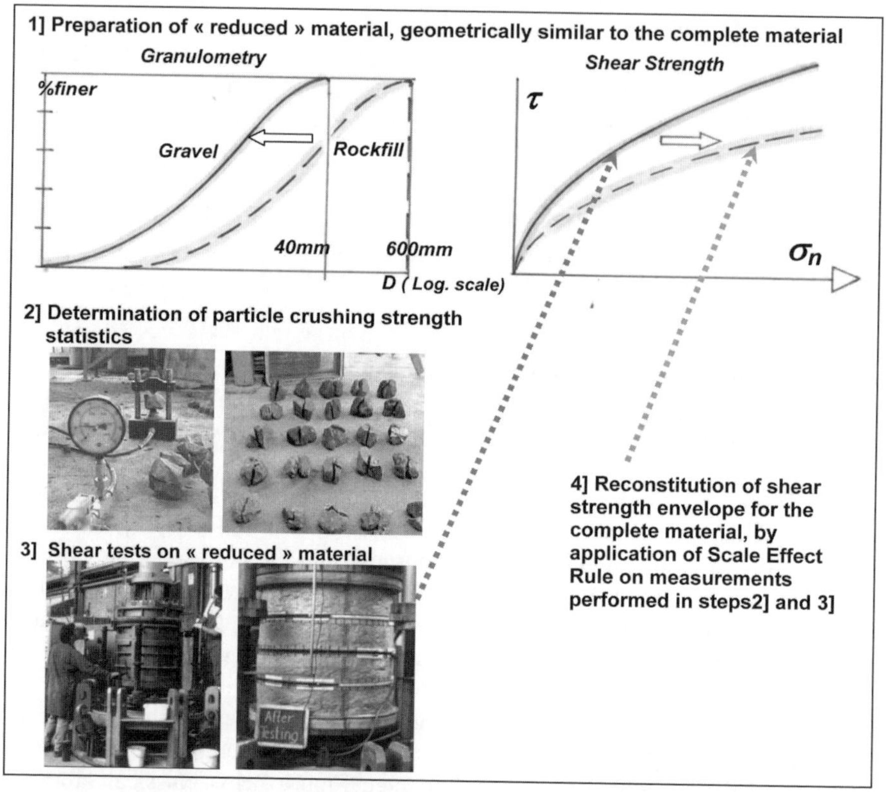

Figure 3.13. *New method for rockfill shear strength assessment*

Cautiously limited to a reduction factor of D_1/D_2 = 15 here, this method is summarized in Figure 3.13. It allows us to predict the shear strength envelope of coarse materials with grain sizes of up to D_{Max}= 600 mm, which can be considered representative of most practical situations, from triaxial test results on gravels with maximum grain sizes up to D_{Max}= 40mm, which can be tested in 250 mm triaxial cells.

This procedure, based on the new physical similarity rule resulting from laws of fracture mechanics governing grain breakage, allows us to reduce the material mass of a representative sample by a factor of $(1/15)^3 \approx 1/3{,}400$.

In order to cover the range of usual stresses $\{\sigma(G_1)\}$ applied on the rockfill within the embankment, the range of stresses $\{\sigma(G_2)\}$ applied on the similarly reduced material in the laboratory must be extended up to the value given by eqiuation [3.29], i.e. ≈ 4 times $\{\sigma(G_1)\}$ if we consider the central trend of the statistic distribution, and a reduction factor equal to 15.

3.3.2. *Incidence of scale effect on rockfill slope stability*

3.3.2.1. *The question of stability assessment*

In recent constructions of large embankments, and especially rockfill dams, the economic interests have directed the design practices towards:

– coarser grain size distributions in order to minimize the quarrying costs;

– steeper slopes in order to minimize the required volumes of material (or at least slopes identical to the ones of mid-sized embankments).

The stability of such constructions is mostly estimated by extrapolating the values of their shear strength from typical values for granular materials, without real measurements for reasons presented above (see section 3.3.1). Therefore, the impact of such design practices – coarser grain sizes, steeper slopes and higher embankments – deserves to be analyzed.

Figure 3.14 summarizes the question arising from these combined effects: starting from the safety factor of a mid-sized embankment of height H_0 and slope β_0, built with a rockfill material of grain size distribution $G(D_0)$, can we determine the safety factor of an embankment having a higher height H and steeper slope β, built with a coarser granular material $G(D)$?

Such a method, if available, would be a useful alternative to the design extrapolation practices described above. Provided that the materials have geometrically similar grain size distributions, come from the same mineral stock, and are prepared at the same density, the scale effects analyzed above bring the answer, as far as the shear strength is concerned, which is the key element of a slope stability analysis.

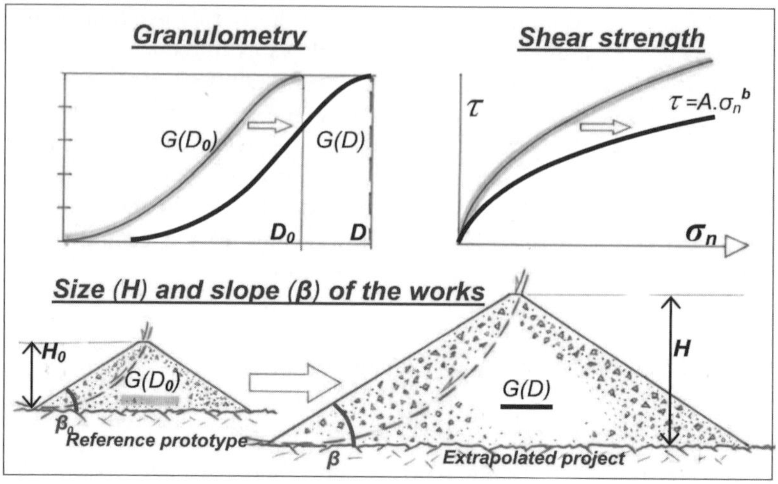

Figure 3.14. *Combined impacts on rockfill embankments safety factors, of size effects and other key parameters: rockfill granulometric size, embankment height, and slopes*

The results published by Charles and Soares [CHA 84] on rockfill slope stability using Bishop's method for shallow circular failure lines including a power law shear strength relation can be combined with our results on scale effects in order to establish such a method.

3.3.2.2. *Explicit scale effect in safety factors*

Charles and Soares have demonstrated that the mathematical structure of the static equilibrium equations leads to the following expression of the safety factor, defined as the ratio between the available shear strength and the shear stresses mobilized at equilibrium along the potential failure line:

$$F_S = \frac{A.\Gamma_{(b,\beta)}}{(\gamma.H)^{(1-b)}} \qquad [3.36]$$

This expression put in separate factors, height *H*, coefficient *A* of the power law expression of the shear strength envelope (see Figure 3.14) and an adimensional stability coefficient Γ, depending on the slope β and on exponent *b* of the power law.

Furthermore, the computed values of coefficient Γ appear to be almost linearly dependent on the slope coefficient $cot\beta$, as shown in Charles and Soares' diagram presented in Figure 3.15, at least for usual slopes between 1.2 h/1v and 2.0 h/1v (outlined zone in the diagram).

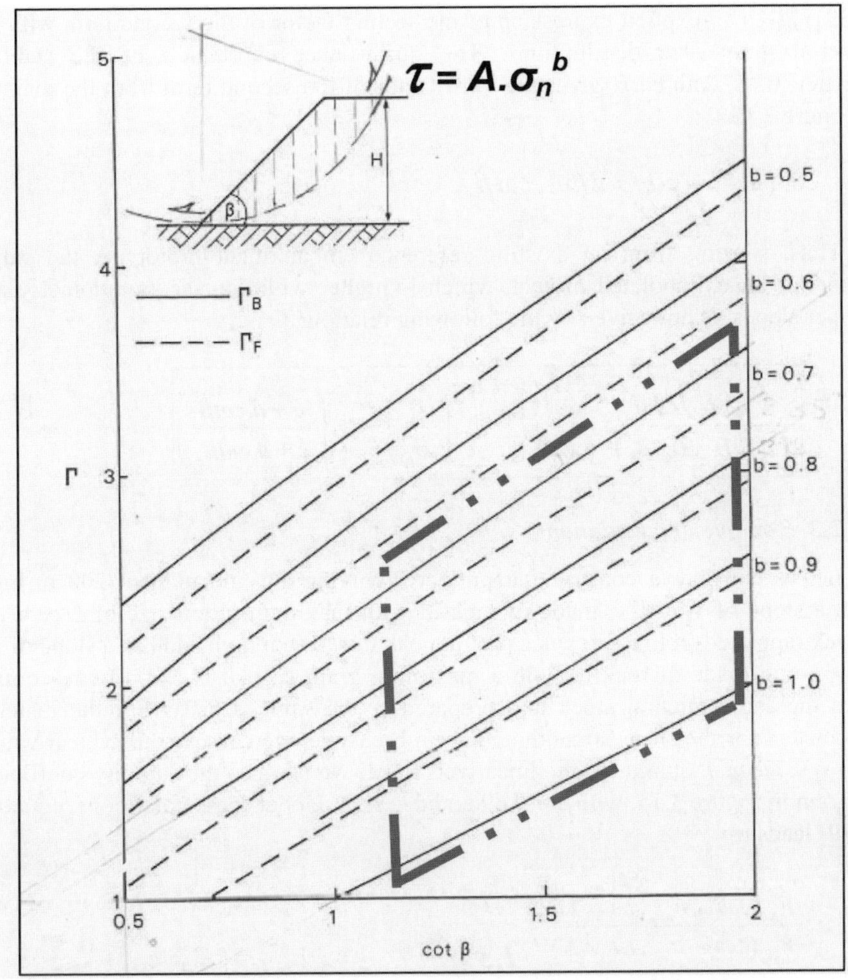

Figure 3.15. *Stability coefficient for rockfill slopes (from Charles & Soares 1984)*

As a consequence, the safety factor for a projected slope compared to a reference prototype is given by the relation:

$$\frac{F_S}{F_{S_0}} = \left(\frac{H}{H_0}\right)^{(b-1)} \times \frac{A}{A_0} \times \frac{\Gamma(b,\beta)}{\Gamma(b_0,\beta_0)} \tag{3.37}$$

The ratio in the first term of the equation directly gives the scale effect due to the embankment height. Our scale effect relationship on the shear strength (equation

[3.32]) gives an explicit expression of the second factor of the second term with the material grain size distributions. The quasi-linear expression of the stability coefficient, Γ, with $cot\beta$, gives the third factor of the second term from the diagram in Figure 3.15:

$$\Gamma(b,\beta) \approx c(b) + d(b) \, . \, cot\beta \qquad [3.38]$$

Thus, starting from an existing reference embankment prototype, the safety factor for an extrapolated project, which is higher with coarser granulometry and steeper slopes, is now given by the following relation:

$$\frac{F_S(H,D,\beta)}{F_{S_0}(H_0,D_0,\beta_0)} \approx \left(\frac{H}{H_0}\right)^{(b-1)} \bullet \left(\frac{D}{D_0}\right)^{\frac{3(b-1)}{m}} \bullet \left(\frac{c+d.cot\beta}{c+d.cot\beta_0}\right) \qquad [3.39]$$

3.3.2.3. *Example of calculation*

Let us consider a construction project involving an embankment 200 m high, with a slope of 1.3 h/1v, made of rockfill with maximum grain size of D_{Max}= 100 cm, extrapolated from a reference prototype that is 100 m high and has a slope of 1.4 h/1v, being made of rockfill with a maximum grain size of D_{Max}= 40 cm, coming from the same mineral stock and prepared at the same density. For the material parameters for the shear strength and grain breakage, we can use our typical values of $b = 0.77$, $m = 6$; and for the linearized slope, we can use the stability coefficient diagram in Figure 3.15, with c = 0.69 and d = 1.28. Under these conditions, equation [3.39] leads to:

$$\frac{F_S(200\,m, D_{100cm}, 1.3_h/1_v)}{F_{S_*}(100\,m, D_{40cm}, 1.4_h/1_v)} \approx \frac{1}{1.37} \qquad [3.40]$$

We can see that the safety factor of the extrapolated project is strongly reduced, when compared to the one obtained for the prototype.

To go further in the analysis, some orders of magnitude may be determined by using the central trend of the shear strengths measured for rockfills with D_{Max} = 150 mm (see section 3.2.2.2) together with the corrections in grain size distribution given by the scale effect rule:

– for shear strength, the central trend for:

$$D_{MAX} = 15\text{cm:} \quad \tau \approx 4\sigma_n^{0.77} \quad \text{(KPa)} \qquad [3.41a]$$

– with correction of grain sizes, central trend for D_{MAX}= 40 cm:

$$\tau \approx 3.57.\ \sigma_n^{0.77} \ \text{(KPa)} \qquad [3.41b]$$

– with Charles and Soares' diagram for stability coefficient Γ, taking into account common values of unit weight γ:

$$\Gamma(0.77\ ;\ 1.4\ \text{h/1v}) \approx 2.46 \qquad \gamma \approx 22\ \text{KN/m}^3$$

– the safety factors corresponding to the end of the construction without water effect are determined for the considered embankment works:

$$[3.42]$$

$$F_{S_0}\left(100m, D_{40cm}, 1.4\ h/1V\right) = \frac{A.\Gamma(b,\beta)}{(\gamma.H)^{(1\text{-}b)}} = \frac{3.57x2.46}{(22x100)^{0,23}} = \mathbf{1.50}$$

$$F_S\left(200m, D_{100cm}, 1.3\ h/1V\right) = \frac{1.50}{1.37} = \mathbf{1.10}$$

The safety factor found for the 100 m high prototype has exactly the usually recommended value (1.5) for this kind of construction. Conversely, due to the combined influences of scale and slope effects, the value of 1.1 found for the extrapolated 200 m high embankment project appears to be far too small.

3.3.2.4. *Scale effect compensation*

Used in a reverse way (see Figure 3.16), the above relations for the safety factors may also be used to determine the constraints on grain size distributions and slopes in the construction projects in order to maintain the value of the safety factor at a given level.

As shown in Figure 3.16, the extrapolation to a 200 m high embankment project starting from our reference prototype with a height of 100 m and a slope of 1.4 h/1v built with rockfill with a maximum grain size equal to 400 mm leads to the following results:

– the 1.3 h/1v slope is definitely too steep, unless the maximum grain size is significantly reduced;

– if the initial maximum grain size equal to 400 mm is kept, the slope of the 200 m high embankment should be decreased to a value close to 1.73 h/1v in order to keep a safety factor value equal to 1.5;

– should a coarser rockfill be chosen, for example with D_{max} = 1,000 mm, then the slope should be flattened further.

This whole set of results gives a detailed analytical view of the influence of scale effects on the stability and design of a rockfill embankment, which confirms and clarifies the results of some previous numerical studies [TEI 99].

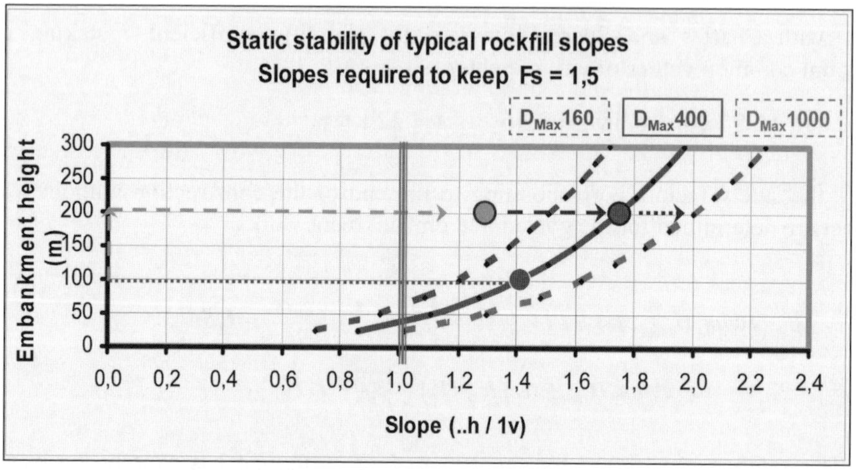

Figure 3.16. *Compensation of scale effects resulting from embankment heights and rockfill size*

3.3.2.5. *Case of small rockfill embankments*

Another consequence of the scale effect is shown in Figure 3.16. For small embankments ($H < 50$ to 60 m), the diagram shows that steeper slopes can be acceptable if the material density is kept constant by means of a confinement system. This is the case for many old small-to-medium-sized rockfill dams built at the beginning of the hydropower development. These dams have a particularly steep upstream slope, confined by hand-placed dry masonry. A remarkable example of such engineering work is the Ghrib Dam in Algeria (built in 1934, which is 65 m high and still in operation).

3.3.3. *Scale effects on deformation features*

3.3.3.1. *Scale effects on strains corresponding to service and exceptional loads*

After having evaluated the scale effects on the equilibrium condition between applied forces and shear strength that can be mobilized within the embankment work and expressed by the safety factor, we can also wonder what such effects may have upon strains. The above section highlighted the fact that the safety margin is certainly lower in high rockfill dams if the slope adopted for lower dams has been kept the same or made steeper. The principle of a large enough safety factor (1.5 for the usual loads) is also to keep strains within a reasonable range. For the same rockfill embankment, the smaller the safety factor is, the larger the strains will be in order to mobilize the resisting forces needed to reach equilibrium.

Keeping the example analyzed above, let us consider two corresponding points: one in the prototype, the other in the extrapolated project (see Figure 3.17). We can calculate the deformations needed to mobilize the strength required for the static equilibrium with safety factors of 1.5 for the prototype and 1.1 for the extrapolated project. From the stress–strain curves measured from large dimension testing on rockfills of different maximum grain sizes [MAR 69], typical stress–strain curves can be drawn, including the scale effect on the grain size distribution and the confining stress influence on the maximum strength and on the strain amplitude at peak.

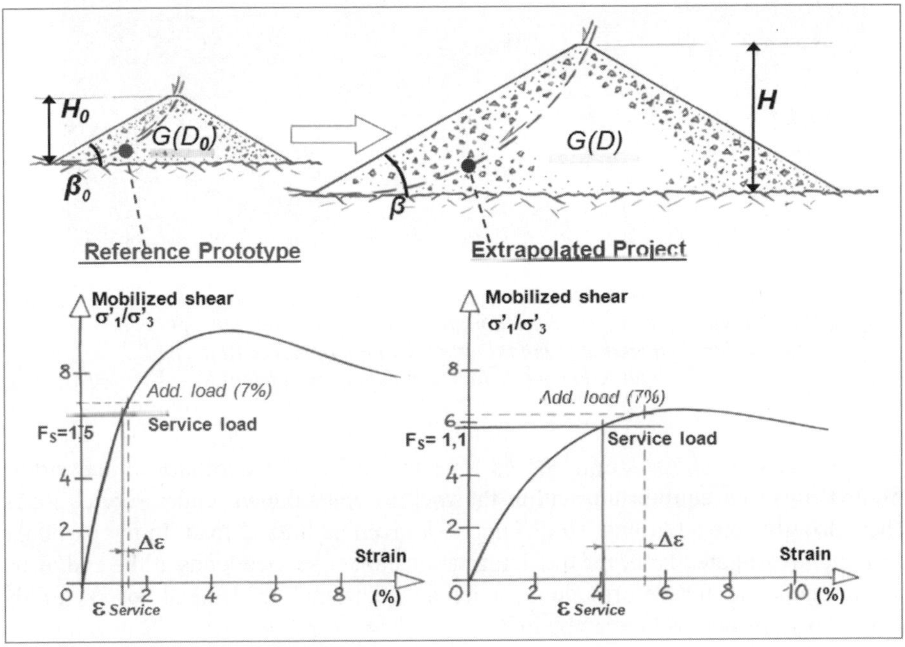

Figure 3.17. *Scale effects on strains*

In our extrapolated project, the coarser material subjected to higher confining stresses will exhibit a softer mechanical behavior than our prototype, due to a larger amount of particle breakage.

Figure 3.18. *Campos Novos rockfill dam in Brazil (height 202 m, volume 13 hm3 of rockfill): a) general view at end of construction; b) view of upstream watertight concrete facing damaged at impounding; c) detail of damages (failure of reinforced concrete facing slab in compression)*

These typical stress–strain curves give the order of magnitude of the strains required to reach equilibrium within the rockfill embankment under service loads. They also give the additional strains due to a given additional load. In the particular example investigated here, for the deformations under service loads at the end of the construction and for deformations under an additional load equal to 7% of the service load, respectively, we obtain

$$\frac{\varepsilon_{\text{Service}}(\text{Project})}{\varepsilon_{\text{Service}}(\text{Prototype})} \approx \mathbf{3} \quad \text{and} \quad \frac{\Delta\varepsilon_{\text{Add. load}}(\text{Project})}{\Delta\varepsilon_{\text{Add. load}}(\text{Prototype})} \approx \mathbf{6} \qquad [3.43]$$

These orders of magnitude show that large strains shall be expected in our extrapolated project compared to those in the prototype under service loads, and even more so, in case of a given extra load.

These results can explain the accident that occurred in the Campos Novos Dam in Brazil when it entered into operation in 2006. The main characteristics of this dam correspond to the ones used in the above example for the extrapolated project: $H = 202$ m, slope 1.3 /1, $D_{Max} = 1$ m. The problem involved a set of fractures by compression that appeared in the watertight upstream concrete slab during impounding, and led to a considerable increase in water flow through the dam body of up to 1,450 l/s. These fractures were discovered after an accidental emptying of the reservoir (see Figure 3.18). They were due to excessive deformations of the material constituting the dam body under service loads incompatible with the rigidity of the concrete slab. Other accidents of this type have occured recently in Brazil, but also in China, Lesotho, etc. (see [PIN 07]).

3.3.3.2. *Scale effects in rockfill apparent rigidity modulus*

The deformability of embankment works made of granular materials, such as rockfill dams, is a key parameter deserving careful monitoring during construction. As they explain most of the compressibility of rockfill under increasing loads, the laws governing grain breakage are applicable here. The stiffness – *or rigidity modulus* – of granular materials coming from the same mineral stock, with similar grain size distributions and the same density, should be inversely proportional to the square root of D_{Max} or of any other characteristic size, $D_{X\%}$.

Figure 3.19 illustrates the statistical correlation proposed by Hunter and Fell [HUN 03, FEL 05] between the apparent rigidity modulus during construction and the coarse fraction characteristics D_{80}. It results from a detailed analysis of data coming from a set of 35 well-documented rockfill dams. The dotted lines added in the diagram correspond to the trend in behavior obtained by the grain breakage laws. They demonstrate that

– the scale effect due to grain breakage appears to explain most of the negative correlation obtained by Hunter and Fell between the apparent rigidity modulus and the coarse grading fraction;

– for rockfills with large blocks (right-hand side of the diagram), the decay of the modulus with block size appears to be more pronounced than predicted by the scale effect. This may be attributed to a loss of efficiency in the compaction techniques when dealing with large blocks of rock.

If in this figure we plot the points corresponding to four high dams having suffered from excessive deformations leading to significant problems during the impounding phase (Barra-Grande, 185 m and Campos-Novos, 202 m in Brazil, Mohale, 145m in Lesotho and Tian Sheng Qiao I, 179 m in China), we can observe that they are located to the far right of the diagram. They were all built with very coarse rockfills ($D_{80} > 800$ mm) and their particularly small rigidity modulus, evaluated through back calculations on strains measured during operation to lie between 30 and 45 MPa, correspond fairly well with their location in the diagram.

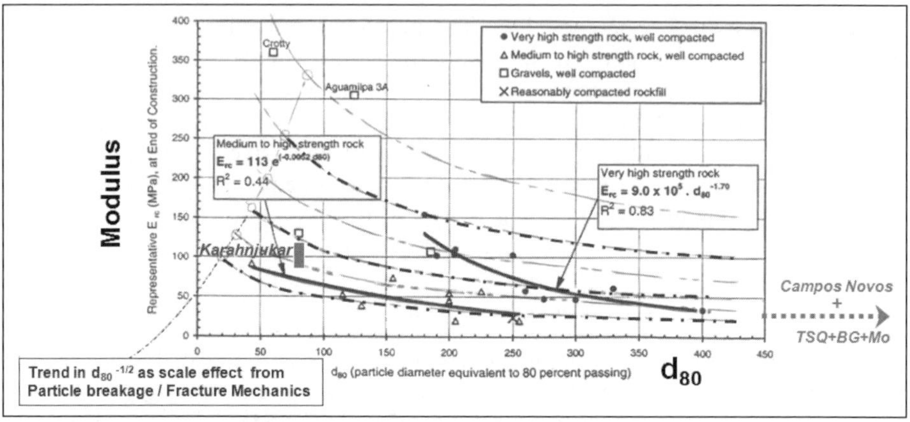

Figure 3.19. *Scale effects in rockfill Rigidity Modulus at end of construction – correlation with grain size (from Hunter & Fell, 2003)*

As a counter-example, we can cite the Karahnjukar Dam in Iceland at a height of 198 m which did not suffer any excessive deformations during the commissioning phase. This can be explained by its characteristics, corresponding to a well compacted granular material with D_{80} < 80 mm, which induced a higher rigidity modulus value measured between 80 and 120 MPa, in fairly good correspondence with the trend displayed in Figure 3.19.

Figure 3.20, published by Johannesson in 2007, gives a practical correspondence between the apparent rigidity modulus measured in various rockfill dams and layer thickness during construction, [JOH 07]. The layer thickness usually being chosen to equal the size of the biggest blocks, the scale effect due to grain breakage may also play a role in this diagram. If we now consider the thickness of the layer rather than the value of D_{Max}, the corresponding trend lines display a similar pattern to those in Figure 3.19, leading to the same observations:

– most of the negative correlation pointed out by Johannesson between the apparent rigidity modulus and the layer thickness can be explained by the scale effect;

– for materials set in place in very thick layers (right-hand side of the diagram), the decay of the rigidity with the increase of layer thickness appears to be more pronounced than predicted by the scale effect rule. This difference can be attributed to the loss of efficiency of the usual compacting techniques when the layer thickness becomes too big.

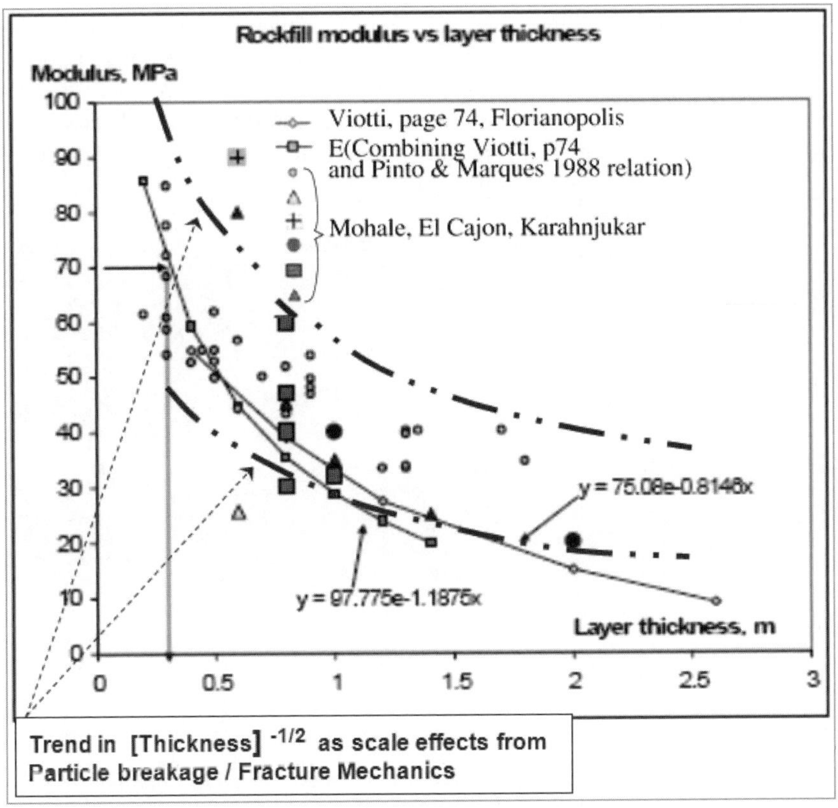

Figure 3.20. *Scale effects in rockfill Rigidity Modulus at end of construction – correlation with layer thickness (from Johannesson, 2007)*

3.3.3.3. *Scale effect on micromechanims inducing settlements – application to rockfill dams*

3.3.3.3.1. Micromecanisms inducing settlement

Several rockfill dams of the CFRD (concrete faced rockfill dams) type have presented disorders at impounding, such as excessive opening or shear displacements at the perimetric joint, especially on steep sloped abutments. The roots of these problems are settlements within the granular fill, which can be schematically explained as follows. In a typical section, the chains of major forces resulting from both impounding and self-weight follow the CC line in Figure 3.21. Under the forces exerted by impounding, settlements resulting from small local shear movements are distributed within the granular assembly, triggered by local stone or block breakage. In a right-to-left bank section, transverse to the valley axis and passing through the CC line, these small shear movements are distributed within the rockfill mass, with mostly random orientation but with some polarization along two characteristic directions. Those shear movements are predominantly clockwise in the granular mass above the left abutment and predominantly anticlockwise above the right abutment. They are in mixed directions in the central part of the valley.

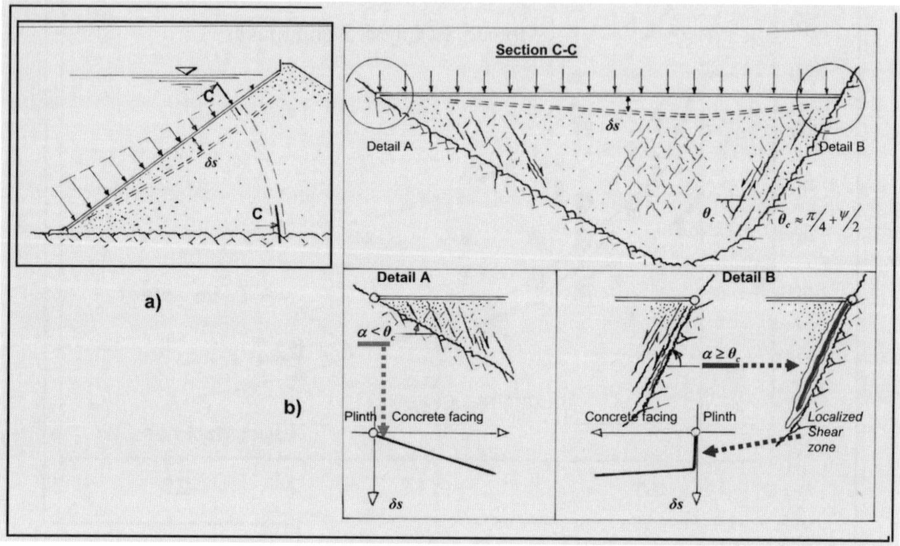

Figure 3.21. *Settlement micromechanisms in concrete faced rockfill dam body*

In the vicinity of the perimetric joint:

– if the abutment slope is sufficiently flat (detail A in Figure 3.21b), the slip lines resulting from the shear movements are intercepted at short distance by the foundation. Therefore, the associated shearing cannot extend over a long distance and, as a consequence, the deflection line of the concrete slab near the perimetric joint will be regular and progressive;

– if the abutment slope is very steep (detail B in Figure 3.21b), the slip lines are no longer intercepted at a short distance by the foundation; they can extend over a long distance and, therefore, develop a localized shear band at the point of contact with the abutment. The result is a concentrated differential settlement of the concrete slab relative to the plinth, localized at the perimetric joint and resulting in a step in the deflexion line of the slab, right at the perimetric joint.

For usual values of friction between the blocks in the rockfill, the corresponding critical abutment slope is about 60° to 65° (without a safety margin). Mitigation measures for steep abutments can be either to provide a flatter slope by excavations or to build a zone of low-compressibility fill at the point of contact with the steep abutment.

3.3.3.3.2. Scale effect on the horizontal contraction under the upstream slab

Another effect of the micromechanical settlements is the horizontal contractive strains resulting from the wedging of the dam body between the abutments. Horizontal contractive strains are induced by friction in the upstream concrete slab, which may reach failure, as in the recent cases in Brazil, Lesotho, and China (section 3.3.3.2).

For steep abutments inducing significant shearing at contact, the order of magnitude of these shear strains at mid-height can be estimated from a basic 3D analysis of the kinematics. Under pressure loading on the upstream face, the dam will settle perpendicularly to this face and the central part will be confined between the two converging abutments. The key parameter controlling the amplitude of this phenomenon is the apparent rigidity modulus of the dam during reservoir filling. This modulus can be related to the apparent rigidity modulus during construction by use of statistical correlations [HUN 03]. The resulting contractive strains are easy to estimate at mid-height, if we assume that initially horizontal directions in the dam remain horizontal during straining.

This leads to the practical relation in Figure 3.22a, which links these contractive strains to two adimensional variables:

– a dam deformability ratio, expressed here as a function of the apparent rigidity modulus during construction, which is the only parameter usually known before impounding;

– a valley shape ratio, expressed here as a function of the geometrical characteristics of a vertical middle section of the valley, schematically represented by a trapezoid.

Knowing that reinforced concrete reaches failure during uniaxial compression at strain amplitude $\varepsilon \sim 0.2\%$, the relation in Figure 3.22a allows us to construct a diagram giving the risk of failure of the concrete slab. In this diagram, we have plotted the corresponding locations of the characteristics of high dams having recently suffered this type of damage (Barra-Grande, H = 185 m, and Campos-Novos, H = 202 m, in Brazil; and Mohale, H = 145 m in Lesotho), as well as those of dams without damage (Karahnjukar, H = 198 m, in Iceland; and El Cajon, H = 189 m, in Mexico). The corresponding points are in fairly good agreement with the observed behavior of these dams, which underlines the interest of such a diagram.

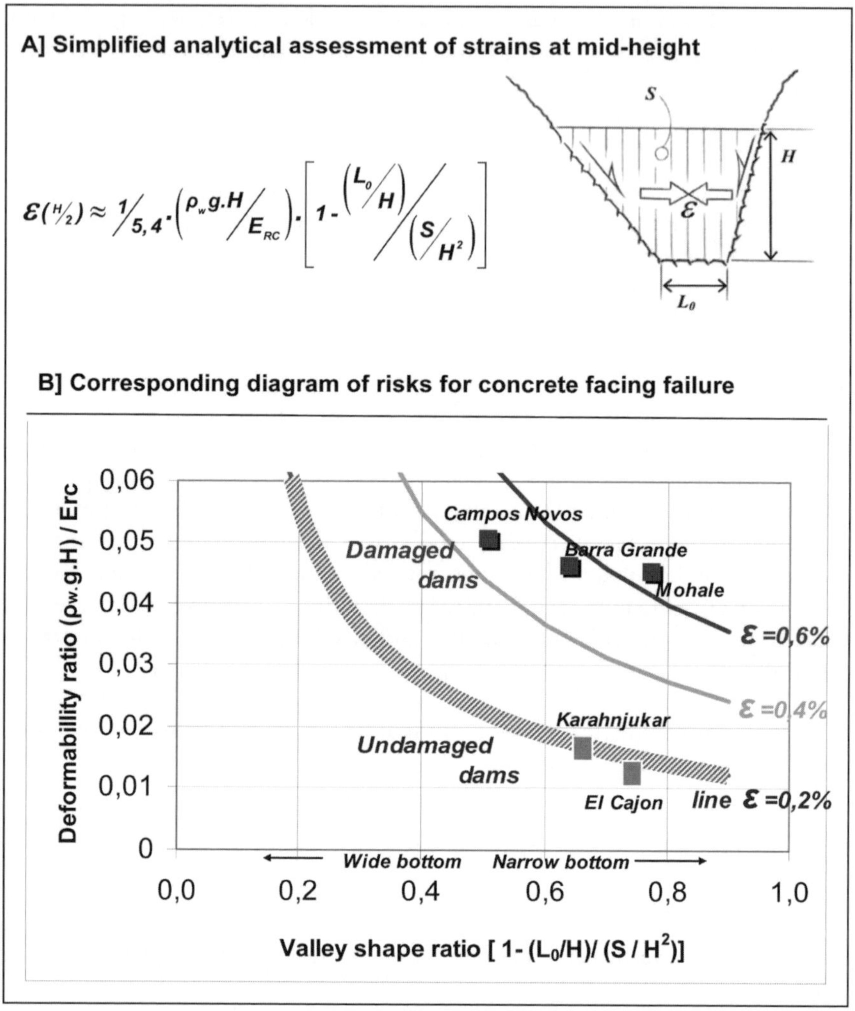

$$\varepsilon(H/2) \approx \frac{1}{5,4} \cdot \left(\frac{\rho_w . g . H}{E_{RC}} \right) \cdot \left[1 - \frac{\left(L_0 / H \right)}{\left(S / H^2 \right)} \right]$$

Figure 3.22. *Scale effects in horizontal compression strains under concrete facing*

3.4. Conclusions

A better understanding of the behavior of granular materials through a physical multiscale approach appears to be very fruitful for the rational use of these materials in civil engineering works. Local phenomena within the material, such as friction at grain contact and grain breakage, are integrated, first, at the scale of granular assembly and, second, at the scale of the whole engineering work. This approach allows us to solve difficult problems, such as the development of shear bands and the description of their internal structure, leading to the formation of failure lines. This astonishing physics of granular assemblies opens new fields of research and has already produced many results that are useful for practical applications. This is particularly the case for the material scale effects and their implication on structural stability in terms of both static equilibrium and the kinetics of deformation, as observed in major civil engineering works.

Several key elements of practical applications have been highlighted in this chapter, such as the central tendency in the shear strength (section 3.2.2.2), that deserve wider investigation in order to enlarge the data bank, to confirm the statistical tendencies and to determine their limits. This is one of the objectives of the joint ECHO (scale effects in civil engineering embankment works) research project that is currently in progress and is financed by the ANR (the French National Research Agency).

These results also show the major relevance of introducing a better representation of the physics of these materials within constitutive models. In the present study, these results were obtained through a multiscale approach starting from the local scale rather than through the usual macroscopic phenomenological approaches. The latter have produced numerous constitutive models that often contain multiple parameters and sometimes lack clear ground.

Along this path, we expect to be able to develop simple, clear and efficient representations of some sort of "ideal granular material", with a small number of physical parameters. Such a representation would be a continuation of the typical models of standard material behaviors in engineering sciences, which have been proved to be very useful in hydraulics, the physics of gas, structural analysis, etc.

The sequence of results obtained on the energy dissipation by friction (section 3.1), grain breakage (section 3.2), and their practical applications (section 3.3) demonstrate that such a representation is now possible.

3.5. Bibliography

[BAR 81] BARTON N., KJAERNSLI B., "Shear strength of rockfill", *J. Geotechn. Eng. Div.*, vol. 107, no. Gt 7, pp. 873-891, 1981.

[BOL 98] BOLTON M.R., MCDOWELL G.R., "On the micromechanics of crushable aggregates", *Géotechnique*, vol. 48, no. 5, pp. 667-679, 1998.

[CHA 80] CHARLES J.A., WATTS K.S., "The influence of confining pressure on the shear strength of compacted rockfill", *Géotechnique*, vol. 30, no. 4, pp. 353-367, 1980.

[CHA 84] CHARLES J.A., SOARES M.M., "Stability of compacted rockfill slopes", *Géotechnique*, vol. 34, no. 1, pp. 61-70, 1984. (see also CHARLES J.A., "Laboratory shear strength tests and the stability of rockfill slopes", in: MARANHA DAS NEVES E. (ed.), *Advances in Rockfill Structures,* Kluwer Publishing, pp. 53-72, 1990).

[DEM 77] DE MELLO V., "Reflections on design decisions of practical significance to embankment dams", *Géotechnique*, vol. 27, pp. 281-355, 1977.

[DES 96] DESRUES J., CHAMBON R., MOKNI M., MAZEROLLE F., "Void ratio evolution inside shear bands in triaxial sand specimens studied by computed tomography", *Géotechnique*, vol. 46, no. 3, pp. 529-546, 1996.

[DUN 04] DUNCAN J.M., "Friction angles for sand, gravel and rockfill", *Lecture at the Kenneth Lee Memorial Seminar*, Long Beach, California, USA, April 28, 2004.

[FEL 05] FELL R., MC GREGOR P., STAPLEDON D., BELL G., *Geotechnical Engineering of Dams*, pp. 589-662, Baalkema Publishing, Stuttgart, Germany, 2005.

[FRO 79] FROSSARD E., "Effect of sand grain shape on interparticle friction", *Géotechnique*, vol. 29, no. 3, pp. 341-350, 1979.

[FRO 83] FROSSARD E., "Une règle d'écoulement simple pour les milieux granulaires", *Géotechnique*, vol. 33, no. 1, pp. 21-29, 1983.

[FRO 86] FROSSARD E., "Dilatance, dissipation d'énergie et critère de rupture dans les milieux granulaires", *Revue Française de Géotechnique*, vol. 34, pp. 17-31, 1986.

[FRO 01] FROSSARD E., "L'approche énergétique en mécanique des milieux granulaires", *Poudres & Grains*, vol. NS2, pp. 1-56, 2001 (see also FROSSARD E., "A micro-mechanical constitutive law resulting from energy dissipation by friction", *Proc. V° Int. Conf. Powders Grains*, Baalkema Publishing, Stuttgart, Germany, vol. 1, pp. 141-145, July 2005.

[FRO 04] FROSSARD E., "Localisation des déformations et chaînons de contraintes en mécanique des milieux granulaires", *Poudres & Grains*, vol. NS3, pp. 1-119, 2004.

[FRO 05] FROSSARD E., Comportement macroscopique des matériaux granulaires mis en œuvre dans les barrages, Report, Coyne & Bellier pour le Projet de Recherches Micromécanique des Barrages en Enrochements (Projet MICROBE-cofinancé par le MRT), Paris, France, 2005.

[FRO 09] FROSSARD E., "On the structural safety of large rockfill dams", *Proc XXIII° Intl. Congress on Large Dams*, Q. 91-R.39, Brasilia, Brazil, May 2009.

[HOE 80] HOEK E., BROWN E.T., *Underground Excavations in Rock*, p. 527, Ed. Instn. Min. Metall., London, 1980 (see also HOEK E., BROWN E.T., "The Hoek-Brown failure criterion – a 1988 update", *Rock Engineering for Underground Excavations, Proc. 15th Canadian Rock Mech. Symp.*, pp. 31-38, 1988.

[HUN 03] HUNTER G., FELL R., "Rockfill modulus and settlement of concrete faced rockfill dams", *J. Geotech. Eng.*, vol. 129, no. 10, pp. 909-917, 2003.

[ICO 02] ICOLD, *Concrete Face Rockfill Dams – Concepts for Design and Construction*, ICOLD Committee on Materials for Fill Dams 2002, (Draft document).

[JOH 07] JOHANNESSON P., "Design improvements of high CFRD's constructed of low modulus rock", *III °Symposium on CFRD Dams Honoring Barry Cooke*, Florianopolis Brazil, October 2007.

[LEP 70] LEPS T.M., "Review of the shearing strength of rockfill", *J. Soil Mech. Found. Div.*, vol. 96, no. Sm 4, pp. 1159-1170, 1970.

[LOB 06] LOBO-GUERRERO S., VALLEJO L., "Application of Weibull statistics to the tensile strength of rock aggregates", *J. Geotech. Geoenviron. Eng.*, vol. 132, no. 6, pp. 786-790, June 2006.

[MAR 69] MARACHI N.D., CHAN C.K., BOLTON SEED H., DUNCAN J.M., Strength and deformation of rockfill, Report no. Te-69-5 Dept. Civil Eng./Geotechn. Eng., University of California, Berkeley, USA, 1969.

[MAR 72] MARSAL R., "Mechanical properties of rockfill", in: HIRSCHFELD R.C. and POULOS S.J. (eds), *Embankment Dam Engineering – Casagrande Volume*, pp. 109-200, John Wiley & Sons, New York, USA, 1972.

[NEM 01] NEMAT-NASSER S., OKADA N., "Radiographic and microscopic observation of shear bands in granular materials", *Géotechnique*, vol. 51, no. 9, pp. 753-765, 2001.

[NOU 05] NOUGUIER-LEHON C., FROSSARD E., "Influence of particle shape on rotations and rolling movements in granular media", *Proc. V° Int. Conf. Powders Grains*, Baalkema, Stuttgart, Germany, July 2005.

[ODA 98] ODA M., KAZAMA H., "Microstructure of shear bands and its relation to the mechanisms of dilatancy and failure of dense granular soils", *Géotechnique*, vol. 48, no. 4, pp. 465-481, 1998.

[PIN 07] PINTO N.L., "Very high CFRD dams – behavior and design features", *Proceedings Symposium on CFRD Dams Honoring Barry Cooke – CBDB/ICOLD*, pp. 3-16, Florianopolis, Brazil, October 2007.

[PRI 77] PRIGOGINE I., NICOLIS G., *Self-Organization in Non Equilibrium Systems*, Wiley Intersciences, New York, USA, 1977 (see also: GLANSDORF P., PRIGOGINE I., *Structure, Stabilité, et Fluctuations*, Masson, Paris, France 1971).

[TEI 99] TEIXEIRA DA CRUZ P., "Sobre a estabilidade de barragens de enrocamento", *II° Simposio Sobre Barragens De Enrocamento Com Face De Concreto – Comite Brasileiro De Barragens*, pp. 91-98, Florianopolis, Brazil, 1999.

Chapter 4

Waste Rock Behavior at High Pressures: Dimensioning High Waste Rock Dumps

Currently, the dumps housing waste rock constitute the highest structures ever built by man, playing an indispensable role during the lifetime of mines. This chapter presents recent experiences related to the design and operation of very high waste rock dumps (HWRDs) with special emphasis on the characterization of the mechanical behavior of certain waste rocks subjected to high confining stresses and *in situ* forced leaching. To achieve this characterization, a large-sized triaxial apparatus was recently built in Chile, allowing confining pressures of up to 3 MPa to be applied on samples 1 m in diameter with a maximum particle size of up to 0.2 m (8"). The waste rock was characterized by means of triaxial compression tests in drained and undrained paths, as well as in oedometric tests. The results obtained were used to conceive and dimension HWRDs; adopt dumping strategies; analyze the possible occurrence of static liquefaction and earthquake effects; and, finally, facilitate eventual closure of these facilities through appropriate methods.

4.1. Introduction

As the production in many of the world's largest open-pit metallic mines has increased dramatically, the demand for waste rock dumps of greater capacity has become high and pressing. Production levels of over 150,000 tons of treated mineral per day or more than 200,000 tons per day are no longer uncommon and, consequently, 300,000 tons to more than 400,000 tons of waste rock per day need to

---

Chapter written by Edgar BARD, María Eugenia ANABALÓN and José CAMPAÑA.

be disposed of. Wherever these high levels of production concern mines in mountainous regions, the only available space for stockpiling these materials is in narrow valleys with steep slopes and, therefore, very HWRDs may be required. In the past two decades, some dumps reaching heights of 400 m to 500 m have been constructed in Canada, Chile and some other countries where mining takes place.

In Chile, for example, some dumps in operation will reach final heights of more than 500 m. Moreover, some of these mines are currently considering expanding their productions, which would mean the building of dumps of 700 m or more in height in the near future. This type of structure would be the highest ever built by man, requiring a meticulous consideration of all aspects related to long-term behavior. These structures should be absolutely stable in order to be environmentally sustainable as well as economically viable.

Thus, the mining industry, geotechnicians and researchers face a major challenge that calls for a significant joint effort to find appropriate solutions to the many problems involved in conceiving and building this type of structure. It is, therefore, of particular interest to analyze the mechanical behavior of waste rock under high confining pressures induced by the very weight of the stockpiled materials. Indeed, high pressures may modify the compressibility, permeability, and shear strength of these granular materials in a significant way.

4.2. Development of new laboratory equipment for testing coarse materials

Mined waste rock has proved to be an extremely heterogeneous granular material, with particles coming from various lithologies or from the same type of rock with different mechanical qualities according to the more or less important weathering degrees. The dimensions of the particles are also very variable, ranging from finely grained soils to meter-sized blocks. The laboratory equipment and the test procedures for these materials have often been limited by the size of the particles and the relatively low levels of the confining pressures applied. Consequently, the geotechnical characterizations available have corresponded mainly to those of the finest parts of these materials and for confining pressures that are generally limited to 1 MPa. Having the proper equipment to permit laboratory tests on samples of greater dimensions and representative *in situ* tests is a foremost necessity.

Reasonable doubts can be had on the representativity of the existing constitutive laws and the failure criteria of very coarse materials, deposited in a loose state and subjected to high confining stresses [VAL 07]. Furthermore, experimental data related to the behavior of these materials are scarce, as is suitable monitoring in any stringent way. Follow-up procedures related to the real behavior of these deposits are also needed that are rigorous enough to inspire confidence in the security conditions related to the operation of these structures.

4.2.1. *Triaxial and oedometric equipment at the IDIEM*

Recently, the *Instituto de Investigaciones y de Ensayos de Materiales* (IDIEM), based at the University of Chile, developed a controlled load triaxial apparatus that can reach confining pressures up to 3 MPa with large-sized samples of up to 1.0 m in diameter and 1.8 m in height. According to these dimensions, the materials tested can present a maximum of size up to 0.2 m (8"). The equipment (see Figure 4.1) has a reaction frame designed for a maximum working load of 20 MN and a testing chamber in the shape of a bottle with an interior diameter of 1.74 m and height of 2.66 m. It has six lower entrances for high-pressure hoses and eight connectors for sensors. The volumetric control is executed with the use of a high-pressure tank located on a support frame with a loading cell that records the volumetric change of the sample by weight. The air and water pressures are regulated through a master panel, while the control is due to a double system of precision manometers and pressure sensors. The load on the sample is applied with four hydraulic jacks with control performed through a loading cell of 5 MN capacity.

Deformation control is managed through three displacement sensors attached to the loading piston and located equidistant from each other. Furthermore, three pressure sensors are used, one to control the confining pressure and the remaining two for the control of the pore pressures inside the sample. Finally, data acquisition during the test is carried out with an analog-digital system.

Figure 4.1. *Loading frame (left) and a sample pf D=1.0 m and H=1.8 m (right) after the application of large-scale triaxial equipment of controlled load. IDIEM, University of Chile [BAR 07a]*

The testing chamber of this triaxial apparatus was adapted to receive an oedometric mould, 0.95 m in internal diameter, permitting the application of vertical pressures greater than 12 MPa and the assembly of samples of 1.0 m high. It is also possible to perform permeability tests under a constant head method with this mould.

4.3. Mining rock waste

Determination of the *in situ* grain size distribution (GSD) for mining waste rock is a difficult task for many practical reasons, given:

– the size of the grains, the heterogeneity of deposited materials;

– the segregation that occurs during dumping;

– the increase in the contents of finer particles due to particle degradation induced by the high pressures;

– the increase in finer particle content by migration due to natural or forced infiltrations.

During construction of the dump, the dumping process produces the segregation of these materials along the advancing slope according to the size of the particles; fines accumulate close to the surface, while the larger particles roll towards the foot of the layer. Thus, segregation leads to a stratification of the particles along the profile of the dump and to the formation of heterogeneous layers, parallel to the advancing face. In complicated situations, when challenges to the stability or environment are serious enough, the most appropriate solution in most cases requires the construction procedure to be controlled as tightly as possible, with dumping in layers of predetermined thicknesses and, on occasion, selecting the type of waste rock to build different sectors of the dump. Modeling this heterogeneity for posterior analyses is generally rather difficult.

4.3.1. *In situ grain size distribution*

To characterize the waste rock, it is recommended to perform *in situ* macro grain size distribution tests at different levels of the dump. The large-sized blocks and particles are separated, sorted by size, measured by hand, and finally weighed at the site. Materials that are smaller than 8 cm (3") should be analyzed by laboratory tests. A complementary approach is to estimate the grain size distribution of the materials exposed on the slopes from digital image-processing. With the latter technique, however, it is not possible to determine the content of the fine particles.

4.3.2. *Analyzed waste rock*

The waste rock analyzed in this section corresponds to large run of mine (ROM) samples obtained from the open pits of two mines located in high mountainous areas in the central region of Chile. These materials are coarse, with angular to sub-angular shape, obtained by the blasting of hard porphyritic rocks. These rocks have a compressive strength of about 120 MPa, and a specific gravity of solids G_s of 2.7 to 2.8. In general, the waste rock is loaded by mine trucks of large tonnage, transported and then dumped in different sectors of the dump, in layers of limited thickness (about 50 m) according to a pre-established placement plan.

The results shown in Figure 4.2 correspond to those obtained with ROM waste rock (ROM WR) and crushed and leached waste rock. ROM WR was obtained from a sampling of about 50 tons directly from a dump. A portion of these materials was crushed to create a second sample. The third sample was taken from a trench dug in a waste rock platform subjected to forced leaching over a period of two years.

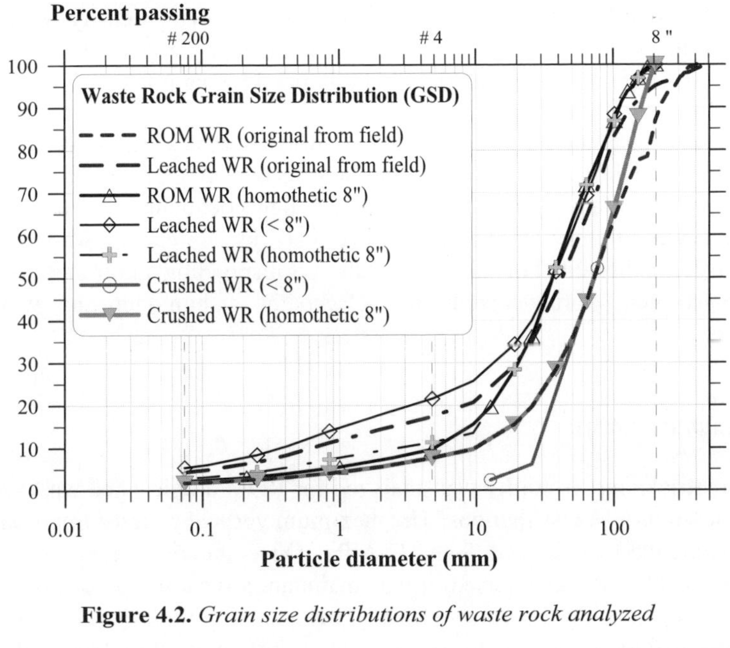

Figure 4.2. *Grain size distributions of waste rock analyzed*

The macrogranulometric analyses (integral grain size distribution) shown in Figure 4.2 made it possible to determine the maximal sizes of the particles, which vary in general between 30 cm and 100 cm (12" and 40"). Below 8 cm (3"), these materials present fine contents (particles passing through the 200 ASTM sieve) making up less than 10% of the sample. The grain size distribution curves of the waste rock analyzed are represented in this same figure. However, in the remolded samples tested, the maximal grain size was limited to 20 cm (8"), to produce grain-size distributions compatible with dimensions of triaxial and oedometric equipment. With the purpose of analyzing the influence of grain size distribution on compressibility and particle crushing in these waste rocks, the remolded samples were prepared with two grain size distributions: homothetic to the one obtained *in situ* with a maximum grain size of 20 cm; and one cut under 20 cm, by eliminating particles of sizes larger than the latter dimension.

4.4. Characterization of mechanical behavior of the waste rock

The mechanical behavior of the three waste rock samples was characterized by the use of oedometric drained tests and compression triaxial tests, in drained (CID) and undrained (CIU) paths. All these tests were carried out on large samples, compacted manually with an initial dry density of $\gamma_d = 18.15$ kN/m^3 (initial void ratio of about 0.46), with a water content of about 5%. This density represents the average of the densities measured *in situ*, obtained by replacing the excavated volumes with water once the materials were dumped by the trucks.

Oedometric tests were carried out on cylindrical samples 0.95 m in diameter and 1.0 m in height, while the triaxial compression tests were carried out on cylindrical samples 1.0 m in diameter by 1.8 m in height. Each oedometric or triaxial sample was prepared into four or eight layers, respectively, by compacting waste rock to the initial density indicated. For each layer, the corresponding grain size distribution was reconstituted, homogenized, and placed in a homogeneous way before compaction.

4.4.1. *Oedometric tests*

Oedometric compressibility was analyzed for ROM and leached waste rock with homothetic grain-size distributions. The maximum vertical stresses imposed in these tests corresponded to $\sigma_v' = 4$, 8 and 12 MPa. Once the maximum vertical stress planned for each test was reached, it was maintained for a period of 30 hours until the vertical deformation became stable. Then, always under a constant vertical stress, these materials were saturated, in counter-current, by imposing low hydraulic gradients. The deformations induced by saturation were also followed for a period of 30 hours. After this last period, permeability tests with the constant head method were carried out. Finally, the resulting grain size distribution was determined after each test.

4.4.2. *Triaxial tests*

Waste rock tested on compressional triaxial paths, with initial confining pressures of up to 2.5 MPa, are indicated on Table 4.1. The samples were completely saturated before each of the tests.

The effective mean stress [p'] and vertical stress [σ_V'], reached at the end of the different tests, performed are equivalent to overburden heights of about 300 m to 560 m, as indicated in Figure 4.3.

Material	Grain size distribution (GSD) tested	Initial confining pressure, σ'_3 (MPa)
ROM waste rock	Homothetic MGS 20 cm (8")	CID: 0.1; 0.2; 0.5; 1.0; 2.0
Leached waste rock	Homothetic MGS 20 cm (8")	CID: 0.2; 1.0; 1.5; 2.5
	Cut < 20 cm (8")	CID: 1.2 (initially consolidated at 2.0); 2.0. CIU: 1.7; 2.5
Crushed waste rock	MGS 20 cm (8")	CID: 1.0; 1.6; 2.5

MGS: maximum grain size; CID: compressional triaxial drained; CIU: compressional triaxial undrained

Table 4.1. *Detail of triaxial tests performed*

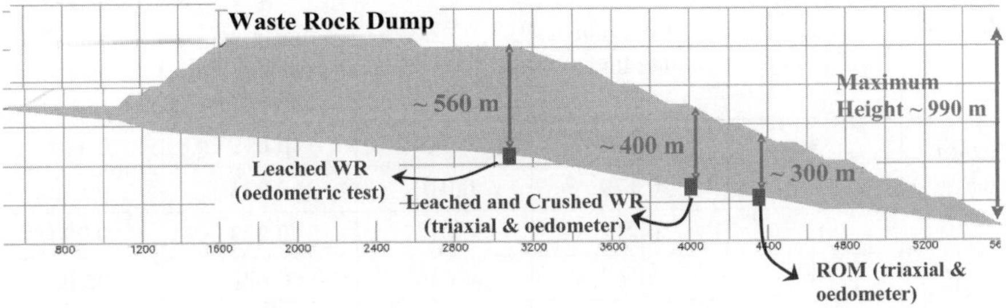

Figure 4.3. *General layout of a high waste rock dump. Equivalent overburden loads to the final mean stresses reached in tests*

4.4.3. *Oedometric test results*

The oedometric behavior of the ROM and the leached waste rocks is indicated in Figure 4.4, where a good reproduction of the various test results for the stresses applied can be noted.

4.4.3.1. *Behavior under monotonic loading*

Initially, for vertical stresses lower than 0.9 MPa, waste rock presents an overconsolidated behavior with a low void ratio decrease under the applied stresses. For vertical stresses higher than 0.9 MPa, a significant void ratio decrease takes place, reflecting an increase in the granular skeleton compressibility and then, finally, a void ratio variation proportional to the logarithm of the imposed vertical stresses is obtained, similar to the normally consolidated behavior. This behavior of granular materials has already been pointed out by many researchers on the subject of gravels, sands and petroleum cokes [BAR 93, BIA 71, BIA 92, COL 86, ROB 58]. Void ratios obtained after monotonous loading are reported in Table 4.2.

4.4.3.2. *Behavior under constant vertical stresses*

For each test, the maximum vertical stress applied was kept constant after monotonous loading, such as after the saturation stage, until the vertical deformation became stable. For both conditions, it could be noted that the compressibility of the granular skeleton continues to develop and that saturation leads to a reactivation of the deformations stabilized in the previous load increment (Table 4.2).

Material	Maximum vertical stress. σ'_v (MPa)	Δ_e	Δ_e (one period of 30 hours)		Δ_e total
		Monotonic loading	σ'_v keep up constant	σ'_v keep up constant and saturation	
ROM waste rock	4	0.144	0.014	0.019	0.177
	8	0.179	0.010	0.023	0.212
	12	0.219	0.012	0.023	0.254
Leached waste rock	4	0.162	0.019	0.022	0.203
	8	0.210	0.015	0.019	0.244
	12	0.251	0.017	0.016	0.284

Table 4.2. *Oedometric tests: void ratios variation*

The effect of time on the compressibility of granular materials has also been analyzed by several researchers. For the waste rock tested, vertical deformations resulting from the period of maintaining the vertical stress constant varied from 5% to 10% of the quasi-instantaneous initial settlement. Similarly, the effect of water on grain crushing was pointed out by [LEE 67] as well as [CLE 81, MAR 77] and others. They showed that the crushing of saturated particles required less stress than it did for dry particles.

4.4.3.3. *Crushing of particles*

For vertical stresses higher than 0.9 MPa, the increase in compressibility of the granular skeleton of waste rock, as shown by the oedometric tests, is related to the particle crushing phenomenon. Several authors have held this phenomenon responsible for the contracting behavior observed.

Thus, the high pressures induced by the very weight of the materials stockpiled in waste rock dumps with heights over 100 m generate significant particle crushing, modifying the initial grain size distribution of the material and leading to an increase in the amount of sands and fines in the granular matrix. This situation modifies the void ratio and, consequently, the density and permeability.

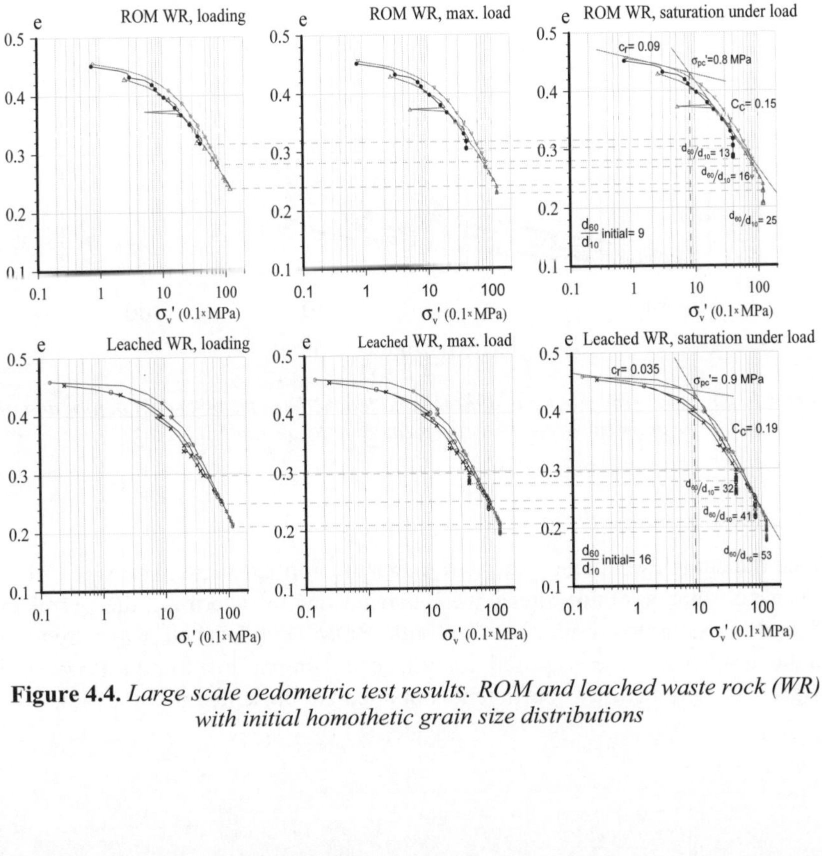

Figure 4.4. *Large scale oedometric test results. ROM and leached waste rock (WR) with initial homothetic grain size distributions*

To evaluate this phenomenon, the grain size of the waste rock was systematically analyzed after each test. The grain size distribution curves of ROM and leached waste rock, before and after oedometric tests, are shown in Figure 4.5 for the different vertical stresses applied and summarized in terms of evolution of d_{10} and d_{60}. These diameters correspond to the openings of the sieves through which 10% and 60% in weight of material pass, respectively, in Table 4.3.

For the two grain size distributions of the analyzed waste rock, the results show that the leached waste rock, with an initial grain size distribution finer than that of the ROM, presents a stronger grain size evolution. This is probably related to the effect of forced leaching on the individual strength of the particles.

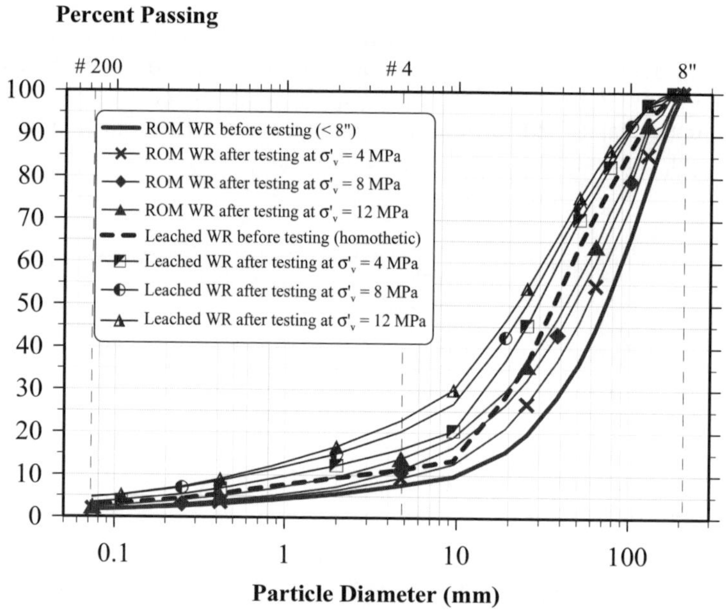

Figure 4.5. *Grain size distributions of ROM and leached waste rock (WR) before and after oedometric tests. Tests performed on large-scale samples*

4.4.3.4. *Saturated permeability*

Once the samples became saturated under the applied vertical stress, infiltration tests at a constant hydraulic head were carried out to determine the permeability coefficient in saturated condition for both ROM and leached waste rock. Low hydraulic gradients were imposed to induce a laminar infiltration flow in these materials. The permeability coefficients obtained are presented in Table 4.4.

According to the results, the saturated permeability coefficients obtained for the ROM WR, at the different dry densities resulting from the imposed vertical stresses, are the same. For the leached waste rock, on the other hand, these results clearly show a lower saturated permeability coefficient as the vertical stress, or the equivalent overburden height of the dump, increases.

These results are particularly interesting in light of the influence of the waste rock characteristics and its initial grain size distribution on the evolution of the coefficient of permeability with the imposed vertical stress (resulting density).

Material	Vertical stress σ_v' (MPa)	% passing < sieve # 4 ASTM mesh		% passing < sieve # 200 ASTM mesh		d_{60} (mm)		d_{10} (mm)		d_{60}/d_{10}	
		Initial	After test	Initial	After test	Initial	After test	Initial	After test	Initial	After test
ROM waste rock	4	7.7	9.3	1.8	1.9	90.6	72.8	10.0	5.7	9	13
	8		11.3		1.9		63.3		3.9		16
	12		13.9		2.4		56.9		2.3		25
Leached waste rock	4	11.3	16.2	2.9	3.4	46.5	38.2	2.9	1.2	16	32
	8		20.3		4.6		34.6		0.6		41
	12		22.8		4.8		31.5		0.6		53

Table 4.3. *Grain size distributions before and after oedometric tests*

Material	Maximum vertical stress. σ_v' (MPa)	Dry density γ_d (kN/m^3)	Saturated permeability coefficient k_s (cm/s)
ROM waste rock	4	20.6	$5 \cdot 10^{-1}$
	8	21.2	$3 \cdot 10^{-1}$
	12	22.0	$3 \cdot 10^{-1}$
Leached waste rock	4	21.1	$2 \cdot 10^{-1}$
	8	21.8	$5 \cdot 10^{-2}$
	12	22.6	$1 \cdot 10^{-2}$

Table 4.4. *Saturated permeability coefficient under constant vertical stress*

4.4.4. *Triaxial test results*

The results and interpretation of the compressional triaxial tests [BAR 07a] are presented in Figures 4.6 to 4.9, in the $[q - \varepsilon_1]$, $[q - p']$ plots with the strength envelope suggested for each waste rock analyzed, as well as the evolution of the void ratio with the axial deformation and effective mean stress, $[e - \varepsilon_1]$, $[e - p']$ plots, respectively. For the particular case of the ROM waste rock (see Figure 4.6), the important strength reduction observed for the confining pressure of σ'_3=2 MPa at an axial deformation of 12% is probably related to a localization of deformations (kinematic discontinuity), a situation that can also be observed in the $[q - \varepsilon_1]$ and $[e - \varepsilon_1]$ plots of this same figure.

4.4.4.1. *Drained tests*

Given the relatively loose initial state of the waste rock, these materials present a contractive behavior (tendency towards volume decrease) for confining pressures above 0.2 MPa, which is accentuated for higher pressures. Indeed, as has been highlighted in the $[e - p']$ plots, the compressibility of these materials increases significantly for confining pressures higher than 1 MPa for the ROM as well as the leached waste rock. This increase in compressibility is associated with the phenomenon of particle crushing, as evidenced by the more widely spread grain size distributions obtained after tests (see Figure 4.12 and Table 4.6).

For the three waste rocks tested at low confining pressures, the results of compressional triaxial tests (CID) show that the strength envelope, presented in the $[q - p']$ plots of Figures 4.6 to 4.9, is characterized by a high internal friction angle and a null cohesion. As the confining pressure increases, the slope of this envelope decreases progressively ($[q - p']$ plots), which gives rise to its curvature.

This behavior, common to granular materials under high stresses, is mainly due to a greater compressibility of the granular matrix associated with the phenomenon of particle crushing, which implies a decrease in the internal friction angle. Indeed, this is accompanied by an evolution of the initial grain size to a finer grain size distribution.

The tendency towards volumetric dilation at lower pressures disappears when the confining pressures are higher than 0.2 MPa.

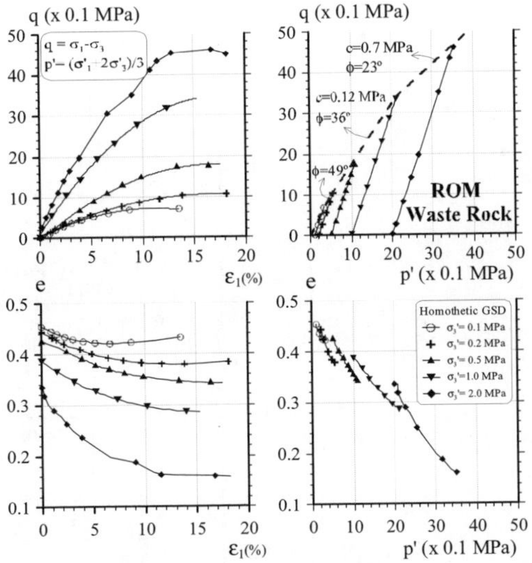

Figure 4.6. *CID triaxial tests results with ROM waste rock, initial homothetic grain size distribution [BAR 07a]*

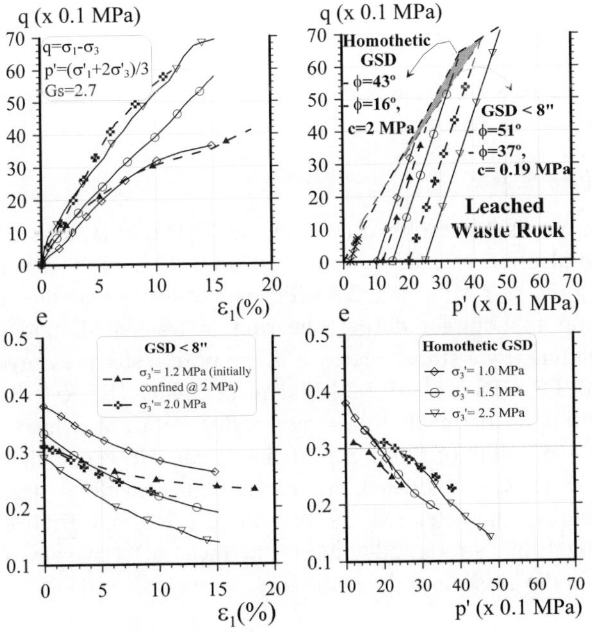

Figure 4.7. *CID triaxial test results with leached waste rock (homothetic and cut initial grain size distributions)*

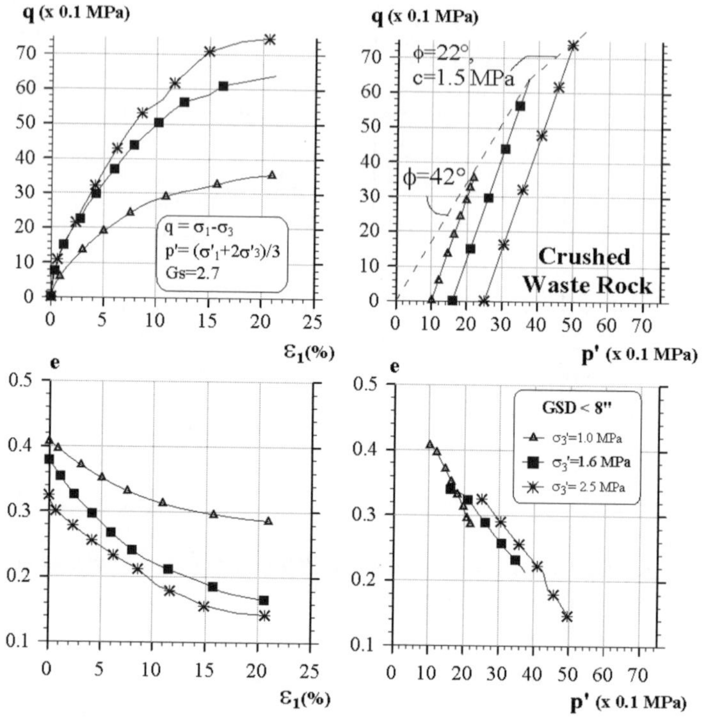

Figure 4.8. *CID triaxial test results with crushed waste rock*

4.4.4.2. *Undrained tests*

Figure 4.9 presents the results of the CIU and CID triaxial tests on leached waste rock for an initial grain size distribution cut under 0.2 m (8"). For both imposed confining pressures, 1.7 MPa and 2.5 MPa, we observe that the tendency of the granular skeleton to contract during shearing is translated, under an undrained compression loading, by a strong increase in the pore water pressure (ΔU), reaching values of about 1.5 MPa and of 2.3 MPa, respectively [BAR 07b]. Compared to the drained condition, this leads to a strong reduction of the shear strength, as the $[q - \varepsilon_1]$ and $[q - p']$ plots of this figure makes clear. However, the shear strengths reached in these tests are situated on the strength envelope determined at low confining pressures, characterized by the biggest internal friction angle of the material. Moreover, this strong reduction of the mean effective stress also leads to a weaker change in the grading of the material.

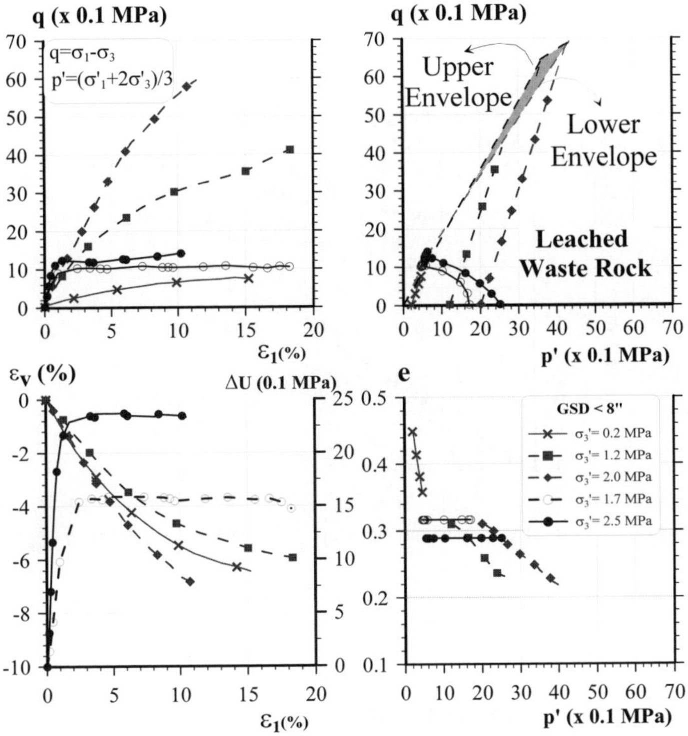

Figure 4.9. *CID and CIU triaxial tests on leached waste rock, cut grain size distribution [BAR 07b]*

It is interesting to note that the undrained strength obtained on these materials (C_u/σ'_{30}=0.28-0.29) is similar to the results obtained on the fine matrix of leached copper waste dump material [VAL 04], presented in Figure 4.10.

4.4.4.3. *Shear strength envelopes*

Figure 4.11 shows the shear strength envelopes, deduced from the tests performed, plotted in the Mohr plane [τ, σ'_n], in addition to those reported in the technical literature for coarse granular materials coming from hard rocks [MAR 75]. As observed in this figure, the results obtained from the waste rocks agree rather well with the shear strength envelopes, suggested for other granular materials, especially in the range of pressures below σ'_n = 2 MPa.

At higher levels of stress, the strength envelopes start to curve due to the greater compressibility of the granular matrix associated with the crushing of particles [BIA 92].

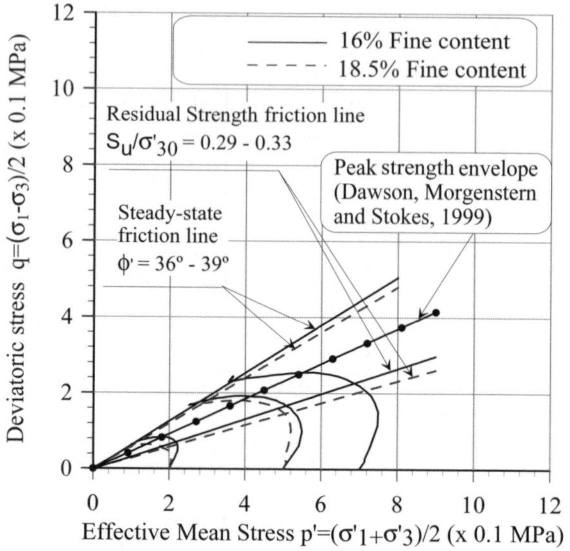

Figure 4.10. *Stress paths derived from CIU triaxial test run on the fine matrix leached waste rock [VAL 04]*

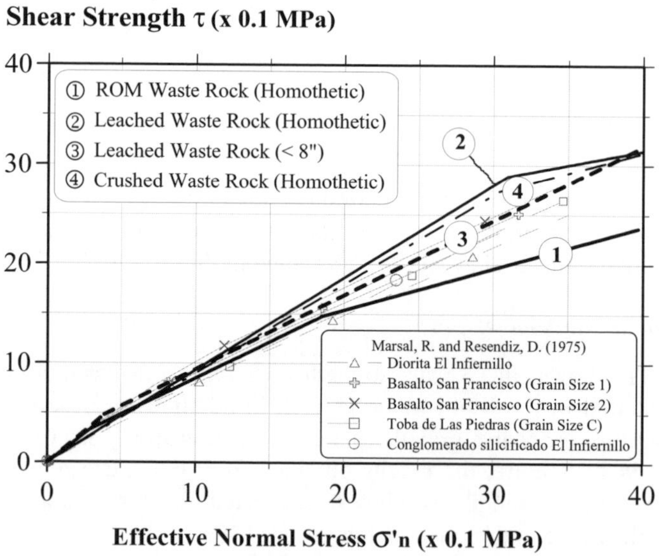

Figure 4.11. *Envelopes of the waste rock in the Mohr plane. Copper mine samples [BAR 07a]*

4.4.4.4. *Deformation modulus*

The results obtained from the different waste rocks analyzed were used to establish stress–strain relations, presented in Table 4.5.

In particular, the hyperbolic model was used to determine the secant modulus representative of deformations of about 10^{-3}, through the initial tangent modulus deduced from the adjustment of the hyperbole, as well as its evolution with the confining pressure [BIA 93].

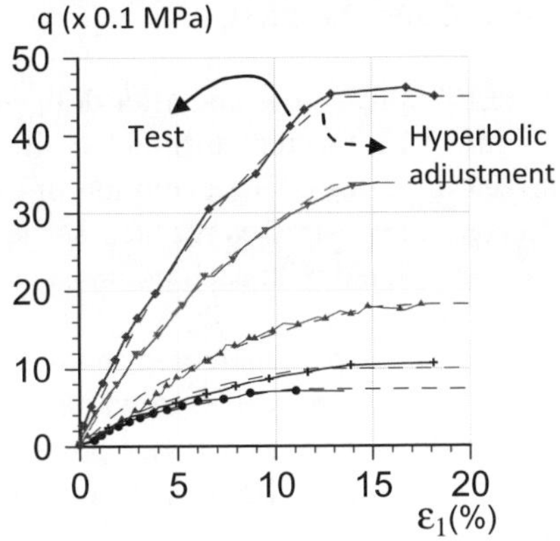

$$E_{th} = k \cdot P_a \cdot \left(\frac{\sigma_3'}{P_a} \right)^n$$

where:

k: modulus factor;

Pa: atmospheric pressure (0.1 MPa);

σ'_3: effective confinement stress; and

n: power.

Materials	k	n
ROM waste rock, homothetic grain size distribution	170	0.43
Leached waste rock, homothetic grain size distribution	120	0.51
Leached waste rock, scalped grain size distribution	90	0.81

Table 4.5. *Deformation moduli deduced from the hyperbolic adjustment*

The *k* values appear lower than the typical values for other granular materials, particularly for ripraps. It is necessary to note that these tests were carried out under controlled loading and information related to mining waste rock, dumped with a low initial density, is hard to come by. However, the results of the compressional triaxial tests agree with those of the oedometric tests.

4.4.4.5. *Particle crushing*

Just as for the oedometric tests, to evaluate the particle crushing on compressional triaxial tests, the grain size distributions of the waste rocks tested were systematically determined after each test.

The grading curves of ROM WR, before and after drained compressional triaxial tests, are presented in Figure 4.12 for the different confining pressures applied. These results are summarized, in terms of the evolution of d_{10} and d_{60}, in Table 4.6. In this same table, the effective mean stresses reached at the end of the test are also indicated, as are the stresses associated with each grain size distribution obtained.

As the final effective mean stress p' is increased, particle crushing is manifested as the initial grain size of the sample evolves towards finer grain sizes, clearly increasing the content of the sands, and, to a lesser proportion, the fines. In other words, this evolution depends on the confining pressure and, significantly, on the deviatoric stress and imposed axial deformation.

For the different initial grain size distribution tested, it was observed that particle crushing was more significant in the coarser grain sizes, except for the leached waste rock. Indeed, the latter materials with a finer initial graduation experienced a greater amount of particle crushing than the remainder of the waste rock tested. However, this effect is more progressive when angular to subangular particles are at stake (see Figure 4.13). Furthermore, the larger sized particles tend to present more defects and, therefore, a greater probability of being crushed when subjected to high tensions (see Figure 4.14). It could be noted in Figure 4.13 that the compressional triaxial path produces a greater amount of particle crushing than the oedometric path, a result which is not surprising.

For larger confining pressures and the same stress path, the final grain size curves, originating from both cut grain size distribution and homothetic grain size, evolve to a single grain size curve. This result is important, since it indicates that waste rock coming from rocks with similar mechanical characteristics but with different initial grain sizes resulting from blasting, loading, segregation, and/or process conditions, will evolve to a similar grain size when subjected to high confining pressures. This result has already been observed in materials with friable grains and sands [BAR 93].

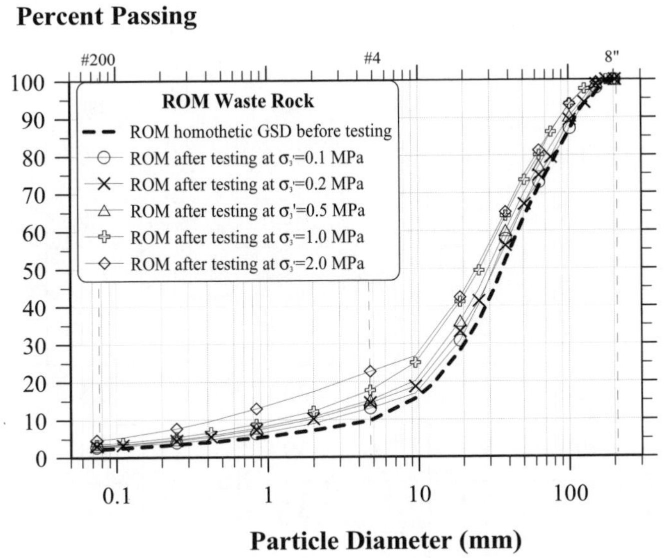

Figure 4.12. *Grain size distribution before and after triaxial tests on ROM waste rock (homothetic GSD)*

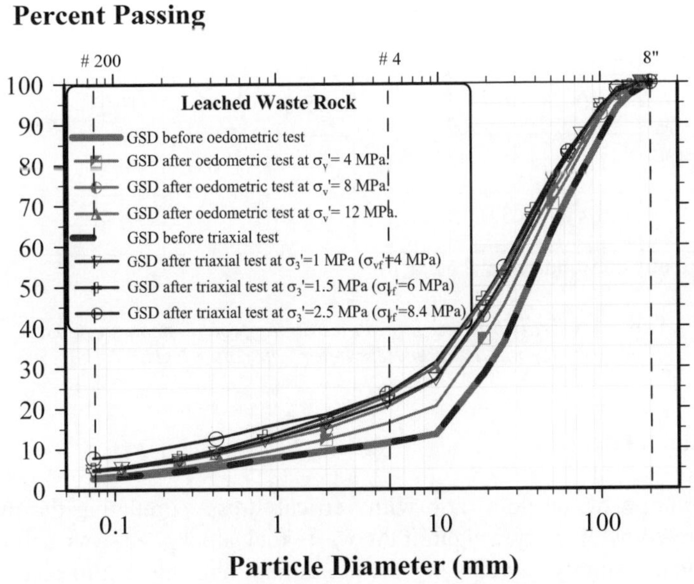

Figure 4.13. *Grain size distributions before and after oedometric and triaxial tests on leached waste rock (homothetic GSD)*

Material		p' (MPa) Initial	p' (MPa) Final	d_{60} (mm) Before test	d_{60} (mm) After test	d_{10} (mm) Before test	d_{10} (mm) After test	d_{60}/d_{10} Before test	d_{60}/d_{10} After test
ROM Waste Rock (homothetic GSD)		0.1	0.3	46.5	41.0	5.0	2.7	9	15
		0.2	0.6		42.6		1.9		22
		0.2	1.1		38.1		1.7		23
		1.0	2.1		34.7		1.3		28
		2.0	3.5		33.6		0.5		73
Leached waste rock	Homothetic GSD	0.2	0.5	46.5	39.8	2.9	1.8	16	22
		1.0	2.2		32.7		0.6		57
		1.5	3.4		30.1		0.4		68
		2.5	4.8		30.4		0.2		152
	GSD < 20 cm (8")	1.2(*)	2.6	49.3	26.6	0.4	0.2	130	140
		2.0	4.0		25.8		0.2		136
Crushed waste rock (< 20 cm or 8")		1.0	2.2	90.6	55.3	10.0	3.3	9	17
		1.6	3.7		45.1		0.9		53
		2.5	5.0		38.5		0.8		46

(*) Initially consolidated at 2 MPa

Table 4.6. *Triaxial tests. Initial and final effective mean stresses, d_{60} and d_{10} diameters before and after tests*

4.5. Evolution of density

The evolution of the void ratio with vertical stress, simulating the overburden effect generated by the very weight of the waste rock stockpiled, was estimated from laboratory tests roughly reproducing the stress path and the *in situ* conditions. Far away from the areas of influence of the slopes, the oedometric path acceptably represents the stress state inside the deposit by supposing that this state is close to the geostatic state, i.e. that $\sigma'_H = k_o \cdot \sigma'_v$, where k_o is the coefficient of lateral earth pressure at rest.

Figure 4.14. *Detail of particle crushing after the compression triaxial test [BAR 07b]*

As for the different waste rock analyzed, with large-scale samples on oedometric and drained compressional triaxial paths, a comparison of the evolution in void ratio with the increase of the effective mean stress can be seen in Figure 4.15.

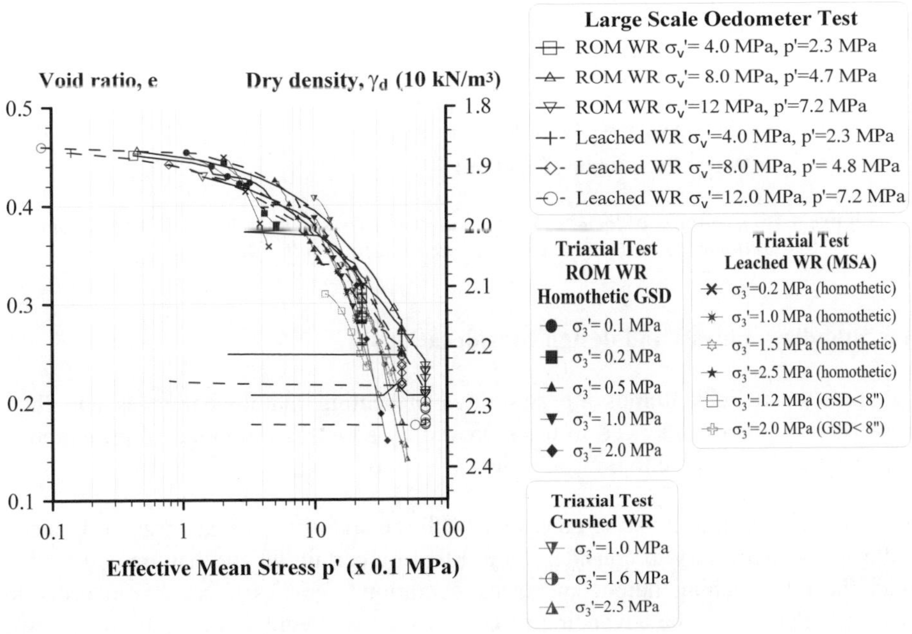

Figure 4.15. *Comparison of the waste rock compressibility on oedometric and compressional triaxial paths using large samples [ARC 06-07]*

These results were also plotted in Figure 4.16 and show the suggested evolution of dry density with the effective mean stress p', deduced from the vertical stress applied σ'_v, associated with the equivalent overburden height of the dump. The values plotted correspond to the dry densities obtained at the end of the tests or from *each* loading step. The results obtained from drained compressional triaxial tests reaching effective mean stresses higher than p' = 5 MPa, roughly equivalent to a vertical stress of σ'_v = 8.5 MPa, are also included in this figure. It is important to mention that for the results of the compressional triaxial tests, the dump height is correlated with the final mean stress of the test.

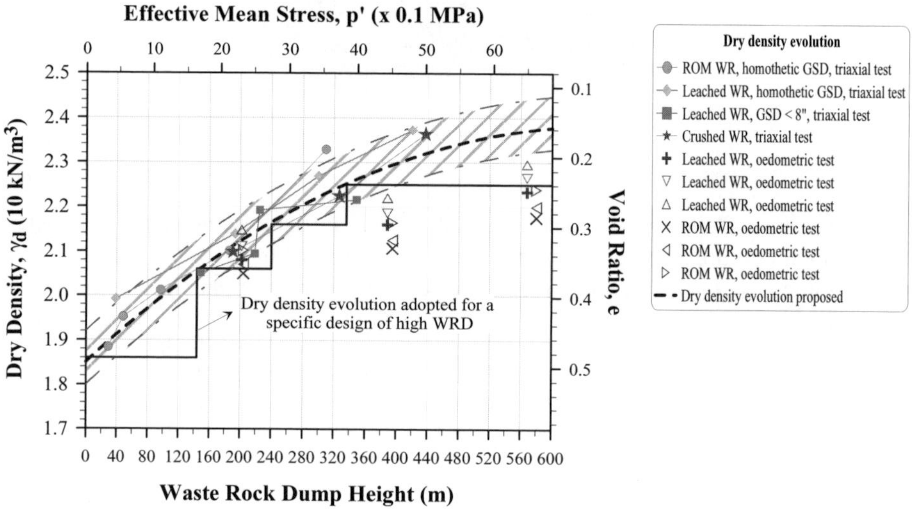

Figure 4.16. *Evolution of the dry density with waste rock dump height, estimated from oedometric and compressional triaxial tests results [ARC 06-07]*

4.6. Stability analysis and design considerations

High waste rock dumps are structures presenting major challenges and their construction procedures need to be improved by adopting a number of pre-emptive measures to reduce potential instabilities.

The dump design must be consistent with the stability criteria adopted. In steep valleys, it is necessary to consider dump lifts of a certain height that are compatible with the pit benching height and truck operations. This is necessary in order to reduce the height of the advancing face and excessive segregation of the waste, two conditions that lead to potential instabilities. The design criteria should also be in accordance with the closure plan, to avoid a large and almost impracticable waste management at a later stage. Indeed, the design criteria have to avoid the creation of unnecessarily restrictive requirements at the closure stage.

Considering the mechanical behavior of the waste rock at high pressures, subjected to overburden stresses well beyond any precedent, the front slope of a dump should be designed with a smooth global angle, much flatter than 37°, with the objective of generating a smaller loaded zone destined to constitute a high-strength and high-permeability buttress. The decrease in stresses in the waste rock forming the buttress will contribute to reduce degradation effects on the materials due to both mechanical and leaching (natural or forced) actions [VAL 07].

Slopes with a low overall angle can be *built on* the waste rock dump by constructing layers of limited thickness, instead of dumping the waste rock regardless of height restrictions. The different layers, each with a local slope of about 37°, should include intermediate berms to control any local instabilities (affecting one layer). To control potential instabilities affecting more than one layer, it is recommended that a large berm is included at a certain number of lifts, in accordance with the overall stability.

An example of the application of these recommendations is presented in Figure 4.17, showing the final geometry of a HWRD located in a narrow mountainous valley in a highly seismic area. Figure 4.18 shows an example of a stability analysis using the limit equilibrium method (LEM). These analyses were complemented with a dynamic analysis and three-dimensional stability analysis based on LEM.

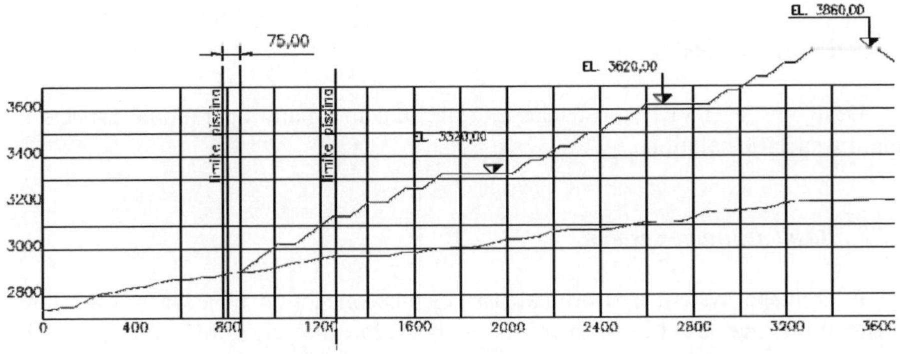

Figure 4.17. *Final layout of a high waste rock dump located in a mountainous environment with high seismicity [ARC 06-07]*

Wherever possible, the advancing faces of each lift should attempt to develop a convex configuration to encourage three-dimensional restraint effects. In addition, the advancing slope should constitute a buttress to provide the appropriate containment and stability of waste rock. These buttresses must also be constructed of high geotechnical quality waste rock, with low fines waste. Fines waste with high water content or low-resistance must be encapsulated in cells located behind the buttress. In order to manage the waste materials properly, sand and fines content of the waste should be checked periodically [MOR 06].

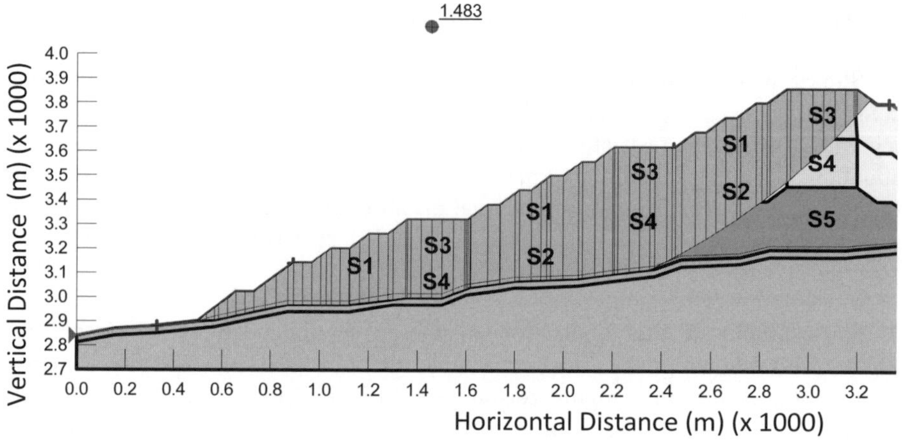

Figure 4.18. *Example of the stability analysis of a high waste rock dump by limit equilibrium methods – static cases [ARC 06-07]*

4.7. Operation considerations

Designing a HWRD must take into account the many operational aspects that require specific solutions.

4.7.1. *Basal drainage system*

A drainage system is a critical part in the conception of a waste rock dump. General practice, used in most mines, is to exploit the segregation into grain sizes during dumping and to rely on the permeability of the coarser layers deposited at the bottom of the dump lifts, formed mostly by large sized particle and blocks. This practice has proven to be valid for many cases where "normal" conditions prevail. There are also many situations where a drainage system of this type could fail, however, in the sense that it will not allow free drainage of the dump. This could ultimately lead to an increase in pore water pressure at the foundation level or in areas above the contact between two layers.

Such "abnormal" conditions may occur in areas with extremely high rainfall, with deterioration of the blocks or particles constituting the drainage system because of natural or forced weathering and/or the increase in the number of thinner particles due to high stress. The solution in these situations is not simple. Usually it involves the selection of sound and ideally inert rocks to form the drainage system and to build it separately, before the deposit begins to be formed. It may even require the building of artificial drains housed in a trench, always with selected waste rock in order to avoid the effect of high pressures on the behavior of these materials [VAL 08].

4.7.2. *Water management*

A water management strategy must be developed to intercept and divert as much surface runoff as possible away from the dump, both for reasons of physical stability of the structure and for environmental considerations. This strategy must consider solutions designed to avoid excessive watering of the dump. Thus, in periods of high potential infiltration, it is necessary to provide effective drainage for the dump platforms; these platforms sloping gently towards collection works, sealing surficial cracks by scarification and compaction [MOR 01-06]. The traffic of high tonnage trucks and/or the presence of the finest waste on the rolling surfaces in order to preserve tires in good conditions, usually leads to highly compacted and relatively impervious platform surfaces.

Previous situations are propitious to the occurrence of perched groundwater, creating conditions favorable for the development of static liquefaction. To restore adequate infiltration conditions, it is therefore necessary to scarify the surface of these compacted and tightened surficial layers before allowing the following lift to be dumped. In certain situations, geotechnical low-quality waste should be dumped in different areas within the dump to reduce or avoid such problems, which underlines the need to deal with the buttresses constructed with selected waste rock [BAR 07a].

4.7.3. *Foundation conditions*

Local deformations induced by the foundation soils of low bearing capacity may be the mechanism that triggers major landslides. Confronted with this possibility, it would be advisable to avoid the materials constituting unstable foundations by removing the weak materials, if the thicknesses involved allows. Another alternative is to implement a particular construction sequence for the advancement of the deposit, encapsulating the sectors with problematic materials. The option to carry out a foundation treatment is not always practical [VAL 08].

The presence of permafrost requires special attention. Indeed, in Canada some waste rock dump failures have been associated with the melting of permafrost due to stresses imposed by the dump. As mentioned previously, potential sliding triggering mechanisms must be identified, avoided if possible, or treated because they can induce major failures or debris flows.

4.7.4. *Effects of rain and snow*

The main problem with rain is the possibility that the water table can rise well above the foundation level and induce the migration of finer particles to the bottom of the deposit or the layer, reducing the permeability of the drainage layer. In these cases, it is necessary to consider the design of a drainage system.

Snow also deserves particular attention. Indeed, snow tends to remain on the upper surface of each layer in construction and on the advancing slope, its accumulation creating potential local instabilities. The modification of the construction sequence and advancing with a narrow section would allow the snow to be pushed back and thrown into the interior of the dump. To minimize local and global instabilities, both due to the accumulation of snow and the presence of finer waste, the dumping of these materials can be made in the transverse direction to the advancement of the deposit [ARC 06-07].

Moreover, in very HWRDs, the snow deposited on steep slopes could produce avalanches along the slope itself. Gentle slopes, as well as intermediate berms, may help solve this problem.

4.7.5. *Effects of in situ leaching on waste rock*

Some high waste rock dumps have been subjected to a forced leaching to recover the remaining copper from these materials. The irrigation process produces an important increase in the saturation degree of the waste rock, weakens the particles, and leads to an increase in the concentration of fine particles in the deeper layers. Consequently, during irrigation, conditions close to saturation can be generated, creating conditions that favor the start of static liquefaction within one or more leach cells of the deposit. In Chile, significant deformations have already been identified in waste rock dumps that were subjected to the forced leaching associated with this phenomenon.

For a more accurate understanding of the effects of chemical degradation and its consequences on the mechanical behavior of waste rock, it is necessary to carry out more systematic research on waste rock from different lithologies.

4.7.6. *Designing for closure*

To construct a HWRD and then ponder what must be done for the closure of this structure is an unthinkable strategy, given the extraordinary dimensions involved. To reshape the geometry of a slope with a height of more than 500 m is a situation that nobody would undertake, not only because of the costs involved but also for the technical complexities and security issues related to works of this magnitude. The most reasonable solution for high waste rock is to design a dump that meets the closure criteria from the outset. The practical implications of this decision means that the dump must be built with more gentle general slopes, conservative drainage systems, effective effluent collection systems and treatment of the effluents, if necessary. The existence of intermediate berms in the design of the dump will also allow global instabilities to be controlled, and if it is called for, a control of possible snow avalanches [VAL 08].

4.8. Conclusions

The results from triaxial tests performed on samples of large dimensions for three types of waste rock with different initial grain sizes have shown that these materials present adequate shear strength at high pressures. The mechanical behavior of these materials does not differ substantially from that of other granular materials used in the construction of dams. It is, therefore, feasible to construct deposits of great heights with waste rock.

The shear strength envelope of the waste rock curves gradually as the level of pressure is increased, showing a reduction in the shear strength induced by the increase in compressibility of the granular matrix, associated with the crushing of particles. In fact, the compressibility and particle breakage are increased significantly for effective mean pressures above 1 MPa. It is worth noting that the final grain size distributions obtained at high pressures for the different waste rocks tend towards a single grain-size curve, independently of the initial grain size being tested, for materials conformed by particles of similar lithology. Considering the mechanical behavior of waste rock at high pressures, subjected to overburden stresses well beyond any precedent, the front slope of the dump has been designed with a global angle of the order of 22° (2.5:1=H:V), with the objective of generating a smaller load zone destined to create a high-strength and high-permeability buttress.

HWRDs are structures that present exceptional challenges for geotechnical engineers, researchers and operators. There is a great need to improve the present state of the art concerning the real behavior of *in situ* waste rock, subjected to the concomitant effect of high stresses and natural or forced leaching, by taking into account the scale effects related to the particle sizes of these materials. There are uncertainties about the possible failure modes of these structures, which can be more complicated than those analyzed by conventional limit equilibrium methods, as well as about the sliding mechanisms, especially when debris flows may occur.

The design, operation and construction practices of HWRDs should be improved through a number of preventive measures to reduce potential instabilities. Water diversion, snow management, and drainage conditions require very special attention to ensure the physical stability and environmental viability of such structures.

It is necessary to know how these structures will behave in different environments. By monitoring and taking *in situ* measurements, documents of their operation can be gathered and valuable information will be saved. It is also necessary to design and manufacture new instrumentation systems that are more robust and destined to survive under very restrictive environments, subjected to large deformations and stresses induced by heavy rock transport equipment and an aggressive chemical environment.

4.9. Acknowledgements

The authors wish to express their most sincere thanks to División Andina of Codelco Chile and to Anglo American Chile, for supporting most of the applied research that was carried out in relation to high waste rock. We also thank the IDIEM Laboratory and Professor Ramón Verdugo for their kind willingness to accept suggestions and comments during the development of the testing campaign. We also recognize, with gratitude, the help of Professor Norbert Morgenstern who, at the early stages of the HWRDs studies, recommended many of the analyses that are discussed in this chapter. Finally, we extend a special acknowledgement of the support provided by the Arcadis Chile geotechnical team, particularly Roberto Leon and Luis Valenzuela, Director of Arcadis Chile.

4.10. Bibliography

[ARC 06-07] ARCADIS, Internal Report, 2006-07.

[BAR 93] BARD E., Comportement des matériaux granulaires secs et à liant hydrocarboné. Doctorate thesis, Ecole Centrale de Paris, France, no. 93 ECAP 0320, 1993.

[BAR 07a] BARD E., CAMPAÑA J., ANABALÓN MA E., "Waste rock behaviour under high pressures", *XIII Pan-American Conference on Soil Mechanics and Geotechnical Engineering*, Venezuela, 2007.

[BAR 07b] BARD E., ANABALON MA E., CAMPAÑA J., APABLAZA R., GALLARDO M., "Comportamiento de Lastre ROM Sometido a Altas Presiones", *VI Congreso Chileno de Ingeniería Geotécnica*, Valparaíso, Chile, 2007.

[BIA 71] BIAREZ J., CAMBOU B., KEIME F., MARTINET M., "Compressibilité des milieux pulvérulents sur chemins isotrope ou oedométrique et rupture de particules. Le comportement des sols avant rupture", *Bull, de Liaison des Laboratoires des Ponts et Chaussées*, Journées Nationales, Paris, France, May-June 17-19, 1972.

[BIA 92] BIAREZ J., BARD E., "Loi de Comportement d'un agrégat routier. Analogie avec le comportement des argiles", *Volumen Raúl J. Marsal, Sociedad Mexicana de Mecánica de Suelos*, A.C., 1992.

[BIA 93] BIAREZ J., BARD E., "Behaviour of granular materials: The moduli", *3er Congreso Chileno de Ingeniería Geotécnica*, La Serena, Ed. Sociedad Chilena de Geotecnia, vol I, pp. 117-128, May 1993.

[CLE 81] CLEMENTS R.P., The deformation on Rockfill: inter-particle behaviour, bulk properties and behaviour in dams, PhD Thesis, Faculty of Engineering, King's College, London University, 1981.

[COL 88] COLLIAT-DANGUS J.L., DESRUES J., FORAY P., "Triaxial testing of granular soil under elevated cell pressure", *Advanced Triaxial Testing of Soil and Rock*, ASTM STP 977, pp. 290-310, 1988.

[DAW 99] DAWSON R.F., MORGENSTERN N.R., STOKES A.W., "Liquefaction flowslides in Rocky Mountain coal mine waste dumps", *Canadian Geotechnical Journal*, vol. 35, pp. 328-343, 1999.

[LEE 67] LEE K.L., FARHOOMAND I., "Compressibility and crushing of granular soil in anisotropic triaxial compression", *Canadian Geotechnical Journal*, vol.4, no. 1, pp. 68-100, 1967.

[MAR 75] MARSAL R., RESENDIZ D., "Presas de Tierra y Enrocamiento", *Editorial Limusa*, pp. 221-267, 1975.

[MAR 77] MARSAL R.J., "Research on granular materials (rockfill and soil-gravel mixtures)". *Experimental work compiled from the IX International conference of Soil Mechanics and Foundation Engineering (Tokyo)*, E-25, Universidad Nacional Autónoma de México, June 1977.

[MOR 96] MORGENSTERN N., *Conference International Symposium on Seismic and Environmental Aspects of Dams Design. ICOLD-Santiago*, vol. II, Chile (not published).

[MOR 01-06] MORGENSTERN N., Personal communications.

[VAL 04] VALENZUELA L., "Stability issues in natural and man made slopes in mining". *International Conference on Landslides*, Brazil, 2004.

[VAL 07] VALENZUELA L., BARD E., CAMPAÑA J., "High waste rock dumps in the mining industry", *XIII Pan-American Conference on Soil Mechanics and Geotechnical Engineering*, Venezuela, 2007.

[VAL 08] VALENZUELA L., BARD E., CAMPAÑA J., ANABALON M., "High waste rock dumps – challenges and developments", *First International Seminar on the Management of Rock Dumps, Stockpiles and Heap Leach Pads*, Australia, 2008.

Chapter 5

Models by Jean Biarez for the Behavior of Clean Sands and Remolded Clays at Large Strains

5.1. Introduction

In the course of his career, Jean Biarez collected a great number of experimental test results that were used to develop an all-encompassing approach for studying soil behavior. In this chapter, we are especially interested in his work on behavior at large strains, which can be important, for example, in calculating the conditions for running penetrometer or pressuremeter tests. The materials examined by our present topic are remolded clay and clean sand. Even if these materials are solely laboratory soils, it can be said that all post-Glacial natural clays and non-cemented sands have these characteristics in common.

Two groups of parameters can sum up the properties of any given soil. The first group involves identification parameters linked to the mineralogical nature of the grains and can be represented, for example, in the case of clays, not only by their Atterberg limits but also by their equivalent void ratios or by e_{max} and e_{min} in the case of sands. The second group involves mechanical parameters (e or w_{nat}) that express the geometric arrangement of grains and of their evolution.

The soil – a granular and discontinuous medium (DM) – may be regarded as a virtual continuous medium (CM) with statistical correlations presented in the explanatory scheme in Table 5.1. Hence, what we call "mechanical properties of

---

Chapter written by Jean-Louis FAVRE and Mahdia HATTAB.

soils" may be defined as the integration of the grain properties (called "nature of grains") in their spatial configuration (or "arrangement of grains") and their mechanical boundary conditions (which can be represented by the consolidation stress tensor).

Nature of grains (DM)	+	Arrangement and confinement of grains (DM)	⇒	Mechanical properties of soils (CM)

Table 5.1. *Logical framework for statistical connections*

For a long time now it has been possible to compare the behavior of laboratory soils to that of natural soils and the method to do so has been well established thanks to a unified vision of clays, sands and granular soils as granular media. We will argue in favor of this unified vision by referring to the explanatory scheme in Table 5.1 for statistical connections between parameters. This approach will enable better integration of specific natural behaviors, such as anisotropy, cohesion, cracking, etc. We can better understand this method of transition, introduced by Biarez, through the representation in Figure 5.1. The numerical formulation of the models, which Biarez drew in graphic form [BIA 75], is the work of Favre [FAV 80, FAV 02]; similar work can also be found in Biarez and Hicher [BIA 94].

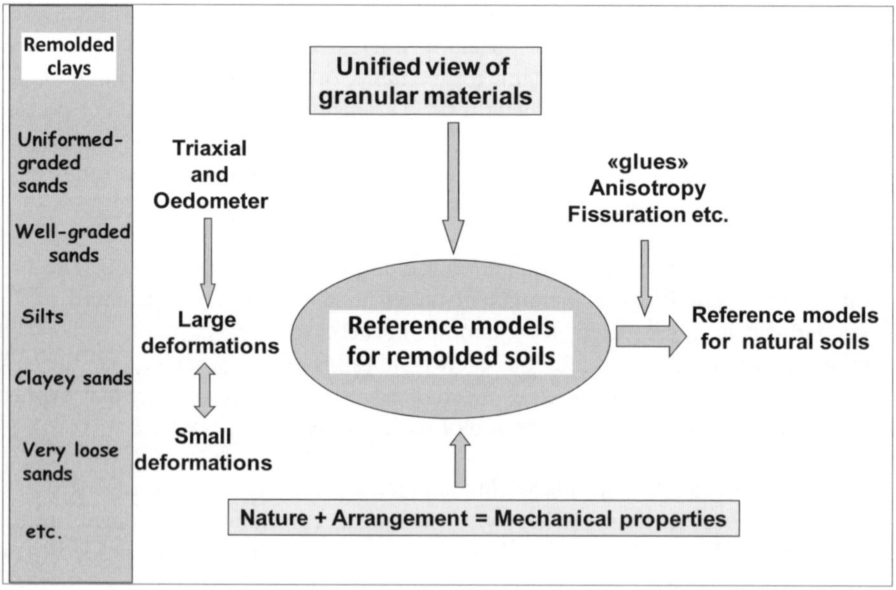

Figure 5.1. *Jean Biarez research approach for behavior modeling*

5.2. Biarez's model for the oedometer test

Grains become bonded or "glued", according to Biarez, due to cementation during the sedimentation and consolidation processes. If this is the case, the clay is defined as "non-remolded" or "intact" and the links between the grains, which can have various origins, form a supplementary parameter for phenomenological behavior [BIA 98], particularly during oedometric loading.

Where the bond between the grains is missing, as in industrial and post-Glacial clays, or when the bond has been destroyed, for example by a high loading, the soil is described as remolded.

It is fundamentally important to study the behavior of material that has been remolded and reconstituted in the laboratory in normally consolidated and overconsolidated conditions [BIA 94] because the "intrinsic" mechanical properties can then be deduced, making it possible to define a fixed reference framework in order to explain the behavior of natural, non-remolded soils.

Biarez [BIA 72, BIA 75, BIA 77], and more recently Burland [BUR 90], introduced the concept of intrinsic compression properties in order to describe, in a normalized plane, the behavior of one-dimensional normally consolidated clay having been remolded and reconsolidated (at w_{sat}=1.5 w_L).

Favre [FAV 72, FAV 80] shows that the Atterberg limits ("nature of grain" parameters) are particularly helpful in explaining the mechanical properties. In mineral clays, these properties are strongly linked by the following relation:

$$I_p = 0.73(w_L - 13) \qquad [5.1]$$

Thus, the model proposed in Figure 5.2a, plotted with the Casagrande A line, defines the mineralogical nature of the clay.

Moreover, a large number of results collected by Favre [FAV 72, FAV 80] show that loadings of 6.5 kPa and 1,000 kPa on the oedometric path correspond statistically on average to w_L and w_P, respectively (or e_L and e_P considering $\gamma_s/\gamma_w = 2.7$, see relation [5.2] and Figure 5.2b). This leads first to equation [5.3] or [5.4] for the compressibility index C_c, then to relation [5.5] when using [5.1] in [5.3].

$$w_{sat} = w_L \text{ for } \sigma'_v = 6.5 \text{ kPa}$$

$$w_{sat} = w_P \text{ for } \sigma'_v = 1 \text{ MPa} \qquad [5.2]$$

$$C_c = \frac{(w_L - w_P)\ 2.7/100}{\log(1000/6.5)} \qquad [5.3]$$

$$C_c = I_P/80 \qquad [5.4]$$

$$C_c = 0.009(w_L - 13) \qquad [5.5]$$

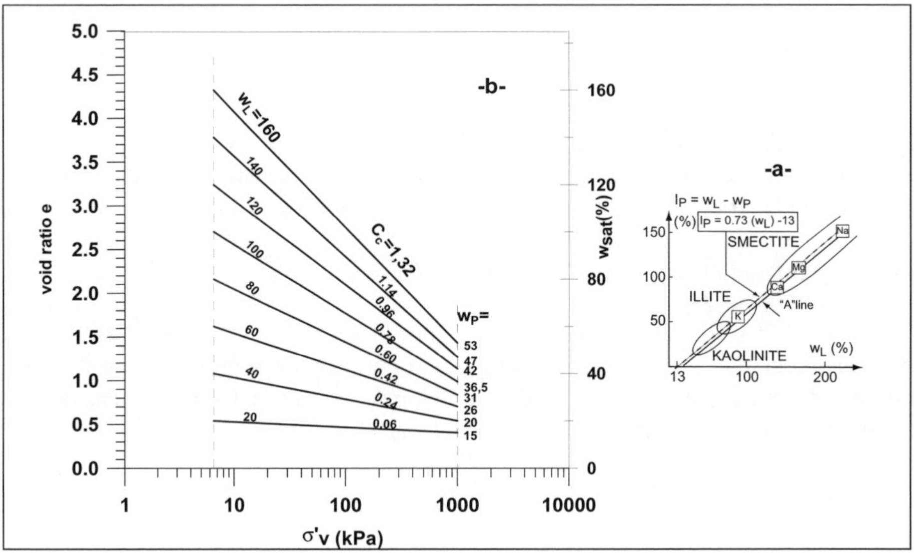

Figure 5.2. *Biarez correlations for the normally consolidated remolded clays*

Compared to relation [5.6] proposed by Skempton in 1944 [SKE 44] for remolded clays, we underestimate the compression index C_c for w_L = 40% by 10% and w_L = 70% by 5%.

$$C_c = 0.007(w_L - 10) \qquad [5.6]$$

Nevertheless, the simple model represented by relations [5.2] and [5.5] was retained for the oedometric test, and hence, by relation [5.1], it uses only one Atterberg limit, w_L. It is a straight line in the $(e, log\sigma'_v)$ or $(w_{sat}, log\sigma'_v)$ plane, passing through the two points given by relation [5.2] with a slope defined by equation [5.5]. A single abacus was then obtained in equation [5.7], σ'_v being expressed in kPa and $I_L = (w - w_P)/(w_L - w_P)$, w_L being between 20 and 160%.

$$I_L = 0.46(3 - \log \sigma'_v) \qquad \sigma'_v \text{ given in kPa} \qquad [5.7]$$

The “arrangement of grains”, therefore, appears to be directly linked to the consolidation stress tensor through the mineralogical properties. This connection corresponds, in the (I_L – σ'$_v$) plane, to a straight line called NCRS (normally consolidated remolded simplified, see Figure 5.3). Thus, a normally consolidated remolded clay will be represented by points located on the NCRS and the overconsolidated remolded clay by points below the NCRS. In this case, the overconsolidation of the material only depends on the clay loading history, expressed by the overconsolidation ratio (OCR) = σ'$_p$/σ'$_v$. Here σ'$_p$ is the maximum effective stress and σ'$_v$ is the overburden effective stress.

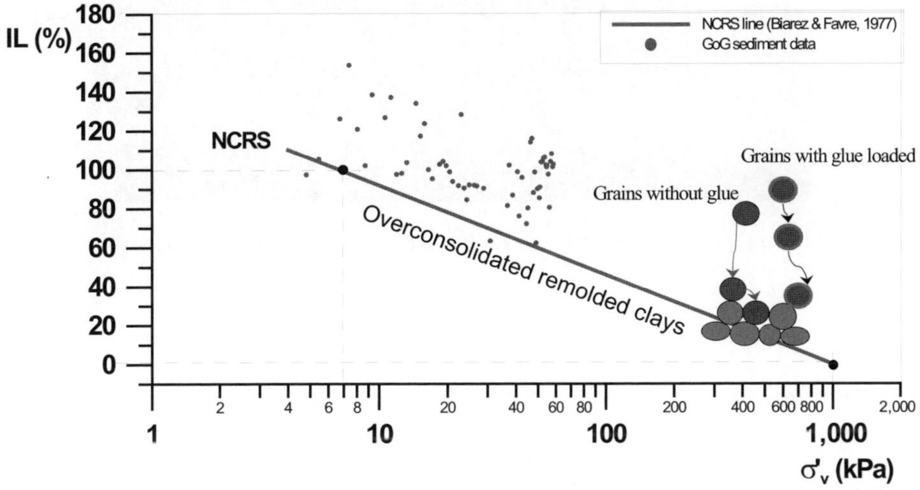

Figure 5.3. *Clayey soils behavior in (I_L-σ'$_v$) plane*

For the natural deepwater GoG (Gulf of Guinea) marine clay, which will be examined below in the Biarez and Burland models, the liquidity indexes calculated at different depths (then, at different overburden stresses) as shown in Figure 5.3, give a cloud of points located above the NCRS line. Greater porosity than that predicted by the NCRS for an equivalent stress σ'$_v$, characterizes this material. Thus, the sediment can be defined, in a first approximation, as a clay constituted of “grains” linked by a physicochemical cementation [FAV 06, HAT 10]. Aggregates seem to have a very complex structure, as seen in Figure 5.4, which shows an organo-mineral aggregate composed of a “foliage” of smectite particles, where some spheroids of framboïdal pyrite are bonded. We could ask whether there is a possible phenomenological analogy between these highly porous deep-water marine sediments and the sandy lumps with dual porosity (inter- and intra-lumps), which remain very loose due to capillary cementation (see section 5.8 and Figure 5.16).

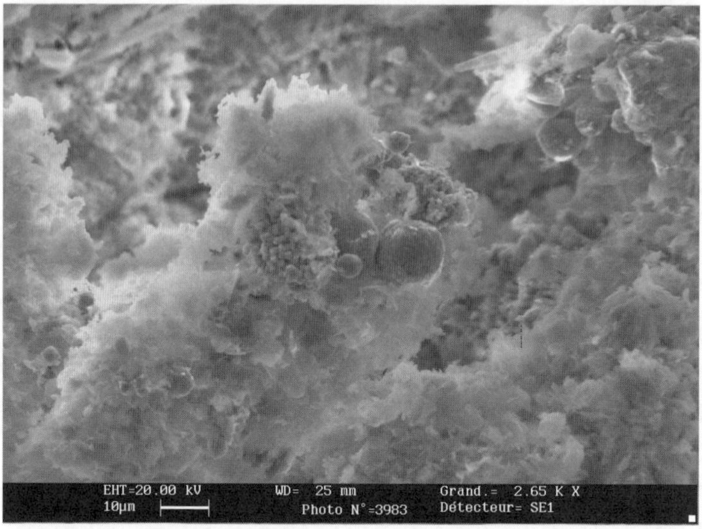

Figure 5.4. *SEM Photo of GoG marine sediment at initial state*

5.3. Perfect plasticity state and critical void ratio

The comparison of the oedometer model with the triaxial test, in particular the isotropic triaxial test, presumes that we know the effective stress ratios σ'_h/σ'_v. This ratio K_0 is considered to be constant by several authors. The following formula by Jacky is generally used, where Φ_{pp} is the friction angle at perfect plastic state:

$$K_0 = 1 - \sin \Phi_{pp} \quad [5.8]$$

Thus, the oedometric loading can be represented in the (e – log p') plane, using the transformation that depends on Φ_{pp}: $p' = \sigma'_v\,(1+2K_0)/3$.

The observation confirms the parallelism of the isotropic and oedometric paths, but does not permit the distance between these two paths to be estimated.

Finally, each path at $q/p' = \eta$ = constant, also has $K = \sigma'_3/\sigma'_1$ = constant, and appears to be parallel with both lines (oedometric and isotropic) in the (e – log p') plane:

– isotropic compression $NC \rightarrow \eta = 0$;

– oedometric loading $\rightarrow \eta = 3(1K)/(1+2K)$.

The perfect plastic state with the critical void ratio is characterized, in the $(q - p')$ plane, by a straight line with the slope $M = 6\sin\Phi/(3 - \sin\Phi)$ or $\sin\Phi = 3M/(6 + M)$.

For remolded clays, the perfect plastic state can be reached easily, starting from the isotropic normally consolidated state in the drained and undrained paths. In the $(e - \log p')$ plane, these paths reach a straight line. Several tests demonstrate that this line passes in the vicinity of two reference points: $p' = 3$ kPa for $w_{sat} = w_L$ and $p' = 0.5$ MPa for $w_{sat} = w_P$. Besides, the same line is reached from overconsolidated states, even if it is more difficult to determine, due to the development of localizations in the large strains.

Given the same ratio $\sigma'_v(w_L)$ and $\sigma'_v(w_P)$ of the oedometric path, which ensures the parallelism, Biarez took relation [5.9] with $p' = 3.5$ kPa (instead of 3 kPa) for $w_{sat} = w_L$, and a slope C_c, defined by equation [5.10], as the model for the critical straight line (CSL), which represents the critical void ratio.

$$w_{sat} = w_L \text{ for } p' = 3.5 \text{ kPa}$$

$$w_{sat} = w_P \text{ for } p' = 500 \text{ kPa} \quad [5.9]$$

$$C_c = 0.009(w_L - 13) \quad [5.10]$$

Equations [5.9] and [5.10] lead to equation [5.11] expressed in the $(I_c - \log p')$ plane with the consistence index $I_c = 1 - I_L$, p' being in kPa:

$$I_c(\text{CSL}) = 0.46(\log p' - 0.54) \quad [5.11]$$

For "glue less" sands, we found very few normally consolidated tests and the overconsolidated tests tend to present localizations before the perfect plastic state. Biarez [SAI 97] selected three tests that reached the perfect plastic state (see Figures 5.5 and 5.6).

The model of the critical void ratio line, represented by CSL for the clean sands, is defined by system [5.12] where the slope is given by [5.13].

$$e = e_{max} \text{ for } p' = 0.1 \text{ MPa}$$

$$e = e_{min} \text{ for } p' = 8 \text{ MPa} \quad [5.12]$$

$$C_c = (e_{max} - e_{min})/1.90 \quad [5.13]$$

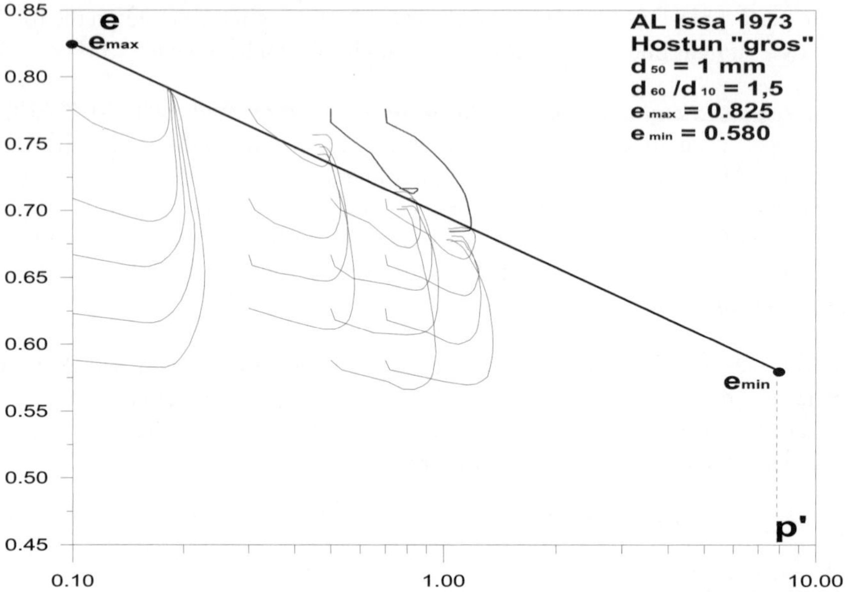

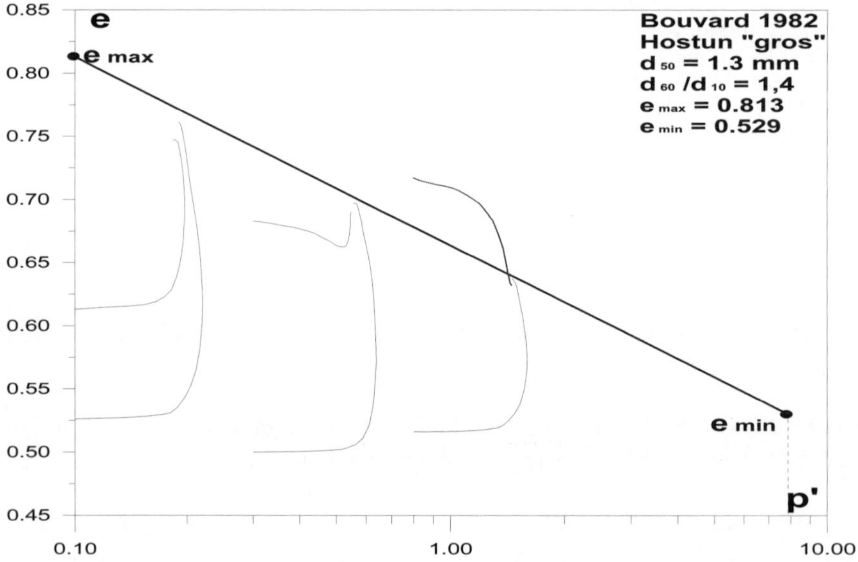

Figure 5.5. *Critical void ratios diagram for Hostun sand (coarse sand) (After Al-Issa tests [ALI 73]; and Bouvard [BOU 82])*

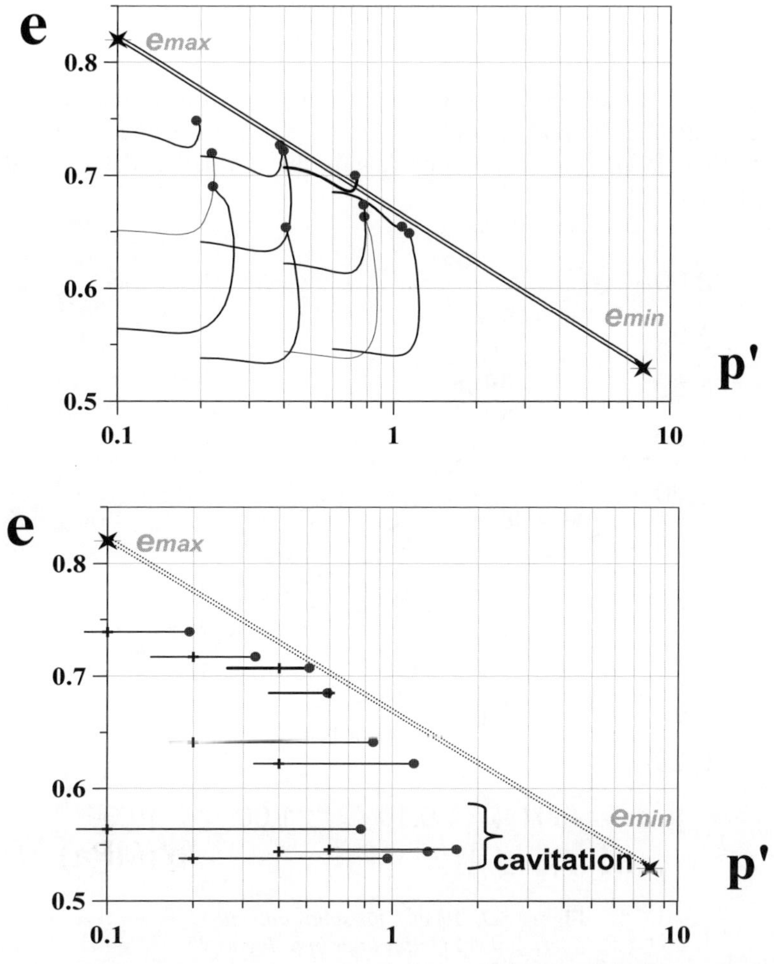

Figure 5.6. *Control of critical void ratio at plastic perfect state. Drained and undrained tests after Mokham [MOK 83]*

Let equation [5.14] be placed in the plane (I_d – log p'), where I_d is the density index (which is what the consistence index is called for sands), p' being in kPa:

$$I_d(\mathrm{CSL}) = 0.52(\log p' - 2) \quad [5.14]$$

The superposing of both perfect plastic models in the same plane shows just how far we can take the analogy between clays, clean sands and granular materials (see Figure 5.7). The very foundation of this analogy resides in the "granular" character of the "glue-less" clays. The granular materials with $d_{60}/d_{10} > 10$ are in the plane $(e - \log \sigma'_v)$ between $w_L = 20\%$ and $w_L = 40\%$.

In Figure 5.7, the three straight lines in bold indicate clean sands: the line of Lee and Seed [LEE 67], the line of Mokham [MOK 83] and the line of Flavigny and Bousquet 1986 [FLA 93]. The dotted line is that of the model of sands with $d_{60}/d_{10} = 10$ [BIA 94].

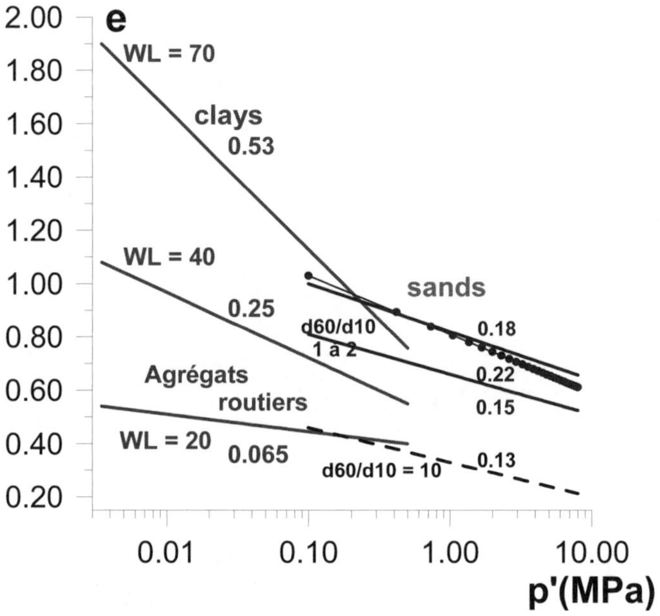

Figure 5.7. *Abacus, for sands and clays, critical void index lines in (e-logp')*

5.4. Normally and overconsolidated isotropic loading

5.4.1. *Analogy between sands and clays*

Experiments have shown that, for sands, a normally consolidated behavior can be had for very large void ratios and for small medium stresses before the grains are crushed (see Figure 5.8, the paths with a pronounced slope before 2 kPa). It is very difficult to obtain a sufficiently large void ratio to observe this NC behavior of sands and granular materials, but with slight moistening, it becomes possible.

The NC domain for these sands is limited: indeed, at around 2 to 3 MPa, we observe grain crushing. The *in situ* sands ($e < 0.8$ or $I_d > 0.3$) are "overconsolidated" (see Figure 5.8, the paths with a slight slope before 2 MPa) and we will *never* find the NC line again, since it is on this side of 2 to 3 MPa, except for sands with very hard grains such as Sacramento sand [LEE 67]. Consequently, we can speak about a fictive consolidation stress p'_{ic} on the extension of the NC line (ISL (See Figure 5.18)) without grain crushing. In this way, for very loosely arranged sands, we again find the same loading NC and unloading-reloading OC, as for clays.

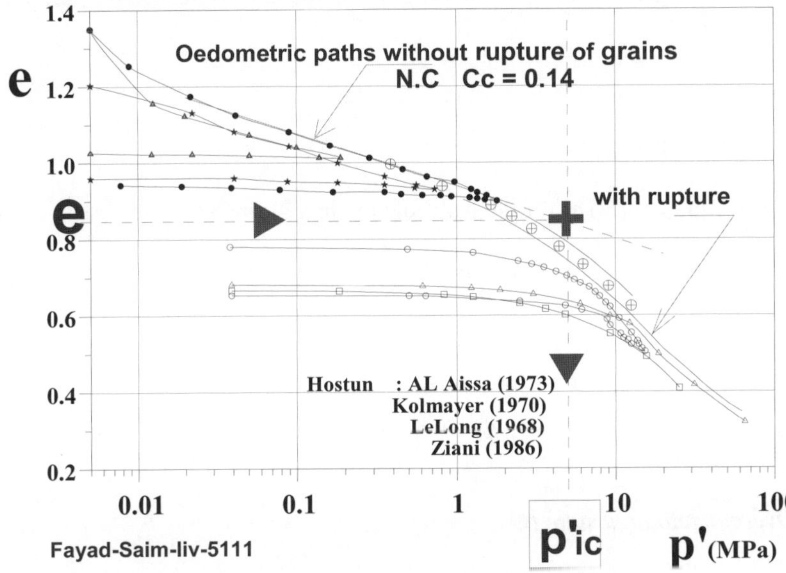

Figure 5.8. *Oedometric tests on laboratory sand [FAY 00]*

5.4.2. *Normally consolidated state (ISL)*

Biarez observed that Roscoe's model [5.15] [ROS 58] represented normally consolidated soils, drained as well as undrained, rather well [DAS 97]:

$$e - e_0 = -C_c \log(\frac{p'}{p'_0}) - Cd \log\left[1 + \left(\frac{q}{Mp'}\right)^2\right] \qquad [5.15]$$

Parameter C_d is easily observed on p' constant paths. Hence, for sands as well as for clays, Biarez observed that the C_d is around 0.3 with a gap between the isotropic path and critical state: $\Delta e = 0.1$. These two values have to be compatible (on the condition that C_d is constant and does not depend on p', which was found to be the case). On p' constant stress paths, we must have relation [5.16] with $\eta = q_{pp}/p' = M$, where the index "$_{pp}$" indicates the perfect plastic state:

$$e_{pp} - e_{NC} = -\Delta e = -C_d \log\left[1+\left(\frac{\eta_{pp}}{M}\right)^2\right] = -0.3\ C_d \qquad [5.16]$$

For loading with p' constant, starting from the NC state, Biarez adopted the following model:

$-\Delta e = 0.1$

$C_d = 1/3 = 0.33$ [5.17]

which gives equation [5.18] for the isotropic loading at the NC state (ISL):

$$e(\text{ISL}) = e_0 + 0.1 - C_c\,(\pi_0 - \log p') \qquad [5.18]$$

where:

– for sands, $e_0 = e_{max}$, $\pi_0 = 2$, $C_c = (e_{max} - e_{min})/1.9$, p' in kPa;

– for clays, $e_0 = (2.7/100)\ w_L$, $\pi_0 = 0.54$, $C_c = 0.009(w_L - 13)$, p' in kPa.

5.4.3. *Overconsolidated state (C_s)*

Let point $(p'_{ic} - e_C)$ be the maximum isotropic stress state and point $(p'_{i0} - e_{OC})$ be the "initial" unloading stress state (see Figure 5.9). p'_{i0} can be noted p'_i or σ'_{30}. The point obtained $(p'_{i0} - e_{OC})$ is easily characterized by its OCR (p'_{ic}/p'_{i0}) for clays, but with great difficulty for sands, as we have just seen above. Therefore, the $(p'_{ic} - e_C)$ point is rarely observable.

Biarez [SAI 97], proposed a new parameter for overconsolidation, or *generalized over-consolidation*, $e_{NC} - e_{OC}$, which could process sands and clays together: instead of the abscissae, he uses the ordinates. This new formulation has the benefit of not only introducing the history of the loading, $\log(p'_{ic}/p'_{i0})$, but moreover the nature of the material, via $(C_c - C_s)$. It is therefore more general than the usual formulation and requires a model for C_c as well as for C_s.

The model of C_c for clays has already been given by [5.4].

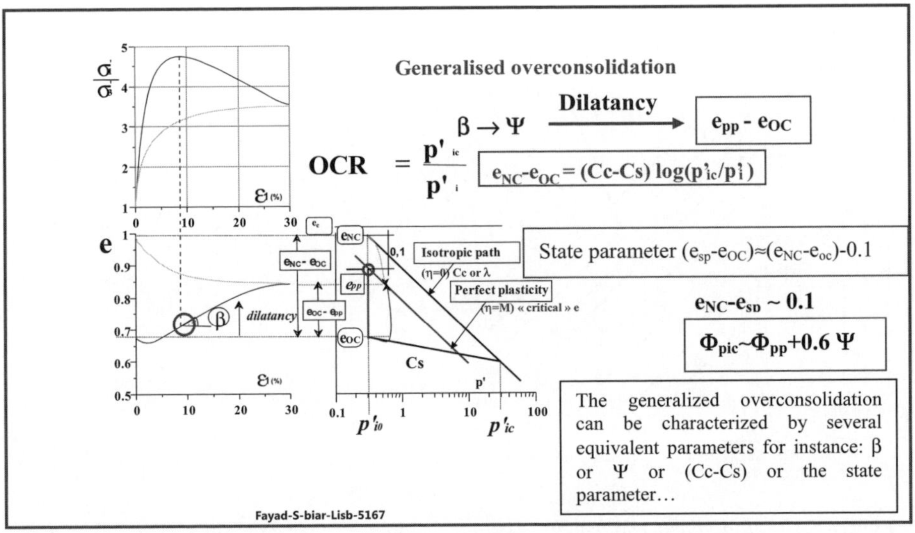

Figure 5.9. *(e_{NC}-e_{OC}) and (e_{sp}-e_{OC}) parameters of the generalized overconsolidated behavior*

There is no well-established model for clean sands. Biarez and Hicher [BIA 94] propose $C_c = 0.20$ for $d_{60}/d_{10} < 2$. This the one we have adopted.

The model for C_s is not well established for clays and even less so for sands. Nevertheless, in accordance with some experimental data [FAV 80, SAI 97], we have adopted the following models:

$C_c = 0.009(w_L - 13)$ $\quad C_c/C_s = 4$ for clays

$C_c = 0.20$ $\quad C_c/C_s = 10$ for clean sands

For clean sands, Saim [SAI 97] researched the correlations between $(e_{NC} - e_{OC})$ and $(e_{sp} - e_{OC})$ and the usual parameters of the dilatancy, Ψ and β_{max}, and also with new measurable parameters, such as $(e_{OC} - e_{pp})$ or Φ_{peak}:

[$\beta_{max} = (d\varepsilon_v/d\varepsilon_1)_{max}$ and $\sin\psi = tg\beta_{max}/(2+tg\beta_{max})$]

Figure 5.10a gives the $(I_D - \log p')$ plane graduated with parallel straight lines having the same Φ_{peak} values, Ψ, and $e_{NC} - e_{OC}$. Favre [MEK 02] graduated it with OCR. In Figure 5.10b, $I_C - \log p'$ gives the graduation with OCR for the clays around wL = 70%. The gap between ISL and CSL is 0.25 on I_d and 0.09 on I_C for these clays.

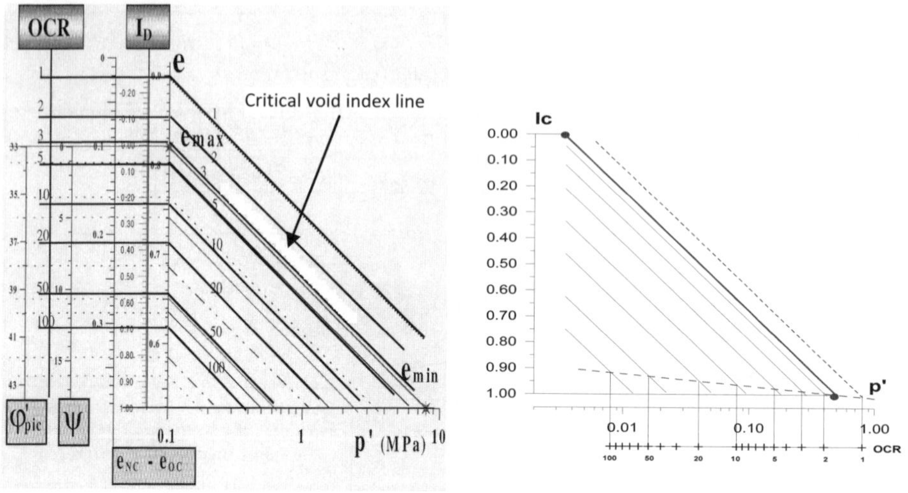

Figure 5.10. *"Relative arrangement of grains". Abacus – "isotropic stress" for sands and clays: a) sands: ($e_{max} - e_{min} \sim 0,4$) after Mekkaoui 2002; b) clays: ($w_L \sim 70\%$)*

5.5. The drained triaxial path for sands and clays

5.5.1. *The reference behavior*

A certain number of criteria were defined by Biarez for the validation of triaxial tests [SAI 97] with the goal of obtaining statistics on the normally consolidated drained behavior of clean sands and clays.

This made it possible for him to establish a set of experimental standards, shown in Figure 5.11, graduated with the available experimental data ($e_{pp} - e_{OC}$ and Ψ).

We note the perfect similarity between the NC "black clay" and the "NC sand". The black clay is a kaolinite with w_L = 70% and M = 0.91. We owe the test to Zervoyannis [ZER 82]. The "NC sand" is a "wet-poured" sand, i.e. it is very loose (w = 5%, M = 1.25). We owe this test to Ziani [ZIA 87].

In making these selection criteria, Biarez chose one of the rarely available drained tests of highly overconsolidated clay without strain localization: the one performed by Ladd [LAD 64]. Further studies could be developed in this area. The test selection criteria ("perfect" tests) can be found in Mekkaoui's thesis [MEK 02].

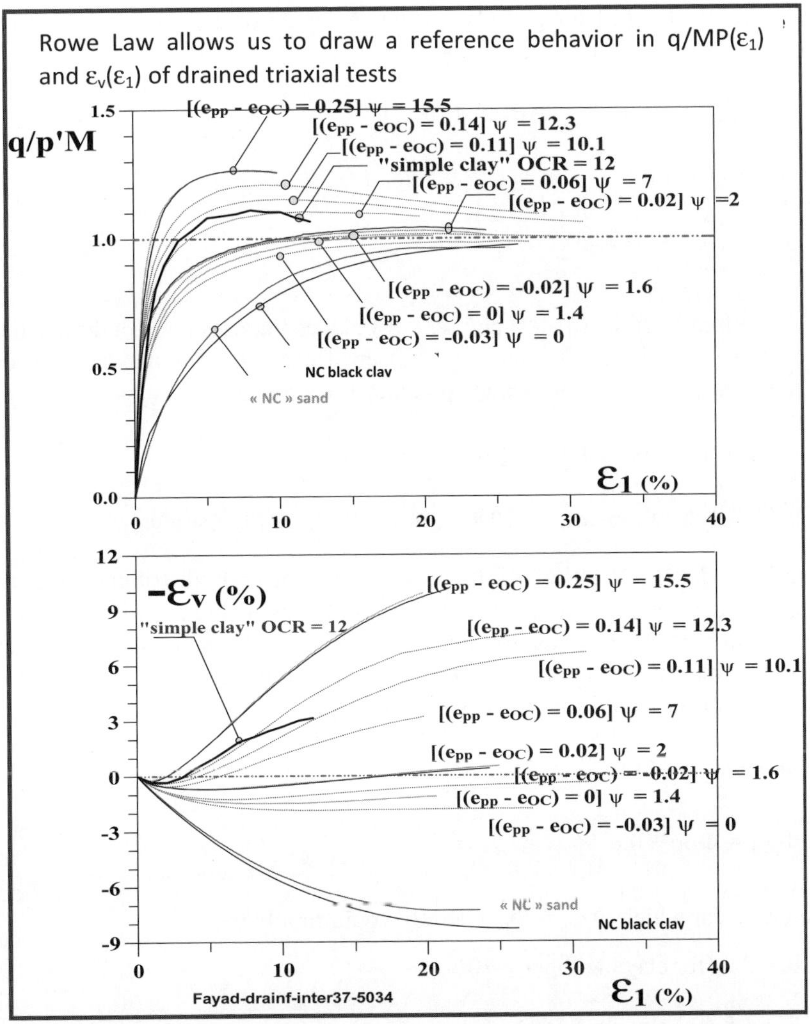

Figure 5.11. *Experimental bundle reference for drained triaxial tests, parametered with dilatancy*

5.5.2. *The mathematical model*

A mathematical model used to represent the set of experimental standards was proposed by Favre and Hachi [HAC 01].

In the $(\varepsilon_v - \varepsilon_1)$ plane, the rather complex model represented by equation [5.19] includes five parameters. Its transcription on the $(q - \varepsilon_1)$ plane comes from Rowe's dilatancy law [ROW 62] for granular media. An additional parameter, the friction angle between grains Φ_f, has been introduced:

$$\varepsilon_v = \theta_0 + \theta_1 \exp(-\theta_2 \varepsilon_1) - \theta_3 \exp(-\theta_4 \varepsilon_1) \quad [5.19]$$

$$\frac{\sigma'_1}{\sigma'_3} = tg^2\left(\frac{\pi}{4} + \frac{\phi_f}{2}\right)\left(1 - \frac{d\varepsilon_v}{d\varepsilon_1}\right) \quad [5.20]$$

Every path in both the $(q - \varepsilon_1)$ and $(\varepsilon_v - \varepsilon_1)$ planes imposes two conditions (at the origin and at the infinite). In this way, the set of experimental standards becomes reduced to a parameter of nature and a parameter of arrangement, such as:

– Φ_{pp} (or M) as "nature of grains" parameter;

– ΦOCR (or $e_{pp} - e_{OC}$, or Ψ) as "grains arrangement" parameter.

Favre's model [FAV 80] gives Φ_{pp} as a function of grain distribution for sands; the Biarez and Hicher's model [BIA 94] gives Φ_{pp} as a function of w_L for clays. Hachi and Favre [HAC 01] quantified the Rowe internal friction angle as a function of the perfect plastic friction angle and of OCR [5.21], in agreement with the model given by Bolton [BOL 86] for the peak friction angle [5.22]:

$$\Phi_f = \Phi_{pp} - 12.4\,(e_{pp} - e_{OC}) \quad [5.21]$$

$$\Phi_{peak} = \Phi pp + 0.6\ \Psi \quad [5.22]$$

and the OCR, for which $\Phi_{peak} = \Phi_{pp}$ ($\Psi = 0$) would finally be:

OCR = 2.6 (for clays with $w_L = 70\%$);

OCR = 5.8 (for clean sands).

5.6. The undrained triaxial path for sands

Biarez became attached to sand models because, unlike clays, sands produce a complex path in the $(q - \varepsilon_1)$ plane. We shall see later that clay paths are actually of the same nature when plotted in the same plane.

5.6.1. *Simplified Roscoe formula for undrained consolidated soils*

The Roscoe surface [5.15] at constant volume becomes:

$$q = p'M\left[\left(\frac{1}{(p'/p'_0)^{C_c/C_d}} - 1\right)\right]^{1/2} \qquad [5.23]$$

This equation defines the bell-shaped curve in the $(q/p'_{ic} - p'/p'_{ic})$ plane in Figure 5.12 and the coordinates for the peak are:

$$\begin{cases} \dfrac{p'}{p'_{ic}} = (1 - C_c/2C_d)^{C_d/C_c} \\ \eta_{pic} = \left(\dfrac{q}{p'}\right)_{pic} = M\left(\dfrac{C_c}{2C_d - C_c}\right)^{1/2} \end{cases} \qquad [5.24]$$

where $C_c = 0.2$ and $C_d = 0.33$ (see section 5.4.2 and 5.4.3).

Hence, the soils present a maximum under the perfect plasticity only if $\eta_{peak} < M$, that is to say $C_c < C_d$.

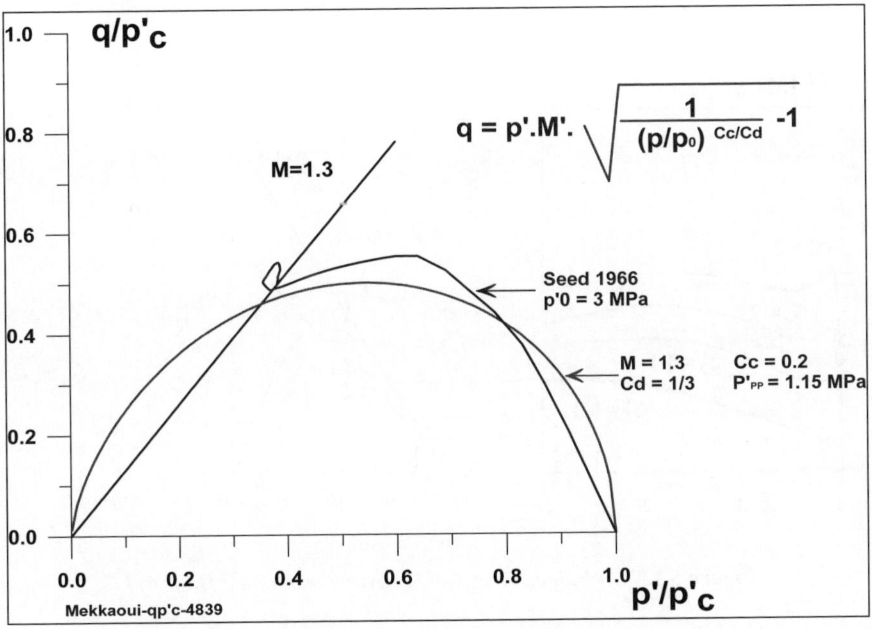

Figure 5.12. *Undrained NC path after the simplified Roscoe law*

Let us note that only the remolded clay–sand mixtures with $w_L < 24$ can be found in this category.

5.6.2. *Modeling of the maxima under the right M on the plan $q - p'$*

Lee and Seed [LEE 67] presented four outstanding tests (see Figure 5.13) on Sacramento sand, three with a peak in the $(q - \varepsilon_1)$ plane for high stresses and maxima under the *M* line.

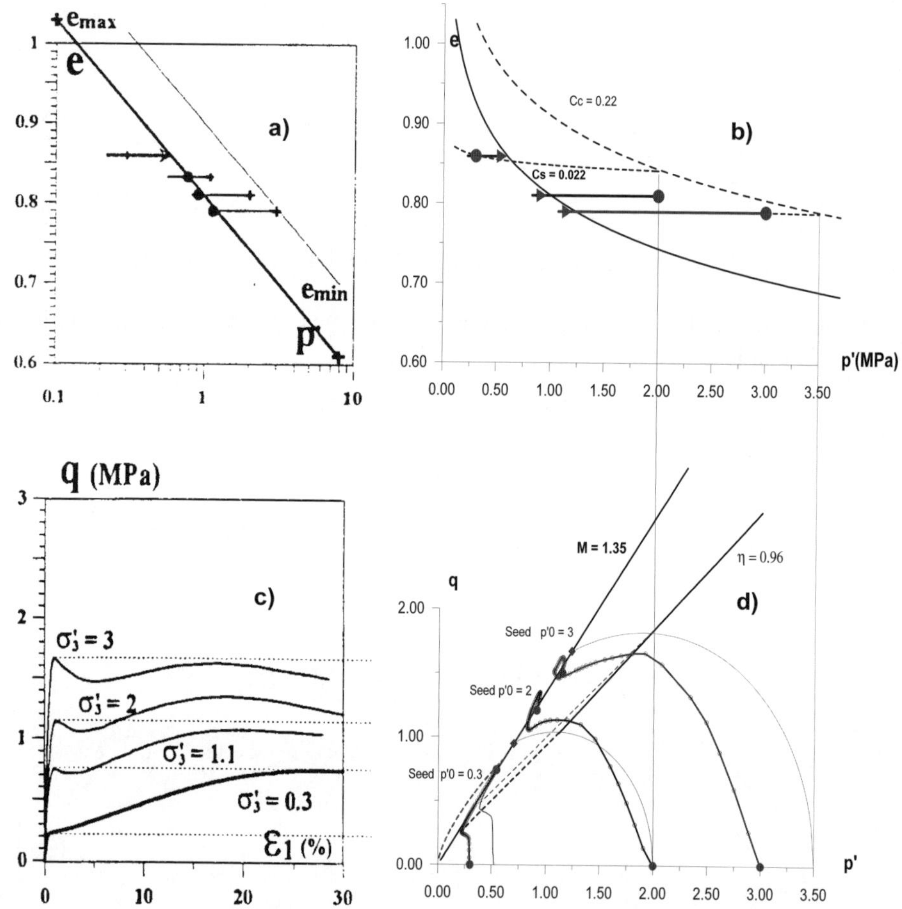

Figure 5.13. *Undrained tests, weakly overconsolidated [LEE 67]*

The isotropic loading with p'_0 = 3 MPa did not provoke grain crushing. The beginning of the triaxial test is under the theoretical NC (ISL) curve, which classes the material in the theoretical overconsolidation category (for sands). Its theoretical overconsolidation stress, following the C_s model, is p'_{iC} = 3.5 MPa. Figure 5.13d represents the results of the simulation by the modified Cam-Clay model of a NC test on this material.

Both of these paths, one real (p'_0 = 3 MPa) and the other fictitious (p'_0 = 3.5 MPa), are two horizontal straight lines in the (e – p') plane, separated by a slight gap until the critical void ratio curve. Thus, at the perfect plasticity state, q_{final} for p'_0 = 3 MPa is a bit smaller than q_{final} would be for p'_0 = 3.5 MPa. The value q_{final} obtained with this Lee and Seed test shows this small gap; furthermore, the curve, after the peak (see Figure 5.13d) descends towards M before moving up towards this q_{final} final value.

The same phenomenon re-appears for both the other tests at high stresses (p' = 2 MPa and p'_0 = 1.1 MPa).

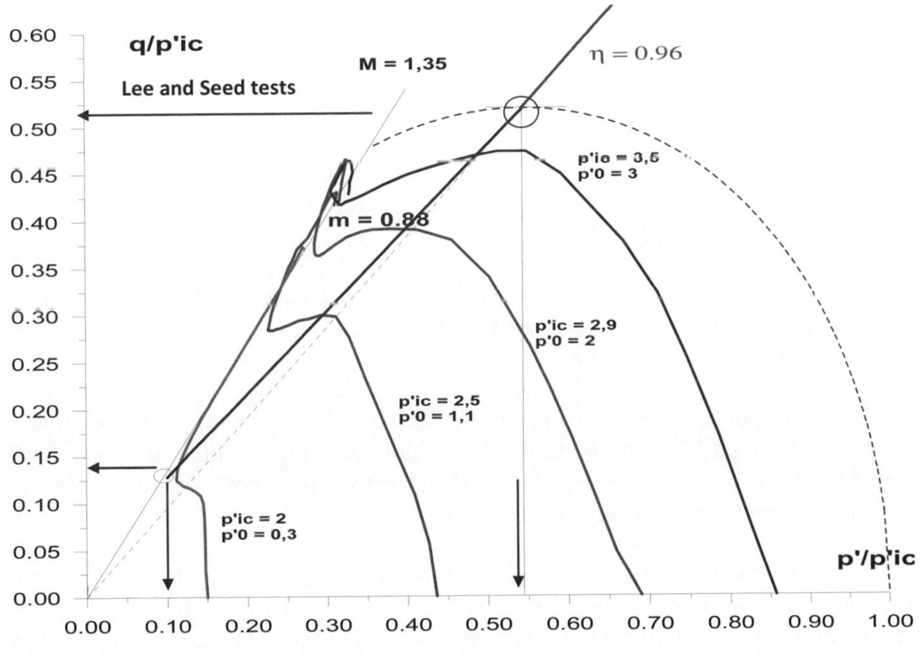

Figure 5.14. *Undrained tests of Lee and Seed [LEE 67] normalized with their theoretical OCR*

On the other hand, for the fourth test – with $p'_0 = 0.3$ MPa – the peak under the straight M line disappears. Its theoretical overconsolidation stress is $p'_{iC} = 2$ MPa, obtained with the C_s model (see Figure 5.13b). The fictive NC test with $p'_{iC} = 2$ MPa is represented in Figure 5.13d. The locus of the peak of overconsolidated paths with a same $p'_{iC} = 2$ MPa could be a straight line (dotted line) that meets up with the M line at the same point as the test performed at 0.3 MPa (OCR = 6.7).

A fictive test can be drawn by considering a homothetic curve with the same OCR = 6.7 for which $p'_{iC} = 3.5$ MPa ($p'_0 = 0.5$). We, therefore, normalized all the tests in relation to their theoretical overconsolidation stress p'_{iC}, as shown in Figure 5.14.

The locus, in the ($q/p'_{iC} - p'/\,p'_{iC}$) plane of the intermediate maxima under the M line is a line passing through the peak-point of the theoretical NC path, for which the coordinates are given by the system of equation [5.24]. Figure 5.14 gives the coordinates of the point on the straight M line: $p'/\,p'_{iC} = 0.1$ and $q/\,p'_{iC} = 0.1$ M.

5.7. Standard behavior for undrained sands

5.7.1. *Normalization by the theoretical overconsolidation stress* p'_{iC}

To be able to compare tests of one material to another, Biarez divides the deviator by M. In Figure 5.15c, the dotted theoretical NC curve is drawn, deduced from equation [5.23] for a clayey material of which the liquidity limit is $w_L = 70\%$, resulting in $C_c = 0.53$. The parameter M would, therefore, be 1.09 and the stress ratio on the peak would be $\eta_{peak} = 1.84\ M = 2$.

This theoretical curve shows a maximum, but placed beyond line M. From an experimental point of view, this would not be possible, since the predictive path has to stop on the perfect plastic line M.

Therefore, for clays with a C_c value in the range $0.33 > C_c > 2C_d$, corresponding to w_L values between 50% and 80%, the maximum stress ratio is located above the M line. For highly plastic clays ($w_L > 80\%$), the effective stress path is characterized by a value of q going to the infinite when p' converges towards zero.

We can determine the theoretical consolidation stress p'_{iC} by drawing the intercepting point between the C_s line passing on the initial point ($I_{d0} - p'_{i0}$) and the ISL line (equation [5.18]) for the sands, which leads to the value (equation [5.25]) of p'_{i0} in kPa.

$$\log p'_{i0} = \frac{\left(1.290 + Id_{i0} - 0.52 \log p'_{i0}\right)}{0.468} \qquad [5.25]$$

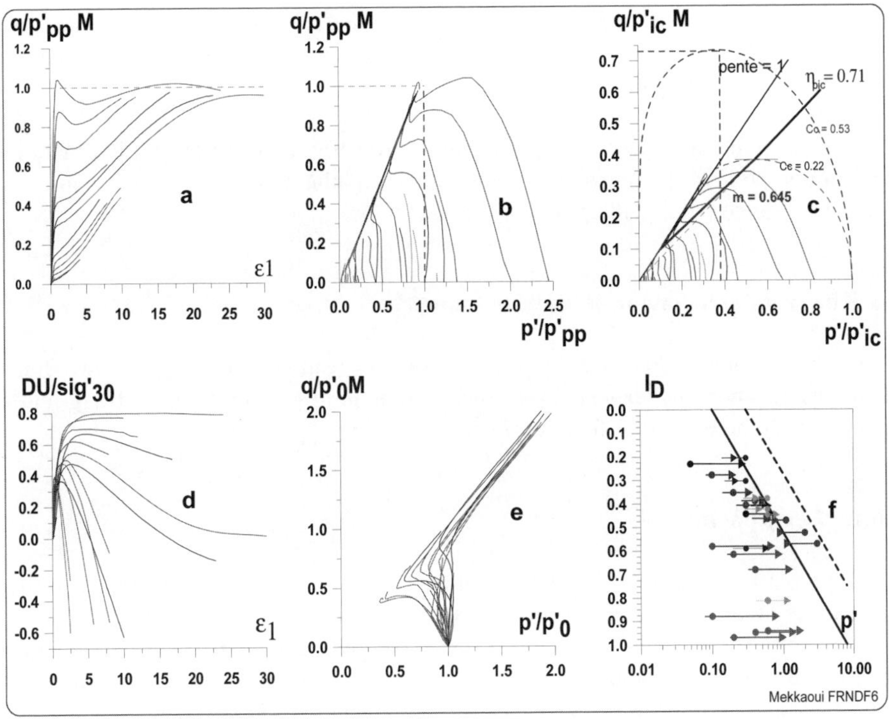

Figure 5.15. *Reference behavior for undrained sands*

5.7.2. *Perfect plasticity normalization of the curves in the $(q - \varepsilon_1)$ plane and pore pressure variation*

All the undrained loading paths in the $(q - \varepsilon_1)$ plane tend towards the horizontal asymptote Mp'_{pp}. A normalization with Mp'_{pp} (Figure 5.15a) keeps the intermediate peak that characterizes the sands with a small theoretical overconsolidation ("loose" behavior). We associate the $(q - \varepsilon_1)$ plane with the $(q/Mp'_{pp} - p'/Mp'_{pp})$ plane, where all the curves tend towards the bissectrice. The $(\Delta U/\sigma'_{30} - \varepsilon_1)$ plane in Figure 5.15d permits the control of cavitation[1] (*the test curves were cut when cavitation appeared*, according to Biarez's criteria [MEK 02]).

5.7.3. *Initial stress p'_0 normalization in the $(q - p)$ plane*

All the paths in the $q/Mp'_0 - p'/p'_0$ plane start from the point with coordinates (1 – 0) and tend towards the first bissectrice (see Figure 5.15e).

1 When water vaporizes under the effect of a strong pore pressure decrease.

Thus, Biarez found a fundamental representation for undrained tests, in this case for sands, but also valid for remolded clays. This is an example of his unified view of materials without "glue".

It is our hope that these six planes can be fruitfully adopted by the international community of soil mechanics and that mathematical modeling of the overconsolidated paths will appear in the near future.

5.8. The triaxial behavior of "lumpy" sands

In the section below we broach the rather peculiar subject of how Biarez attempted to show by *experimental* means that lumpy sand is indeed a granular material with the particular feature of containing "glue", i.e. capillary forces.

5.8.1. *"Lump" sands*

In preparing a sample of sand in moist conditions, we observe that a lumpy structure is created. With an electronic scanning microscope, Benhamed [BEN 01] produced images of sands with a moist deposit, showing a geometric arrangement of grains with lumps and lacunae (see Figure 5.16), i.e. two void ratio classes: an intra-lump class and an inter-lump class.

Figure 5.16. *"Lumpy sands" observations with scanning electron microscope [BEN 01]*

On the drained triaxial path, we can assume that the large strains have effaced the history of the material. Therefore, the same perfect plasticity line as for the dry sand deposits could be obtained, except that in this case the behavior undergoes an intermediate stage of lacunal geometry collapse.

On an undrained path, the collapse effect is replaced by a pore pressure increase. Indeed, since its tendency to re-arrange itself into lumps (intra-lumps class) does not compensate for the tendency of the lacunae (inter-lumps class) to collapse, the lumps "disintegrate" and we are left with the possibility of static liquefaction.

The principal tendency, until a stress of about 1 MPa, could be "collapse" or "liquefaction", not compensated for by the re-arrangement of the relatively consolidated lumps. We could, therefore, talk about a "new" critical state of the "collapsible" arrangements. This may be what is called the "steady state line" or SSL. It is characterized, in the (e – log p') plane by the Castro SSL or by the SSL curve [ISH 93].

5.8.2. *The Roscoe model applied to lump sands*

The Roscoe equation [5.23] on the undrained path (see Figure 5.17) gives the bell-shaped curve for an equivalent C_c of 0.08 (slope of the Castro SSL) and the above curve, for $C_c = 0.20$.

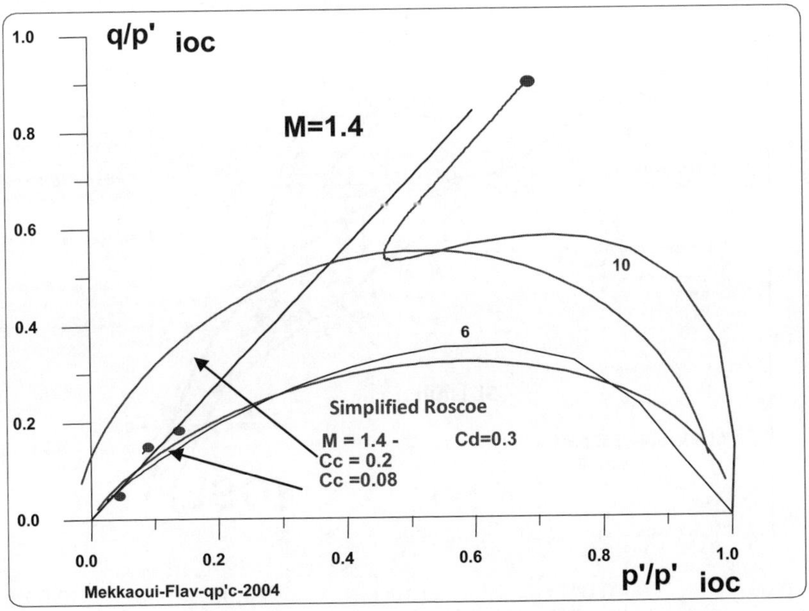

Figure 5.17. *Simplified for Cc = 0.2 and 0.08*

Two tests are drawn in the same figure: test 10 by Bousquet-Flavigny [FLA 93] of a very loose sand with $e_0 = 0.93$ for $p'_0 = 0.3$ MPa prepared under *dry* conditions; and test 6 by Flavigny and Meghachou [FLA 93] of a very loose sand with $e_0 = 1$ for $p'_0 = 0.28$ MPa, prepared under moist conditions.

Test 10 follows the Roscoe curve with $C_c = 0.2$, going through a maximum then descending to the straight line *M*, the path being nearly normally consolidated and the sand without lumps. Test 6, with lumps, follows the Roscoe curve with $C_c = 0.08$ and stops at the straight *M* line.

5.8.3. *Synthesis of several lump sand behaviors*

Biarez decided to redraw a certain number of drained as well as undrained tests (see Figure 5.18) of clean and lumpy badly graded sands in the (e – log p') plane [MEK 02]. The starting point of the test is marked with a large point and the end of the test with an arrow. The latter is situated in the vicinity of the SSL.

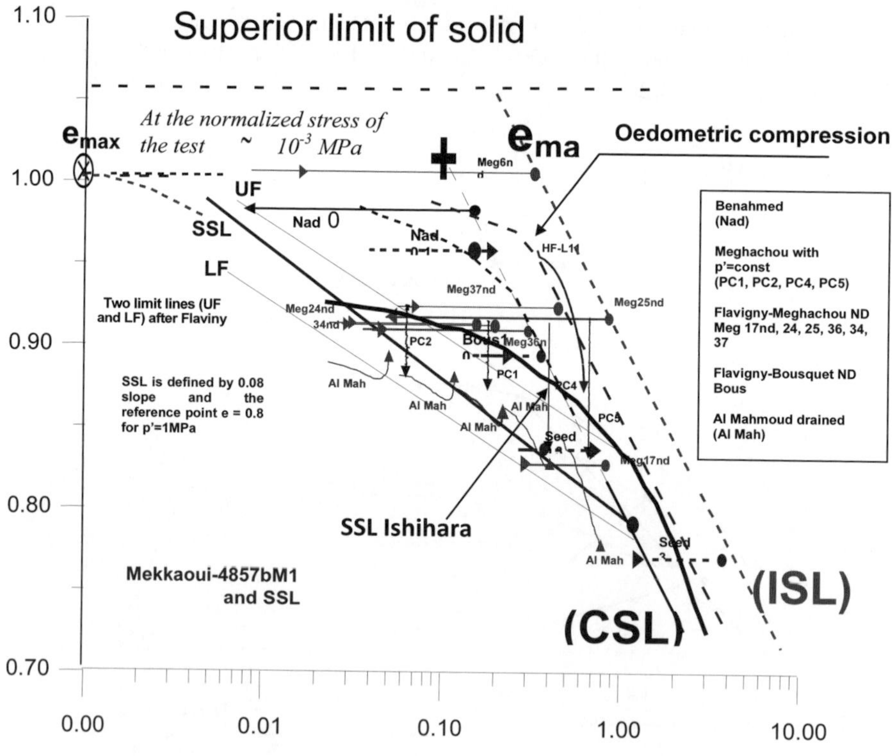

Figure 5.18. *Tests on lumpy sands in the (e–logp') plane with the Castro SSL, the lower and upper limits of Flavigny and the Ishihara SSL*

Figure 5.18, which displays many different nature tests, may be complex. Nevertheless, it is perfectly representative of the investigative methods adopted by Biarez, who, through an experimental approach, sought to arrive at an all-encompassing logical system of explaining the behavior of granular materials (with or without glue).

In this figure, we find a number of standard reference lines, the ISL, the CSL, the Castro SSL, the two frontier lines UF and UL of Flavigny, and the Ishihara SSL curve (Figure 5.18).

Four outstanding undrained tests (thick dotted line) permit the CSL for the lump-less sands to be located: two Lee and Seed tests [LEE 67] with e_0 = 0.78 and e_0 = 0.83 (denoted *Seed 0.3* and *Seed 3*), the Flavigny and Bousquet test 10, with e_0 = 0.93 (denoted *Bous 1*), and a test by Benhamed [BEN 01] with e_0 = 0.93 (denoted *Nad 0.1*).

We remark that the lump tests reach the SSL (*Nad 0*), but at times they return to it after going beyond it (*Meg 17, 24, 34*). Certain tests, before or after reaching the SSL, return to the CSL (*Meg 6, 37, 36*).

Most of the tests, stopped at 20% for axial strain, were not strictly speaking "liquefied"; so the deviator might have slightly moved upwards to produce what could be called an attempt to "reach the perfect plasticity stage" (as a return to *CSL*). Thus, the question of a possible return of the lumpy sands towards CSL seems to be an open affair. We could attribute the abrupt fall towards zero of certain tests in the literature to a phenomenon of large strain localization. Nevertheless, to continue running a test beyond 20% in strain without finding localization would, according to Biarez, be the equivalent of a technical miracle.

Finally, Biarez asked to draw two other lump type tests in this plane:

– the drained tests of Al Mahmoud [ALM 97], which tend towards the Castro SSL into the Flavigny gap UF-UL that, with regard to the SSL, have normally consolidated and overconsolidated behavior;

– the Meghachou tests, with p' constant, (*PC1, PC2, PC4* and *PC5*), which tend towards the SSL.

It was with the utmost care that Biarez verified the current models (lump sands moving towards the SSL line), even if the tests did not accord with his new hypothesis (a return to the CSL). His approach was experimental, his observational method meticulously elaborate and guided by a fundamental intuition of the physics of granular materials.

5.9. A new model to analyze the oedometer's path

This study, partly conducted with Biarez, wished to compare the predictions of two models for the compression of reconstituted normally consolidated clays under oedometric loading. The Biarez model presented in section 5.2 uses the liquidity index I_L, and the more recent Burland model introduces I_v, a normalized void index based on two characteristic void ratios (e^*_{100} and e^*_{1000}), corresponding to the oedometric curve of σ'_v=100 kPa and σ'_v=1,000 kPa.

These models, which represent the intrinsic properties of clays under compression, are compared via two experimental curves: one representing a remolded and reconstituted clay (Kaolinite P300) [HAT 04]; and the other, a deepwater clay sediment (GoG clay) taken from the Gulf of Guinea at a depth of 700 m [FAV 06].

5.9.1. *Burland's model*

Burland [BUR 90] collected and analyzed the oedometric compressibility of several clay sediments with a liquid limit w_L varying from 25% to 159%, remolded and reconstituted in the laboratory. The curves obtained in the $(e - log\sigma'_v)$ plane are slightly concave when σ'_v is located between 10 kPa and 1,000 kPa (see Figure 5.19a). The intrinsic compressibility index C^*_c is, therefore, introduced [5.26] as the difference between e^*_{100} and e^*_{1000}, two characteristic quantities corresponding to the consolidation stresses at 100 kPa and 1,000 kPa, respectively. C^*_c is also the slope of the oedometric compression curve (linear in this interval of stresses), the quantity $\{-[\log(100) - \log(1,000)]\}$ being equal to 1. With the transformation of the variable defined in [5.27], and by introducing the normalized void index I_v, the Burland model requires all experimental curves to pass through two fixed points in the $(I_v \text{-} log\sigma'_v)$ plane, corresponding to $e = e^*_{100}$, $I_V = 0$ and $e = e^*_{1,000}$, $I_V = -1$ for σ'_v = 100 kPa and σ'_v = 1,000 kPa, respectively.

$$C^*_c = e^*_{100} - e^*_{1000} \qquad [5.26]$$

$$I_v = \frac{e - e^*_{100}}{C^*_c} \qquad [5.27]$$

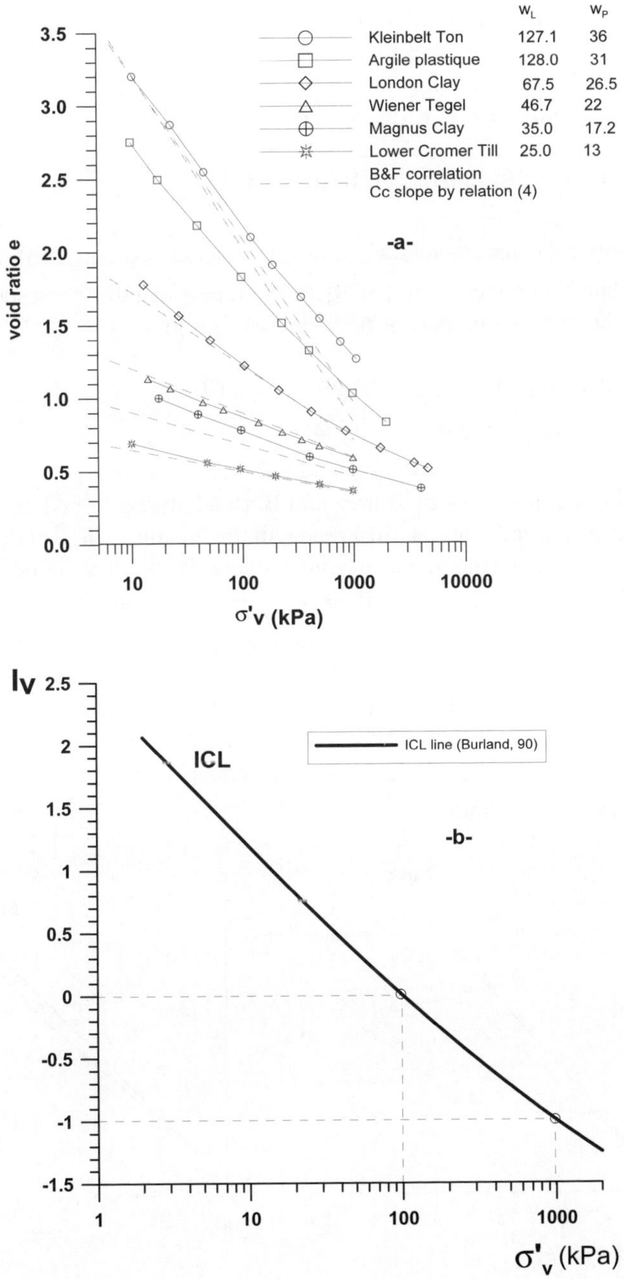

Figure 5.19. *Intrinsic compression line by Burland [BUR 90]*

The ICL (intrinsic compression line) curve, corresponding to equation [5.28] and represented by Figure 5.19b, can then be deduced from a statistical model based on the experimental curves passing through the two specific points (σ'$_v$ = 100 kPa, I_v = 0) and (σ'$_v$ = 1,000 kPa, I_v = -1). σ'$_v$ is expressed in kPa.

$$I_v = 2.45 - 1.285 \log\sigma'_v + 0.015 (\log\sigma'_v)^3 \qquad [5.28]$$

The two particular parameters e^*_{100} and e^*_{1000} for the "arrangement of grains" are, on the other hand, connected by [5.29] to e_L, the void ratio corresponding to the liquid limit – the only parameter for the "nature of grains".

$$\begin{cases} e^*_{100} = 0.109 + 0.679\, e_L - 0.089\, e_L^2 + 0.016\, e_L^3 \\ C^*_c = e^*_{100} - e^*_{1000} = 0.256\, e_L - 0.04 \end{cases} \qquad [5.29]$$

Although both approaches by Biarez and Burland appear to be based on a similar method, there is a fundamental difference in the manner in which the intrinsic characteristics of the "arrangement of grains" were defined on the oedometric path. Burland used two experimental values (e^*_{100} and e^*_{1000}) for the "arrangement of grains". On the other hand, Biarez, by introducing I_L, used two parameters of the "nature of grains" (e_L and e_P). Both models link the "nature of grains" parameters to the "arrangement of grains" parameters, however, using statistical correlations under oedometric loading. This corresponds to relation [5.2] in Biarez's model and to relation [5.29] in Burland's model. We can deduce that e^*_{100} and e^*_{1000} are nature parameters in Biarez's method.

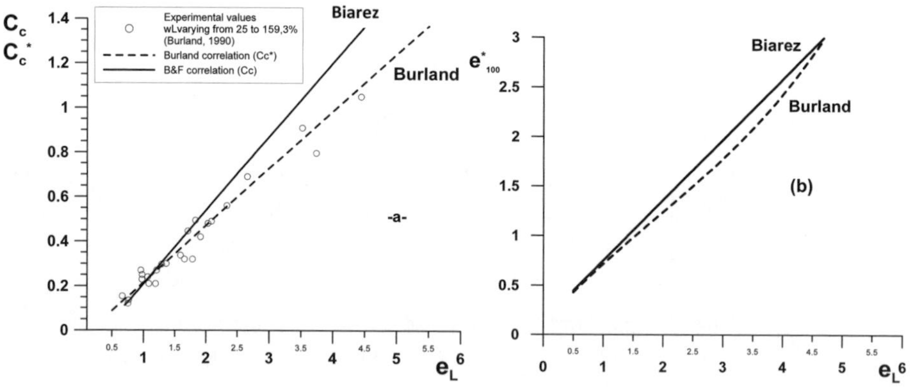

Figure 5.20. *Variation of C_c, C^*_c and e^*_{100} parameters as fonction of e_L*

An examination of Figures 5.19a and 5.20a reveals a difference in the C_c (Biarez's approach) and C^*_c (Burland's approach) slopes, particularly between $\sigma'_v = 100$ and $\sigma'_v = 1,000$, when the liquid limit is high (w_L=100% to 160%). On the other hand, for the lower w_L, C_c and C^*_c are similar, and fit well the experimental data given by Burland for different values of w_L (see Figure 5.20a).

5.9.2. *Comparison of models and mixed model*

5.9.2.1. *Plotting Biarez's model in Burland's space ($I_v - \log\sigma'_v$)*

According to correlation [5.7] we have:

$$I_L = \frac{e - e_P}{e_L - e_P} = 0.46\left(3 - \log\sigma'_v\right) \qquad [5.30]$$

Using [5.27] and [5.30] to express Biarez's model as a function of I_v, we obtain the NCRS equation [5.31] in Burland's space:

$$I_V = 1.38\frac{e_L - e_P}{C^*_c} - \frac{e^*_{100} - e_P}{C^*_c} - 0.46\frac{e_L - e_P}{C^*_c}\log\sigma'_v \qquad [5.31]$$

According to Biarez, hence [5.2] $e_P = e^*_{1000}$ and $e_L = e^*_{6,5}$, equation [5.31] becomes:

$$I_V = \frac{e^*_{6,5} - e^*_{1000}}{C^*_c}\left[1.38 - 0.46\ \log\sigma'_v\right] - 1. \qquad [5.32]$$

Considering a linear variation between Δe and $\Delta\log\sigma'_v$, we obtain:

$$\frac{e^*_{100} - e^*_{1000}}{\log 100 - \log 1,000} = \frac{e^*_{6.5} - e^*_{1000}}{\log 6.5 - \log 1,000}. \qquad [5.33]$$

Thus, Biarez's model given by relation [5.32] becomes:

$$I_V = 2 - \log\sigma'_v. \qquad [5.34]$$

Equation [5.34] is a straight line in the (I_v – logσ'_v) plane, passing through two specific points ($I_v = 0$, $\sigma'_v = 100$ kPa) and ($I_v = -1$, $\sigma'_v = 1,000$ kPa), represented by the NCRS line in Figure 5.21a. A representation of both the NCRS line and the ICL in Burland's plane shows that the two curves converge at a vertical stress range between 40 kPa and 1000 kPa. Normalization by means of the I_V parameter and $e_P = e^*_{1000}$ assumption are, of course, fundamental in this result because they force the NCRS to have the (I_v = -1, σ'$_v$ = 1,000 kPa) point on the ICL. However, in the stress range below 40 kPa, the two lines diverge slightly due to the concavity of the curves (Figure 5.19a), which is taken into account in the ICL expression. Both approaches may, therefore, be summarized in Table 5.2, with the e^*_{100} and e^*_{1000} indices for "Biarez in Burland's space" being deduced from both interpolation [5.33] and relations [5.2] to [5.5], considering $e_P = e^*_{1000}$ and $e_L = e^*_{6.5}$.

	Burland's model	Biarez's model	Mixted model
Equation	$I_V = 2.45 - 1.285 \log \sigma'_v + 0.015 (\log \sigma'_v)^3$	$I_L = 0.46(3 - \log \sigma'_v)$	$I_V = 2 - \log \sigma'_v$
Characteristic points	$\begin{cases} e^*_{100} = 0.109 + 0.679\, e_L - 0.089\, e_L^2 + 0.016\, e_L^3 \\ e^*_{1000} = 0.149 + 0.423\, e_L - 0.089\, e_L^2 + 0.016\, e_L^3 \end{cases}$	$w_{6.5} = w_L$ $w_{1000} = w_P$	$\begin{cases} e^*_{6.5} = e_L \\ e^*_{100} = 0.60\, e_L + 0.14 \\ e^*_{1000} = 0.27\, e_L + 0.26 \end{cases}$
Estimation of compression index	$C^*_c = 0.256\, e_L - 0.04$	$Cc = 0.009(w_L - 13)$	$C^*_c = C_c = 0.33\, e_L - 0.12$

Table 5.2. *Characteristics of the models*

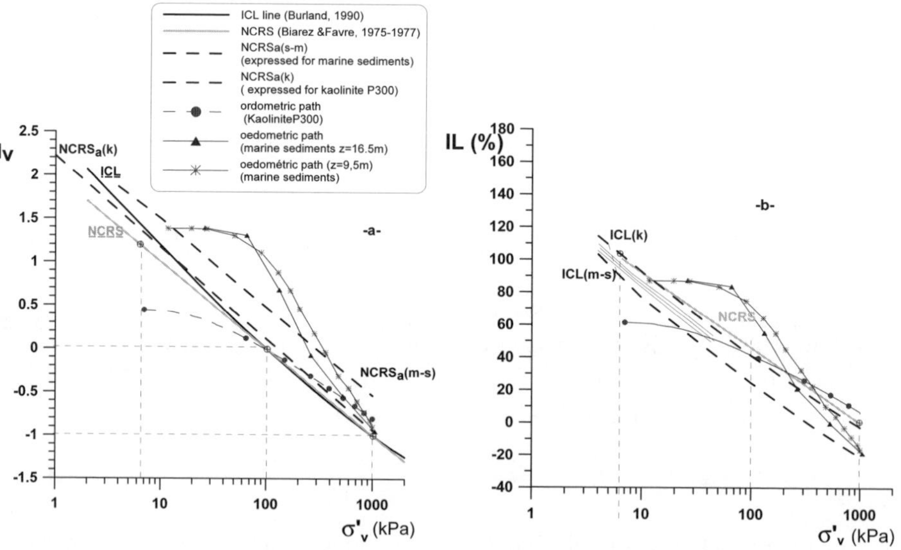

Figure 5.21. *Comparison between the two statistical models for the one dimensional loading in (σ'$_v$,I_v) and (σ'$_v$,I_L) planes*

5.9.2.2. *Biarez's mixed model in Burland's space ($I_v - \log\sigma'_v$)*

In the linear relation [5.31] resulting from Biarez, e^*_{100} and C^*_c are replaced by expressions [5.29] proposed by Burland for the ICL expression [FAV 08]. This gives the "mixed" model [5.35]:

$$I_V = 0.46\ \zeta\ [3 - \log\sigma'_v] - \chi\ . \qquad [5.35]$$

Here ζ and χ are the parameters given by [5.36] as a function of the e_L and e_P limits, defined by their experimental values:

$$\begin{cases} \zeta = \dfrac{e_L - e_P}{0.256\ e_L - 0.04}\ , \\ \chi = \dfrac{0.109 + 0.679\ e_L - 0.089\ e_L^2 + 0.016\ e_L^3 - e_P}{0.256\ e_L - 0.04}\ . \end{cases} \qquad [5.36]$$

Since the assumption of $e_P = e^*_{1000}$ is not considered, the NCRS$_a$ line is not forced to pass through a point on the ICL. The NCRS$_a$ line, however, will depend on the e_L limit.

When equation [5.35] is applied to the kaolinite through the e_L value (w_L = 40% and w_P = 20%), we obtain a curve termed NCRS$_{a/k}$, located very close to the ICL (see Figure 5.21a), whatever the stress domain may be. Both correlations are thus very similar, and compatible with the experimental curve on the oedometric path (see Figure 5.21a), the kaolinite having first been consolidated under one-dimensional conditions under an axial stress of 140 kPa.

The NCRS$_{a/m\text{-}s}$ curve calculated for the GoG marine sediments (the deepwater sediments from the Gulf of Guinea, whose w_L and w_P limits vary according to depth, therefore according to σ'_v, from 110% to 160% for w_L and from 30% to 90% for w_P [FAV 06]) diverges from the ICL, but also from the experimental curve (Figure 5.21a). The experimental path evolves first from one point located above the ICL curve (sensitive clays) and also the NCRS$_{a/m\text{-}s}$. This point (which exhibits a "fictitious consolidation stress" and a supplementary cohesion on the oedometric curve) is associated with the beginning of the damage of the cementation (destructuration according to Leroueil *et al.* [LER 79]), developed *in situ* during the sedimentation and consolidation processes. The experimental curves thus tend to converge at high stresses towards the ICL curve of the remolded state of the material.

The principal difference between the ICL and the NCRS, on one hand, and NCRS$_{a/m\text{-}s}$ on the other hand, seems to be due mainly to the difference between the slopes of the oedometric curves obtained for clays with high values of w_L (100% to 160%), which is the case of the GoG marine sediments. The C_c slopes used in Biarez's model are higher than those used in Burland's model (see Figure 5.20) through the secant C^*_c values taken between $\sigma'_v = 100$ kPa and $\sigma'_v = 1{,}000$ kPa.

5.9.3. *Burland's model in ($I_L - \log\sigma'_v$) Biarez's space*

Writing Burland's model in the (I_L – log σ'$_v$) plane requires us to know both e_L and e_P indices (by their experimental values, for instance), on one hand, and parameters e^*_{100} and C^*_c, on the other hand, defined by relation [5.29]. The Burland relation [5.28] becomes [5.37], ζ and χ being defined by relations [5.36]:

$$I_L = \frac{1}{\zeta}\left[2.45 - 1.285\ \log\sigma'_v + 0.015\ (\log\sigma'_v)^3 + \chi\right] \qquad [5.37]$$

Since the ICL depends on e_L (Figure 5.20b), it is transformed in the (I_L – log σ'$_v$) plane by a bundle of parallel curves. In this plane, the similarity between the two models can be clearly seen, which is also in accordance with the experimental curve for the low w_L values (as is the case of the kaolinite $ICL_{/k}$ (see Figure 5.20b)). As previously seen in Burland's plane, the $ICL_{/m\text{-}s}$ of marine clay, located in the high w_L domain, diverges from the NCRS. The experimental oedometric loading curves converge towards the $ICL_{/m\text{-}s}$ in the domain of high stresses, which shows that it is Burland's correlation that is closer to the experimental results for these high values of w_L.

Biarez assumed that $e_P = e^*_{1000}$ could not be introduced into Burland's model because it would force the $ICL_{/m\text{-}s}$ to converge to the NCRS at the (I_L = 0, σ'$_v$ = 1,000 kPa) point; in this case, the curve would no longer reflect the experimental observation.

5.10. "Destructuration" of clayey sediments

The evolution of the degradation of clayey sediments has been quantified by Cotecchia and Chandler [COT 97, COT 00], using the sensitivity S_t parameter (Skempton and Northey [SKE 52]), corresponding to the σ'_v/σ'^*_{ve} ratio, σ'^*_{ve} being the equivalent vertical effective stress.

In the case of the GoG sediment, a destructuration by oedometric loading is clearly highlighted through the S_t evolution, represented as a function of I_L in Figure 5.22a and as a function of e in Figure 5.22b. We can see that the path evolves towards a peak that shows a maximum sensitivity of $S_t = 8.5$, then gradually tends towards the remolded sediment behavior with $S_t = 1$.

This study shows that Burland's ICL better reflects the experimental results and represents the compressibility for a large range of w_L. On the other hand, the NCRS line in the Biarez model is less well adapted to high values of w_L, but remains valid for representing compressibility in the case of medium and low values of w_L. The NCRS line also has the advantage of being simpler to use because it is based on two current "nature parameters": the Atterberg limits w_L and w_P.

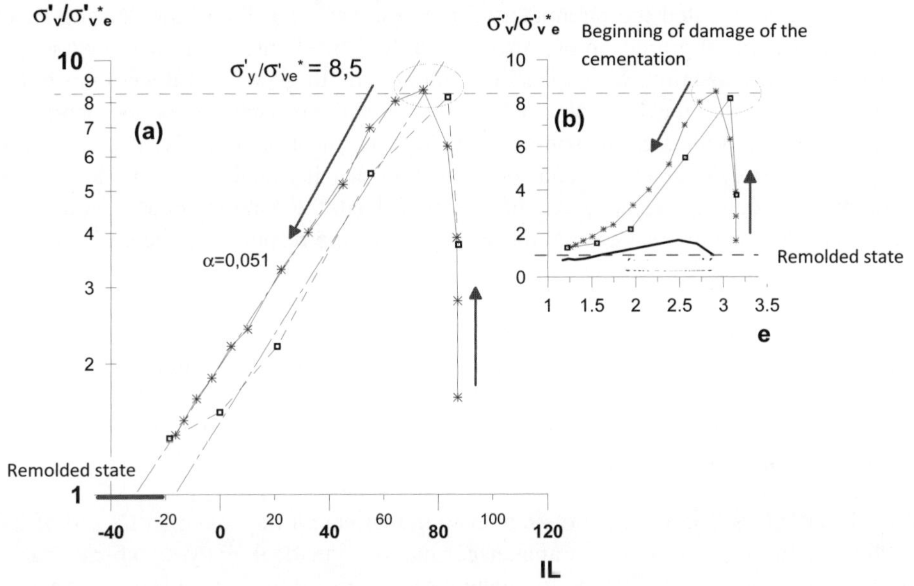

Figure 5.22. *Evolution of cementation damage in the GoG sediment*

5.11. Conclusion

It has been shown by Biarez that non-cohesive granular materials, such as the clean sands and the remolded clays studied in this chapter, behave similarly in the large strain domain. The same can be said for the critical state line, isotropic loading paths, the over-consolidation concept, drained and undrained triaxial test paths. As an analogy to the above, we have also found a similar behavior between loose sands with or without lumps and deep sea marine clays with low physico-chemical cohesion. It was certainly his vision of sands and clays as "granular materials" (Biarez often referred to clay as "*les paquets de plaquettes*", i.e. particle clusters) and his unflinching physical and experimental approach concerning their assemblage that led him to pronounce this sort of sweeping generalization.

His particular vision of grain assembly led him to conceive a system of logic that could subsequently connect the parameters of all soils. Considering grains as many continuous media (the properties of grains) with their boundary conditions in space (the assembly) and in stresses (the confinement), he deduced that the properties of soils (the deformability and rupture parameters of the imaginary "continuous medium") were indeed the combined substance of the mechanical properties of these multiple continuous media, or real "discontinuous medium", with different boundary conditions. He created the parameters for "nature of grains" and the "arrangement and confinement of grains" in order to explain the "mechanical parameters of soils". His system of reasoning, which has been adopted by those who followed him in the quest to resolve mechanical continuous media problems, coupled with an experimental and physical reasoning, is at work in all of his models. This approach is particularly evident in the confrontation between his oedometric modeling and Burland's modeling; indeed, as we have emphasized, the parameters e^*_{100} and e^*_{1000} are simply the parameters of "nature of grains", as the response of the grains (as w_L and w_P) to a particular loading.

We should point out that Biarez's approach was basically experimental and physical, and that he left to others the task of quantifying and digitalizing all the elements that for him were only reference points, reference standards and graphs within the system of axes; all these elements had enormous pedagogical value for understanding granular media behavior.

Nevertheless, Biarez was no stranger to numerical work. Towards the end of the 1960s, he himself promoted parameter correlation statistical analyses which, thanks to the emerging models, helped to create a safety approach. This approach aimed to ensure the safety of earthwork constructions by considering all the uncertain elements that still clouded the engineer's knowledge about the real behavior of granular materials. A real behavior could be compared to a modeled standard reference behavior, from which it would be possible to appreciate the errors and uncertainties contained in the measured parameter values. By taking into account all these elements of uncertainty, academics could propose a global safety approach to earthwork calculations [FAV 95]. Before these scientific developments, it was not possible to treat the uncertain elements confronting real earthwork behavior from the outset, simply because the uncertainties were part and parcel of the predictions of the model. Given the fact that the sources of uncertainties could be distinguished from the beginning, it has become possible to assess their relative importance more efficiently.

The statistical analysis Biarez performed towards the end of his career led to the probabilities calculation and thus to the safety of earthworks [FAV 04]. This particular aspect of his scientific and engineering contribution shows that Biarez was a visionary in the field of soil mechanics.

5.12. Examples of manuscript notes

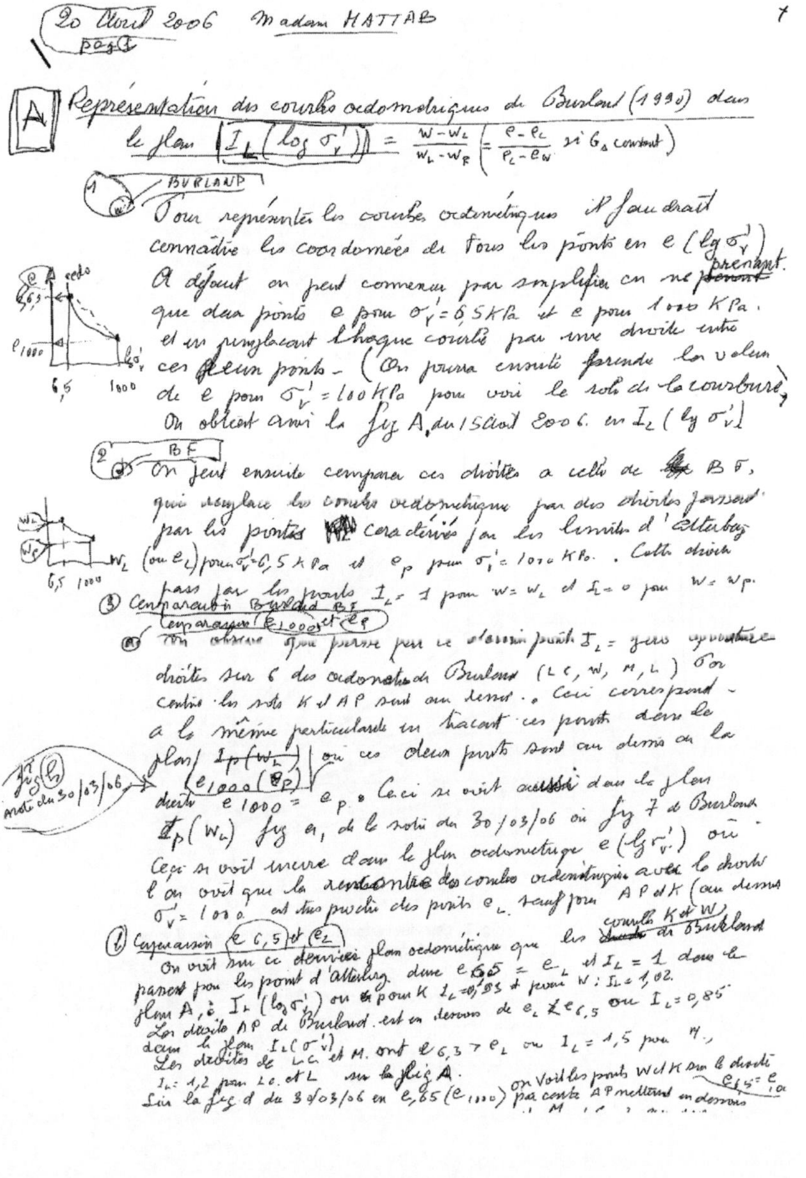
20 Avril 2006 Madam HATTAB

page 1

A Représentation des courbes oedométriques de Burland (1990) dans le plan $I_L(\log \sigma'_v) = \frac{w - w_L}{w_L - w_P} = \frac{e - e_L}{e_L - e_w}$ si G_s constant)

1 BURLAND

Pour représenter les courbes oedométriques il faudrait connaître les coordonnées de tous les points en $e(\lg \sigma'_v)$. A défaut on peut commencer par simplifier en ne prenant que deux points e pour $\sigma'_v = 6,5$ kPa et e pour 1000 kPa et en remplaçant chaque courbe par une droite entre ces deux points. (On pourra ensuite prendre la valeur de e pour $\sigma'_v = 100$ kPa pour voir le rôle de la courbure). On obtient ainsi la fig A du 15 avril 2006 en $I_L(\lg \sigma'_v)$

2 BF

On peut ensuite comparer ces droites à celle de BF, qui remplace les courbes oedométriques par des droites passant par les points caractérisés par les limites d'Atterberg w_L (ou e_L) pour $\sigma'_v = 6,5$ kPa et e_P pour $\sigma'_v = 1000$ kPa. Cette droite passe par les points $I_L = 1$ pour $w = w_L$ et $I_L = 0$ pour $w = w_P$.

3 Comparaison Burland BF

comparaison (e_{1000} et e_P)

On observe que pour ce dernier point I_L = zéro approximativement droites sur 6 des oedomètres de Burland (LC, W, M, L) Par contre les sols K et AP sont au dessus. Ceci correspond à la même particularité en traçant ces points dans le plan $I_P(w_L)$ où ces deux points sont au dessus de la droite $e_{1000} = e_P$. Ceci se voit aussi dans le plan $I_P(w_L)$ fig e, de la note du 30/03/06 ou fig 7 de Burland. Ceci se voit encore dans le plan oedométrique $e(\lg \sigma'_v)$ où l'on voit que la rencontre des courbes oedométriques avec la droite $\sigma'_v = 1000$ est très proche des points e_L sauf pour AP et K (au dessus)

fig (e) note du 30/03/06

1 Comparaison $e_{6,5}$ et e_L

On voit sur ce dernier plan oedométrique que les courbes K et W de Burland passent pour les points d'Atterberg dans $e_{6,5} = e_L$ et $I_L = 1$ dans le plan A, c. $I_L(\log \sigma'_v)$ ou pour K $I_L = 0,85$ et pour W: $I_L = 1,02$. Les droites AP de Burland est en dessous de e_L $e_{6,5}$ ou $I_L = 0,85$ dans le plan $I_L(\sigma'_v)$. Les droites de LC et M ont $e_{6,5} > e_L$ ou $I_L = 1,5$ pour M, $I_L = 1,2$ pour LC et L sur la fig A. On voit les points W et K sur la droite $e_{6,5} = e_{10}$ par contre AP mettant en dessous

Sur la fig d du 30/03/06 en $e_{6,5}(e_{1000})$

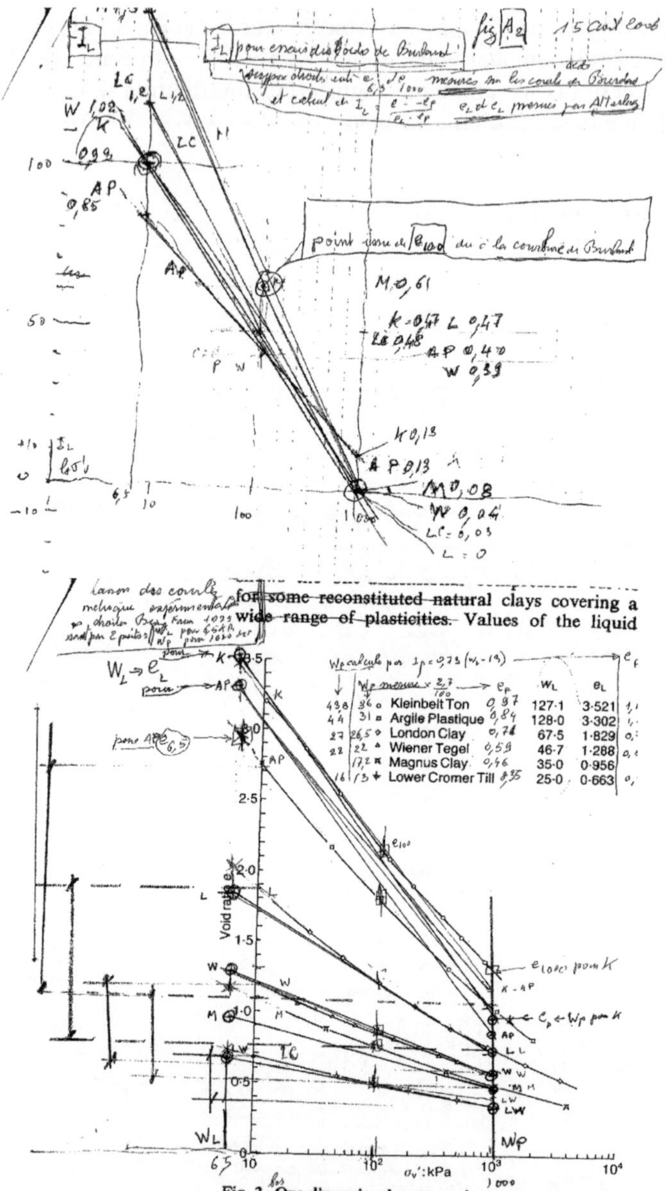

Fig. 2. One-dimensional compression curves for various reconstituted clays

Les droites BF passant par les deux points {W_L pour 6,5 kPa / W_P pour 1000 kPa} (droites rouges) donnent un ordre de grandeur des mesures oedométriques mais la qualité est meilleure pour $W_L < 70$ que pour $W_L = 127-128$ où les valeurs de W_P sont nettement inférieures aux mesures de l'oedomètre pour $\sigma'_v = 1000$ kPa. On note que ces derniers valeurs sont au dessus de la moyenne $I_p = 0,73(W_L - 13)$. Si l'on utilise cette relation pour avoir un W_P à partir de W_L les droites sont plus satisfaisantes. (Vardo)

5.13. Bibliography

[ALI 73] AL ISSA M., Recherche de lois contrainte-déformation des milieux pulvérulents; analyse de la validité des lois hyperboliques; application aux pieux et barrages, Doctorate thesis, University of Grenoble, France, 1973.

[ALM 97] AL MAHMOUD M., Etude en laboratoire du comportement des sables sous faibles contraintes, Doctorate thesis, Ecole Centrale de Lille, France, 1997.

[BEN 01] BENHAMED N., Comportement mécanique d'un sable sous cisaillement monotone et cyclique: Application aux phénomènes de liquéfaction et de mobilité cyclique, Doctorate thesis, Ecole Nationale des Ponts et Chaussées, Paris, France, 2001.

[BIA 72] BIAREZ J., FAVRE J. L., Table ronde sur les corrélations de paramètres en mécanique des sols, Ecole Centrale de Paris, Paris, France, 1972.

[BIA 75] BIAREZ J. FAVRE J.L., "Parameters filing and statistical analysis of data in soils mechanics", *2 Int. Conf. Appl. Stat. Prob.*, vol. 2, pp. 249-264, Aachen, Germany, 1975.

[BIA 77] BIAREZ J., FAVRE J.L., "Estimation statistique et extrapolation à partir d'observations", *Rapport général 6ème Session Spéciale, Proc. 9th Int. Cong. Soils Mech. Found. Eng.*, vol. 3, pp. 505-509, Tokyo, Japan, 1977.

[BIA 94] BIAREZ J., HICHER P.Y., *Elementary Mechanics of Soils Behavior. Saturated Remolded Soils*, A.A. Balkema, Rotterdam and Brookfield, The Netherlands, 1994.

[BIA 98] BIAREZ J., FAYAD T., TAILLEZ S., GOMES CORREIA A., FLAVIGNY E., BRANQUE D., "Argiles et craies du tunnel sous la manche – grains sans et avec colle", *2nd International Conference on Hard Soil and Soft Rocks*, pp. 437-445, Naples, Italy, 1998.

[BOL 86] BOLTON M.D., "The strength and dilatancy of sands", *Géotechnique*, vol. 36, no. 1, pp. 65-78, 1986.

[BOU 82] BOUVARD D., Rhéologie des milieux pulvérulents: Etude expérimentale et identification d'une loi de comportement, Doctorate thesis, University of Grenoble, France, 1982.

[BUR 90] BURLAND J.B., "On the compressibility and shear strength of natural clays", *Géotechnique*, vol. 40, no. 3, pp. 329-378, 1990.

[COT 97] COTECCHIA F., CHANDLER R.J., "The influence of structure on the pre-failure behaviour of a natural clay", *Géotechnique*, vol. 47, no. 3, pp. 523-544, 1997.

[COT 00] COTECCHIA F, CHANDLER R.J., "A general framework for the mechanical behaviour of clay", *Géotechnique*, vol. 50, no. 4, pp. 431-447, 2000.

[DAS 97] DA SILVA I., Simulation of the Menards Pressuremeter Test Using the F.E.M and an Advanced Mohr-Coulomb Model. IST Report, IST, 1997.

[FAV 72a] FAVRE J.L., Pour un traitement par le calcul des probabilités et des statistiques des problèmes de mécanique des sols, Doctorate thesis, University of Grenoble, France, 1972.

[FAV 72b] FAVRE J.L., "Corrélations entre paramètres", in: FAVRE J.L., BIAREZ J. (eds), *Table Ronde sur les Corrélations de Paramètres en Mécanique des Sols*, vol. II, pp. 1-51, Ecole Centrale de Paris, Paris, France, 1972

[FAV 80] FAVRE J.L., Milieu continu et milieu discontinu. Mesure statistique indirecte des paramètres rhéologiques et approche probabiliste de la sécurité, Doctorate thesis, UPMC, Paris VI, France, 1980.

[FAV 95] FAVRE J.L, TOUATI K., "Probabilistic estimations of the errors in finite element calculation code for shallow foundation", *7th Inter. Conf. Applic. Stat. Prob. Soils Struct. Eng., ICASP'7*, vol. 2, pp. 1309-1315, Paris, France, 1995.

[FAV 02] FAVRE J.L., BIAREZ J., MEKKAOUI S., "Modèles de comportement en grandes déformations des sables et argiles remaniées à l'oedomètre et au triaxial", *Symposium International Identification et Détermination des Paramètres des Sols et des Roches pour les Calculs Géotechniques, PARAM*, pp. 369-384, Paris, France, 2002.

[FAV 04] FAVRE J.L., *Sécurité des Ouvrages, Risques – Modélisation de l'Incertain*, Ellipses, Paris, France, 2004.

[FAV 06] FAVRE J.L., HATTAB M., "Propriétés des argiles marines de grandes profondeurs", *Revue Française de Géotechnique*, 116, pp. 3-13, 2006.

[FAV 08] FAVRE J.L., HATTAB M., "Analysis of "Biarez-Favre" and "Burland" models of the compressibility of remoulded clays", *C. R. A. S. Géoscience*, vol. 340, no. 1, pp. 20-27, 2008.

[FAY 00] FAYAD T., Mécanique des grains avec et sans colle pour une comparaison avec des sols naturels et remaniés, Doctorate thesis, Ecole Centrale de Paris, France, 2000.

[FLA 93] FLAVIGNY E., Compilation des essais triaxiaux de révolution sur le sable d'Hostun RF, internal research report, University of Grenoble, France, 1993.

[HAC 01] HACHI F., FAVRE J.L., "Comportement de référence pour la caractérisation des erreurs de modèle et de mesures", *1ère Conf. Int. Albert Caquot*, Paris, France, 2001.

[HAT 04] HATTAB M., HICHER P-Y., "Dilating behavior of over consolidated clay", *Soils and Foundations*, vol. 44, no. 4, pp. 27-40, 2004.

[HAT 10] HATTAB M., FAVRE J.L., "Analysis of experimental compressibility of deep water marine sediments from the Gulf of Guinea", *Marine and Petroleum Geology*, vol. 27, pp. 486-499, 2010.

[ISH 93] ISHIHARA K., "Liquefaction and flow failure during earthquakes", *Géotechnique*, vol. 43, no. 3, pp. 351-415, 1993.

[LAD 64] LADD C.C., "Stress-Strain behavior of saturated clay and basic strength principle", *Rec. Rapp.*, no. 1, Part 1, pp. R64-17, 1964.

[LEE 67] LEE K.H., SEED H.B., "Undrained strength characteristics of cohesion less soils", *Proc. ASCE*, 93 (SM6), pp. 333-360, 1967.

[LER 79] LEROUEIL S., TAVENAS F., BRUCY F., LA ROCHELLE P., ROY M., "Behaviour of destructuted natural clays", *Proc. ASCE*, vol. 105 (GT6), pp. 759-778, 1979.

[MEK 02] MEKKAOUI S., Modélisation de l'incertain sur les essais triaxiaux en grande déformation sur les sables et argiles remaniées, Doctorate thesis, Ecole Centrale de Paris, France, 2002.

[MOK 83] MOKHAM M., Contribution à l'étude expérimentale et théorique du comportement du sable sous chargements cycliques, Doctorate thesis, University of Grenoble, France, 1983.

[ROS 58] ROSCOE K.H, SCHOFIELD A.N., WROTH C.P., "On the yielding of soils", *Géotechnique*, vol. 8, no. 1, pp. 22-53, 1958.

[ROW 62] ROWE P.W., "The stress-dilatancy relationship for static equilibrium of an assembly of particles in contact", *Proceedings of Royal Society*, vol. 269 (A), pp. 527, 1962.

[SAI 97] SAIM R., Des comportements repères des grains sans colle à un exemple de sol réel, Doctorate thesis, Ecole Centrale de Paris, France, 1997.

[SKE 44] SKEMPTON A.W., "Notes on compressibility of clays", *Q.J. Geol. Soc.*, vol. 100, pp. 119-135, 1944.

[SKE 52] SKEMPTON A.W., NORTHEY R.D., "The sensitivity of clays", *Géotechnique*, vol. 3, no. 1, pp. 30-53, 1952.

[ZER 82] ZERVOYANNIS C., Etude synthétique des propriétés mécaniques des argiles saturées et des sables sur chemin œdométrique et triaxial de révolution, Doctorate thesis, Ecole Centrale de Paris, France, 1982.

[ZIA 87] ZIANI F., Contribution à l'étude du compactage des sols – Cas particulier du comportement des sables très peu denses, Doctorate thesis, University of Gembloux, Belgium, 1987.

Chapter 6

The Concept of Effective Stress in Unsaturated Soils

Very small to large strain properties of a silty sand were measured at several water contents ranging from a few percentage points to saturation level by the use of three triaxial devices with different types of transducers. In the analysis of the data, attention was paid to the influence of the water content on the maximum (Young's) modulus and the effect of the confining stress, etc. Isotropic and triaxial tests, with the measurement of capillary pressure (suction), were also performed on the same material. The effective stress concept was used to interpret the results. The predictions of a micromechanical model were compared with the experimental data in the elastic domain and on the failure criterion. The method highlights a simple way to relate the strength and stiffness of the soil to its capillary pressure.

6.1. Introduction

The "effective stress" concept is perhaps one of the most debated notions in unsaturated soil mechanics. It was originally defined as "…a function of the total stress and pore pressure that controls the mechanical effects resulting from a change in stresses…" [BIS 63] and it applies exclusively to the solid skeleton of soil [TER 43]. Effective stress has been used to formulate both elastic and elasto-plastic constitutive laws of the solid skeleton, where it controls stress–strain relationships, the volume changes and strength of the solid skeleton, independently of the pore

---

Chapter written by Said TAIBI, Jean-Marie FLEUREAU, Sigit HADIWARDOYO, Hanène SOULI and António GOMES CORREIA.

pressure changes [BI0 41; COU 95; DEB 90; LOR 00; RIC 76]. At the origin, effective stress was the macroscopic stress of the solid skeleton. In the case of saturated soils, the expression which is generally used is [NUR 71; SKE 61]

$$\sigma' = \sigma - \alpha p_w \, \delta \tag{6.1}$$

where σ' is the effective stress tensor, σ the total stress tensor, δ the unit tensor, p_w the pore fluid pressure and α the Biot coefficient.

In the case of an isotropic material:

$$\alpha = 1 - \frac{K_T}{K_s} \tag{6.2}$$

where K_T is the compressibility modulus of the solid phase and K_s, the compressibility modulus of the solid grains. If the solid constituent is supposed to be incompressible ($K_s \rightarrow \infty$; $\alpha = 1$), the previous relationship becomes Terzaghi's effective stress [TER 36]:

$$\sigma' = \sigma - p_w \, \delta \tag{6.3}$$

In the case of unsaturated soils, effective stress was expressed for the first time by Bishop [BIS 59; BIS 63] as

$$\sigma' = \sigma - p_a \, \delta + \chi \, (p_a - p_w) \, \delta \tag{6.4}$$

Here p_a is the pore-air pressure and χ is an effective stress parameter, called Bishop's parameter, which is a function of the degree of saturation (equal to 1 for saturated soils and 0 for dry soils). Effective stress can also be written as

$$\sigma' = \sigma_{net} + \chi \, p_c \, \delta \tag{6.5}$$

with

$\sigma_{net} = \sigma - p_a \, \delta$: net total stress

$p_c = p_a - p_w$: capillary pressure.

[BIS 61] and [BIS 63] showed, first on the basis of experimental results, that the shear strength and volume change remained the same when the net total stress σ_{net}

and the capillary pressure p_c were kept constant during the test, even though σ, p_a and p_w varied. However, [JEN 62] questioned the validity of the effective stress principle in unsaturated soils because it could not explain the phenomenon of collapse observed during wetting tests under constant total stress.

In recent years, a number of researchers have shown that irreversible strains, such as those in collapse, could be correctly described by a generalized effective stress concept associated with an elasto-plastic law that integrated a yield surface function of capillary pressure [BOL 96; DAN 02; KHA 01; KHA 04; KOH 63; LOR 00; MOD 94].

These studies have shown that the effective stress concept, extended to unsaturated soils, remains an efficient means by which to qualitatively and quantitatively describe the unified mechanical behavior of soils passing from saturated to unsaturated domains and vice versa.

Table 6.1 summarizes the expressions of χ proposed by several researchers.

Today, different approaches are used to characterize the behavior of polyphasic granular media (Figure 6.1). We can mention, as an example, the macroscopic phenomenological, homogenization, and discrete elements methods (DEM).

The extension of the effective stress tensor concept to unsaturated soils, under the hypothesis that the unsaturated soil is a poro-elastic material, in other words, when the solid constituent has a reversible elastic behavior, is presented as

$$\sigma_{ij}' = \sigma_{ij} + \pi_{ij} \qquad [6.6]$$

π_{ij} : isotropic tensor, $\pi_{ij} = \pi\delta_{ij}$

The micro–macro approach of the unsaturated porous media has led some investigators to generalize the Hill's lemma [CHA 95].

This approach, based only on mechanical considerations of statics, gives an expression of π as

$$\pi = S_g p_g + S_l p_l - U \qquad [6.7]$$

Expression of χ	Definition of the parameters	Authors
$\chi = S^w$	S^w : degree of saturation in water	Hassanizadeh & Gray (1980) Lewis & Shrefler (1987) Bolzon & Schrefler (1995) Bolzon *et al.* (1996) Hutter *et al.* (1999) Wheeler *et al.* (2003)
$\chi = 1 \quad \text{for } s < s_e$ $\chi = \dfrac{s_e + \dfrac{a_e}{s - s_e + a_e}(s - s_e)}{s} \quad \text{for } s \geq s_e$	s: suction (kPa) s_e: air entry suction a_e: parameter function of the soil	Kohgo *et al.* (1993)
$\chi = S^w\left(2 - S^w\right)$	S^w : degree of saturation in water	Gudehus (1995)
$\chi = \begin{cases} 1 & \text{for } \dfrac{s}{s_e} \leq 1 \\ \left(\dfrac{s}{s_e}\right)^{-0.55} & \text{for } 1 \leq \dfrac{s}{s_e} \leq 12 \end{cases}$	s: suction (kPa) s_e: air entry suction	Khalili & Khabbaz (1998) Loret & Khalili (2000) Loret & Khalili (2002)
$\psi = \begin{cases} 1 & \text{for } \dfrac{s}{s_e} \leq 1 \\ 0.45\chi & \text{for } 1 \leq \dfrac{s}{s_e} \leq 25 \\ 0 & \text{for } \dfrac{s}{s_e} \succ 25 \end{cases}$	s: suction (kPa) s_e: air entry suction $\psi = \dfrac{d(\chi s)}{ds}$	Khalili *et al.* (2004)
$\chi = \begin{cases} 1 & \text{for } \dfrac{s}{s_e} \leq 1 \\ \left(\dfrac{s}{s_e}\right)^{-0.55} & \text{for } 1 \leq \dfrac{s}{s_e} \leq 25 \\ 25^{0.45}\left(\dfrac{s}{s_e}\right)^{-1} & \text{for } \dfrac{s}{s_e} \succ 25 \end{cases}$	s: suction (kPa) s_e: air entry suction	Russel & Khalili (2006)

$\chi = S^w$ $\frac{e}{e_s} = 1 - a\left[1 - \exp(b\xi)\right]$	*e(p')*: void ratio (isotropic compression test under controlled suction) e_s *(p')*: void ration (isotropic saturated test) $\xi = f(s)\left(1 - S^w\right)$ a=0.369 ; b=1.419 *f(s)*: suction function	Gallipoli *et al.* (2003)
$\chi = \frac{1}{s + a}$	*s*: suction *a*: parameter	Sun *et al.* (2003)
$\chi = S^w \quad or \quad \chi = \sqrt{S^w}$		Sheng *et al.* (2003)

Table 6.1. *Different expressions of the Bishop χ parameter*

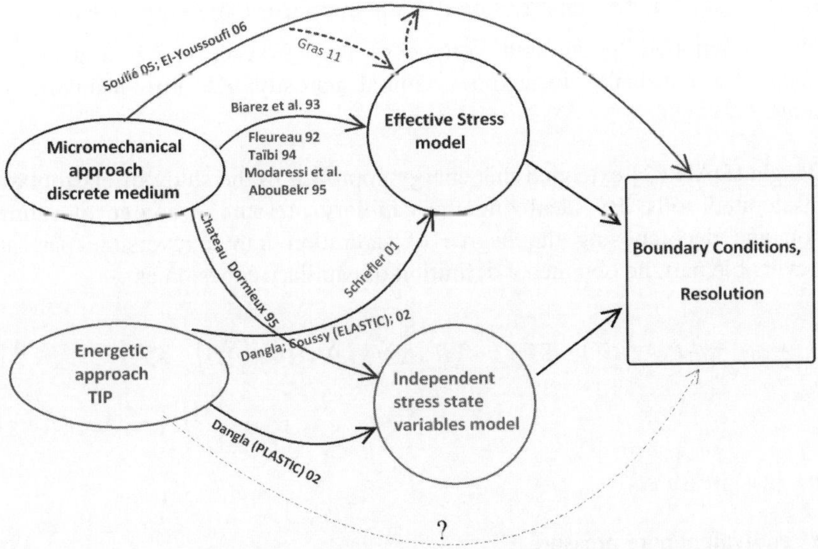

Figure 6.1. *Different approaches that can be used to solve an unsaturated granular material problem*

with

S_g : Saturation in gas (air + water vapor) at pressure p_g

S_l : Saturation in liquid water at pressure p_l

$$S_g = 1 - S_l$$

$$p_c = p_g - p_l = \text{capillary pressure}$$

U is a function derived from the wetting curve and by taking into account the strain of the skeleton. Following an energy approach, Coussy and Dangla [COU 02] showed that the effective properties derived from the analysis at the microscopic scale could be combined to the relations deduced from thermodynamics to interpret the effects of the pore pressure and the scale effects. This approach leads to the same expression of π:

$$\pi = S_g p_g + S_l p_l - U(S_l, T) \qquad [6.8]$$

where $U(S_l, T)$ is the free energy stored in the interfaces (solid–liquid; solid–gas and liquid–gas) reported to the unit volume of porous space $n_0 d\Omega$ in which these interfaces are physically located (n_0: initial porosity; $d\Omega$: infinitesimal volume element).

Dangla [DAN 02] extended this energy approach to the study of elastoplasticity in unsaturated soils. By idealizing the capillary pressure – degree of saturation relation and decomposing the degree of saturation into a reversible part and a non-reversible part, he obtained a definition of capillary cohesion as

$$C_{cap} = (p_g - \pi)\ tg\phi = (p_c(S_l)\ S_l + U(S_l))\ tg\phi \qquad [6.9]$$

with

p_g : gas pressure

π : equivalent pore pressure

ϕ: internal friction angle, assumed independent from capillary pressure

$p_c(S_l)$: capillary pressure curve, only depending on (S_l); for example, in the case of the Brooks and Corey model [BRO 64]:

$$p_c = p_e S_l^{-\alpha} \qquad [6.10]$$

with

p_e : air entry pressure

α: fitting parameter.

Figure 6.2 shows the evolution of the function $C_{cap} / tg\phi\,(p_c)$, with α as parameter for $p_e = 30$ kPa.

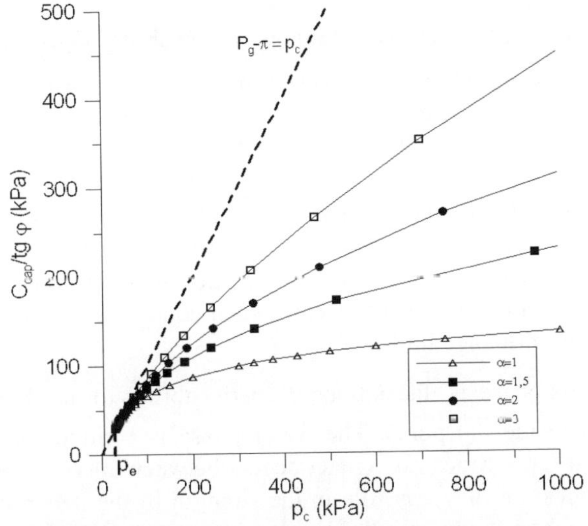

Figure 6.2. *Evolution of the function $c_{cap}/tg\varphi\,(p_c)$, with α as parameter, for p_e = 30 kPa [DAN 02]*

In a somewhat similar approach, Gray and Schrefler [GRA 01] suggested a thermodynamic definition of the effective stress notion in porous media. Their macroscopic approach also takes into account the microscopic boundary conditions, in particular the liquid–solid interfaces. Two cases were studied: in the first one, the wetting phase forms menisci at the points of contact of the solid phase; in the second case, the wetting phase is present as a film enveloping the solid.

In this chapter, the effective stress concept will be compared to the results of an experimental study on a compacted silty sand. The following points are examined:

– the behavior before failure, from very small to large strains;

– the influence of the capillary pressure on the maximum strength;

– the validity of an effective stress approach based on a simple microstructural model to interpret the results.

6.2. Microstructural model for unsaturated porous media

The model is based on a medium composed of elastic spherical isodiametral particles. Gravity is not considered and the pressure in the air phase is atmospheric (p_a = 0). Characteristic behaviors can be brought to light in two cases [BIA 89; BIA 93; BIA 94; FLE 95; TAI 94]:

– when the water phase is discontinuous (permeability to water $k_w = 0$), water forms torical menisci at the points of contact between particles (Figure 6.3). The air phase is continuous and its pressure plays a role, as does the water pressure, in the creation of the capillary strength; the water pressure inside the meniscus is negative with respect to that of the air ($p_w < p_a$). In this case, the intergranular forces due to water are perpendicular to the tangent planes at the contact points between the particles and, therefore, do not produce any rearrangement of the particles or variation in volume. However, these forces play an important role in the strength of the material. From a practical point of view, such conditions are observed for water contents below the shrinkage limit;

– when the air phase is discontinuous (permeability to air $k_a = 0$), air forms isolated bubbles inside the pores. The water phase is continuous and completely wets the solid particles. Since no contact occurs between air and the solid particles, water pressure does not play any role in the strength of the material. However, the presence of air bubbles in the pore fluid makes it more compressible.

When the particles arrange themselves randomly, the inter-granular forces due to external loading are generally not perpendicular to the tangent planes and can, therefore, induce volumetric strains.

The passage from a discontinuous to a continuous medium is made by considering regular arrangements of particles. Four types of arrangements were studied, with densities ranging from 0.83 (tetrahedral) to 1.81 (dodecahedral) (Figure 6.4).

In a representative elementary volume (REV), the "capillary" stress in a direction is expressed as

$$\sigma'_{cap} = \frac{\Sigma F_i}{S_{REV}} \tag{6.11}$$

where ΣF_i is the sum of the capillary forces acting in this direction and S_{REV} the cross-section of the REV normal to that direction.

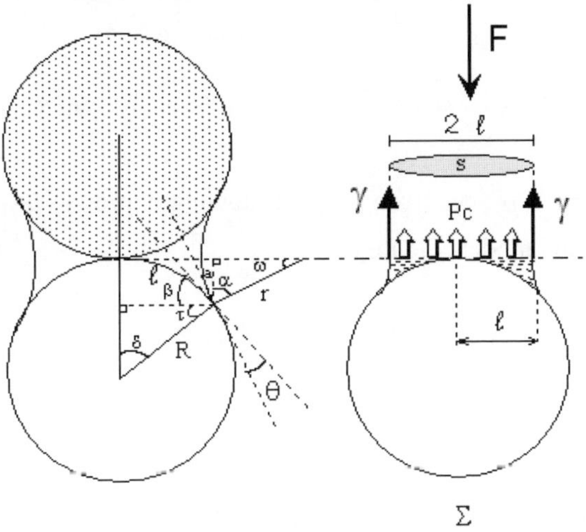

Figure 6.3. *Two spherical particles in contact with a water meniscus*

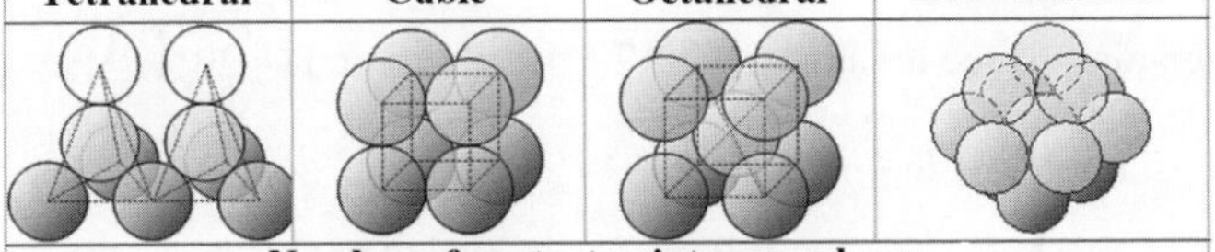

Tetrahedral	Cubic	Octahedral	Dodecahedral
Number of contact points per sphere			
4	6	8	12
Void ratio			
1.95	0.91	0.47	0.35

Figure 6.4. *Different types of regular arrangements of spherical particles*

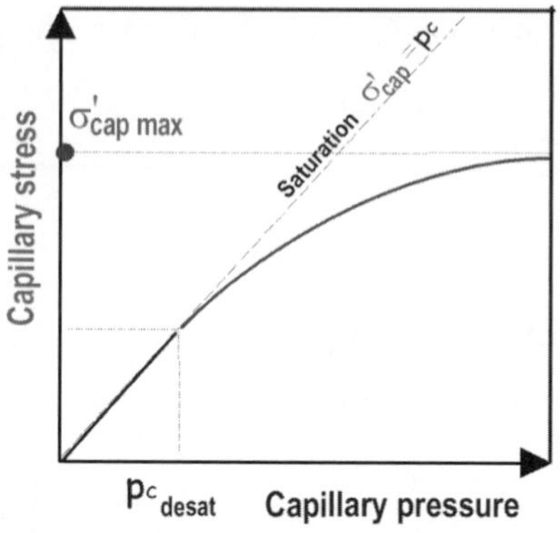

Figure 6.5. *Relation between capillary pressure and capillary stress*

The relation $\sigma'_{cap}(p_c)$ is given by the following expression, within the limits of the calculation of the model (Figure 6.5):

$$\sigma'_{cap} = \frac{\pi\gamma_{lg}}{2K(e)R^2}\left[4R + \frac{3\gamma_{lg} - \sqrt{8\gamma_{lg}p_c R + 9\gamma_{lg}^2}}{p_c}\right] \quad [6.12]$$

and

$$\sigma'_{cap\max} = \frac{2\pi\gamma_{lg}}{K(e)R} = \frac{4\pi\gamma_{lg}}{K(e)d} \quad [6.13]$$

with

γ_{lg} : water-air surface tension ($\gamma_{lg} = 7.5\ 10^{-2}$ N/m at T = 20°)

$d = 2R =$ diameter of the particle

$K(e)$: parameter depending on the packing type, given by the expression:

$$K(e) = 0.32\, e^2 + 4.06\, e + 0.11 \quad [6.14]$$

At large strains, the representation of the critical state of a granular material in Mohr's plane shows the existence of an apparent cohesion. In general, the solid skeleton is submitted to an intergranular (or effective) stress, which is the sum of the total stress and the capillary stress due to capillary pressure:

$$\sigma' = \sigma + \sigma'_{cap} \qquad [6.15]$$

Therefore, Mohr-Coulomb's expression for a granular medium ($c' = 0$) becomes

$$\tau' = \left(\sigma + \sigma'_{cap}\right) \; tg\,\phi' = \sigma \; tg\,\phi' + \sigma'_{cap}\, tg\,\phi' \qquad [6.16]$$

The quantity $C_{cap} = \sigma'_{cap}\, tg\phi$ is called "capillary cohesion". It is dependent on the degree of saturation and becomes zero when $S_r = 0$ or $S_r = 1$.

For a material with a natural cohesion C_{nat} (in the case of cementation), the total cohesion, when the material is unsaturated, is (Figure 6.6)

$$C_{tot} = C_{nat} + C_{cap} \qquad [6.17]$$

Recently, [SOU 06] generalized these calculations to the case of contact between spheres of different diameters and with imperfect wetting (Figure 6.7). They proposed the following analytical form to express the capillary force versus the geometry of the pair of particles:

$$F = \pi \gamma_{lg} \sqrt{R_1 R_2}\left[c + \exp\left(a \frac{D}{R_2} + b\right)\right] \qquad [6.18]$$

where R_1 and R_2 are the particle radii and D is the interparticular distance.

$$a = -1.1 \left(\frac{V}{R_2^3}\right)^{-0.53} \qquad [6.19]$$

$$b = \left[-0.148 \ln\left(\frac{V}{R_2^3}\right) - 0.96\right]\theta^2 - 0.0082 \ln\left(\frac{V}{R_2^3}\right) + 0.48 \qquad [6.20]$$

$$c = 0.0018 \ln\left(\frac{V}{R_2^3}\right) + 0.078 \qquad [6.21]$$

V: volume of the liquid bridge

θ: contact angle

γ_{lg}: water-air surface tension

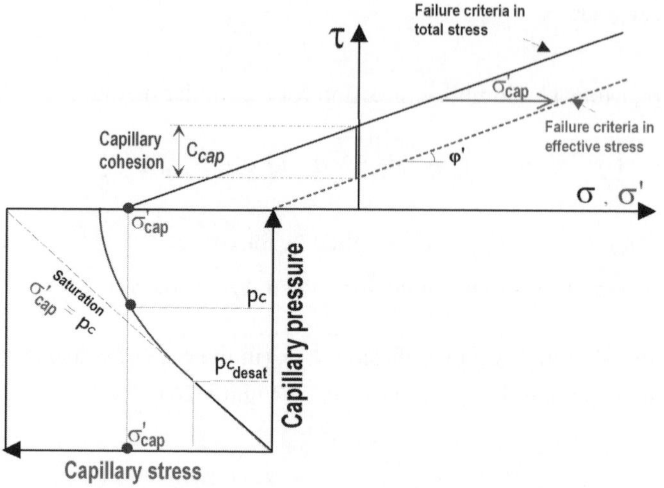

Figure 6.6. *Relation between capillary cohesion and stresses in Mohr plane*

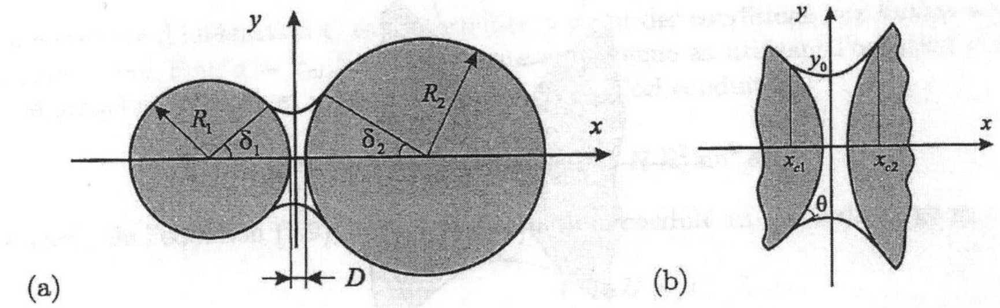

Figure 6.7. *(a) Geometrical representation of the liquid bridge between two grains of different diameters and (b) detail of the liquid bridge [SOU 06]*

6.3. Material and methods

6.3.1. *Material and preparation of samples*

The tests were carried out on a residual silty sand, called "Perafita sand", mainly composed of decomposed granite, which was used for the construction of a road in the north of Portugal. Its geotechnical properties are shown in Table 6.2 and its grain size distribution in Figure 6.8. The soil contains approximately 2% particles smaller than 2 μm and nearly 20% smaller than 80 μm. Standard and modified Proctor curves are plotted in Figure 6.9. Photographs of the material taken with a scanning electron microscope (SEM) are shown in Figure 6.10. The grains seem to be made of several assembled layers that are several micrometers thick. This fabric, which is the characteristic of a residual soil and similar enough to that of a clay, gives a liquid limit of nearly 33% and a few original properties, especially with respect to saturation.

Drying–wetting tests were carried out on specimens compacted to the modified Proctor optimum water content and close to maximum density with tensiometric plates and osmotic devices. Details of the experimental techniques are given, for instance, in [BIA 88] or [VER 02]. The drying and wetting paths starting from the compaction point are shown in Figure 6.11, in the void ratio and water content versus capillary pressure planes, respectively. When the soil is normally consolidated or slightly overconsolidated, these paths exhibit hysteresis [FLE 93]; whereas for a soil compacted under these conditions, the paths are generally reversible [FLE 02b]. The points corresponding to the specimens prepared for the large strain triaxial tests, i.e. compacted to different water contents and densities, were also plotted on the graph. These points are not located near the drying–wetting paths and cannot, therefore, be used to estimate the capillary pressure of compacted specimens.

6.3.2. *Experimental devices and test procedures*

The modules were measured by means of two different triaxial cells with local strain measurements, one (called LPTT, large cell for precision Triaxial tests) at the Instituto Superior Técnico (IST) in Lisbon for the tests on large specimens and the other one (SPTT, small cell for precision Triaxial tests) at the Ecole Centrale de Paris (ECP) for the tests on smaller specimens. Capillary pressure measurements were carried out at ECP in a conventional triaxial cell (TTu, Triaxial tests with control of suction) equipped with a semi-permeable porous stone and an absolute pressure transducer.

w_L %	w_P %	I_P %	< 2µm %	< 80µm %	w_{opt} %	γ_{dmax} kN/m^3	G_s	M_{pp}
32.6	25.0	7.6	2.5	20	Standard: 17.6 Modified: 13.2	Standard: 16.8 Modified: 18.6	2.66	1.1

Table 6.2. *Geotechnical properties of Perafita silty sand*

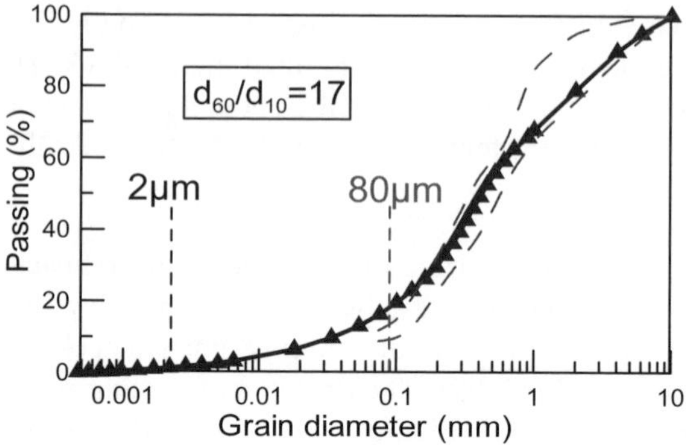

Figure 6.8. *Grain size distribution of Perafita sand*

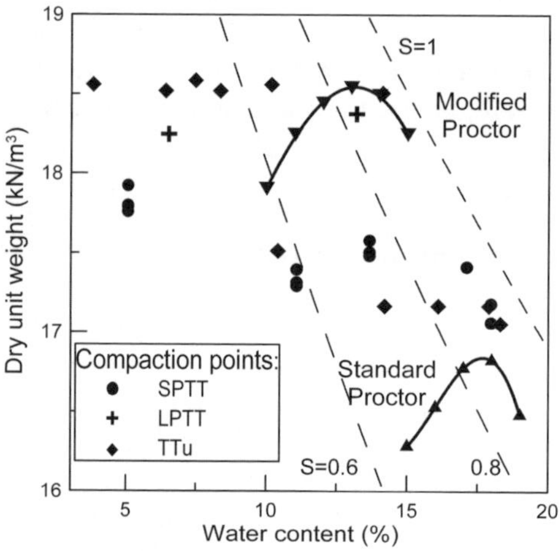

Figure 6.9. *Compaction points of the different specimens and standard and modified Proctor curves*

The procedure of preparing the specimens is the same for all the tests: the soil is passed through a 5 mm sieve to eliminate the largest grains, then it is mixed with the adequate quantity of water and lastly it is placed in a sealed plastic bag for at least 24 hours to obtain a uniform humidity. Two mold dimensions were used to prepare the samples: 70 mm in diameter and 140 mm in height for SPTT and TTu at ECP, 160 mm in diameter and 320 mm in height for LPTT at IST. The small specimens were compacted to the chosen dry density in three layers by a 24 N weight falling from a height of 305 mm, with the required number of blows per layer. The large samples were compacted in seven layers by means of a vibrating hammer with a static weight of approximately 30 N equipped with a 105 mm plate. The vibration time was chosen so as to reach the chosen density. The compaction properties of the samples used in the study are indicated in Tables 6.3 and 6.4, and the corresponding points are shown in Figure 6.11.

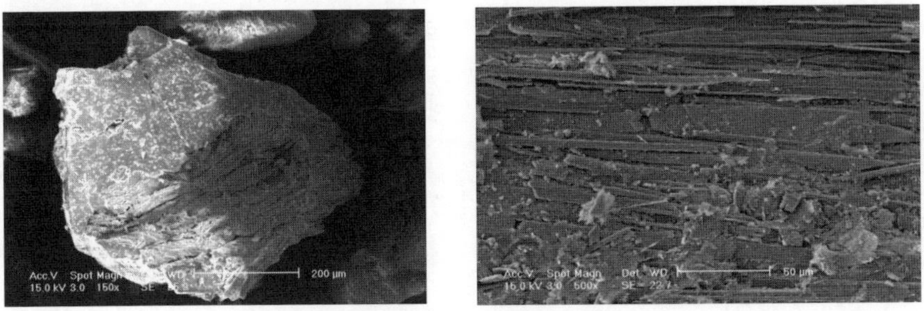

Figure 6.10. *SEM photos of Perafita sand at enlargements of 150 (left) and 500 (right)*

w (%)	5.1	6.5	11.1	13.2	13.7	18.0
D x H mm x mm	70 x 140	160 x 320	70 x 140	160 x 320	70 x 140	70 x 140
e	0.469	0.430	0.509	0.420	0.509	0.529
S_w (%)	28.9	40.2	58.0	83.6	71.6	90.5
Ref.	[DUF 02]	[HAD 02]	[DUF 02]	[HAD 02]	[DUF 02]	

Table 6.3. *Conditions of compaction of the samples for the LPTT and SPTT tests*

w (%)	3.74	6.25	7.5	8.0	9.5	10.4	14.1	14.2	16.1	17.9	18.3
D x H mm x mm	70 x 70	70 x 140	70 x 70	70 x 70	70 x 70	70 x 140	70 x 140	70 x 140	70 x 140	70 x 140	70 x 140
e	0.406	0.409	0.404	0.409	0.401	0.490	0.410	0.520	0.520	0.520	0.530
S_w (%)	24.3	40.5	48.7	54.6	63.2	57.6	93.8	74.1	84.0	95.2	97.4
Ref.	[HAD 02]					[LAN 01]					

Table 6.4. *Conditions of compaction of the samples for the TTu tests*

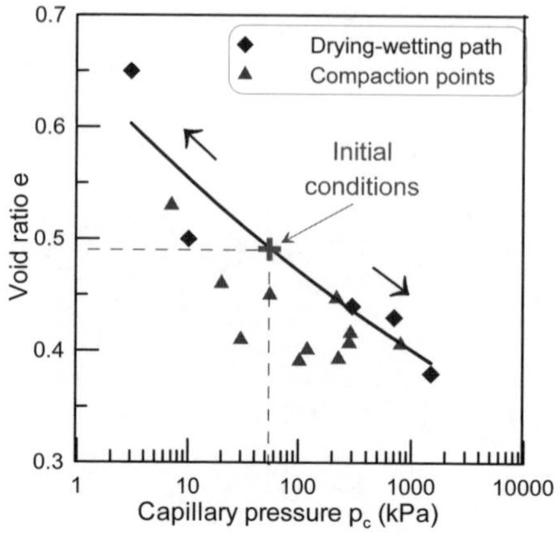

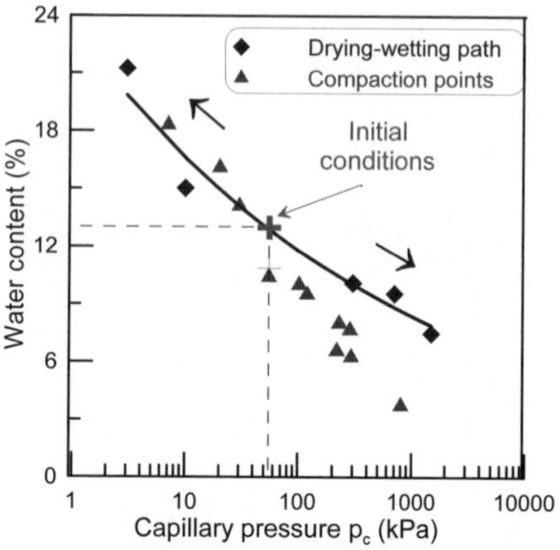

Figure 6.11. *Drying and wetting paths of specimens of Perafita silty sand compacted to modified Proctor optimum water content and maximum density and compaction points of triaxial tests specimens*

6.3.2.1. *Precision triaxial cell for large samples (LPTT)*

The precision triaxial cell for tests on large samples, 320 mm high and 160 mm in diameter, is equipped with three LVDT (Linear Variable Differential Transformer) for measuring axial strains and three LVDT for measuring radial strains. The frame and the armature rod of the vertical LVDT are fixed to target studs placed in the specimen during compaction. The same type of studs is also used to measure radial strains, the frame of the LVDT being supported by a ring mounted in the triaxial cell. The strain resolution is around 10^{-6} with a 16-bit A/D converter. A standard pressure transducer and a sensitive load cell located inside the triaxial cell are used to measure the confining stress and axial force. The specimen is in contact with air at atmospheric pressure through semi-permeable membranes at the top and bottom.

The test procedure uses the multistage technique. For each confining pressure (26 and 52 kPa), the test starts with five unloading–reloading cycles of very small vertical stress amplitude. The amplitude of the cycles is controlled to ensure that the cycles are closed and linear, in order to evaluate the elastic Young's modulus. Then, deviatoric loading is applied up to an axial strain of about 5×10^{-4} to obtain the decay curve of the secant Young's modulus with vertical strain. The strain rate of the tests is approximately 1.4×10^{-7} s^{-1}.

During the unloading process, very small unloading–reloading vertical stress cycles are performed at different steps. Figure 6.12 shows a typical result for a confining pressure of approximately 52 kPa. At the end of all these tests, an isotropic stress path is followed in order to evaluate the anisotropy of the soil.

6.3.2.2. *Precision triaxial cell for small samples (SPTT)*

The precision triaxial cell for tests on small samples, 140 mm high and 70 mm in diameter (SPTT), is fitted with two 80 mm-long Hall effect-based transducers (HETs) for measuring axial strains and one HET for measuring radial strains. After a critical analysis of the possible sources of error, several improvements were made to the standard GDS equipment [DUF 02]. A very high precision voltage source (HP 3245A) and a multimeter (HP 3458A) were used for data acquisition and the tests were carried out under temperature-controlled conditions. Calibration and comparison with other techniques were made on different reference samples. In soils, errors are most likely due to the displacement of the sensor supports, and the accuracy of the strain measurements does not exceed 10^{-6} [FLE 01]. The experimental procedure is the same as that used for the large specimens.

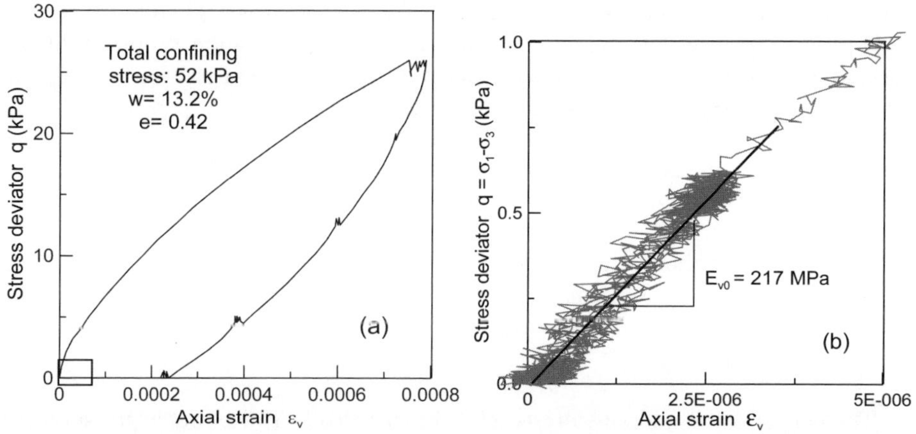

Figure 6.12. *(a) Example of deviatoric stress versus axial strain curve; (b) detail of very small unloading–reloading cycles*

After compaction, the specimen is placed in the triaxial cell and allowed to consolidate under the first isotropic stress for a period of one day, then a deviatoric loading of up to 10^{-4} axial strain is applied, so the sample is not damaged. At this point, the confining pressure is increased to the second consolidation stress level, and so on for the other levels.

For the last confining pressure, i.e. 79 kPa, the specimen stays in consolidation for three days (aging) before the test.

At the end of all the tests, a larger cycle is made, up to a few 10^{-3} axial strain. The drainage of the specimen remains open for the whole test. The strain rate is approximately 10^{-8} s^{-1}.

6.3.2.3. *Triaxial tests with measurement of capillary pressure (TTu)*

In most cases, capillary pressure was measured separately from the modulus measurements.

A usual triaxial cell is equipped with a semi-permeable porous stone (with 1.5 MPa air entry pressure), located within the pedestal in order to prevent air from passing into the water circuit. The pressure in the water phase is measured by means of an absolute pressure transducer, while a relative pressure transducer connected to the upper part of the specimen is used to measure the pressure in the air phase.

For water contents above the modified Proctor optimum water content, corresponding to pore-water pressures higher than -50 kPa, the device is used as a tensiometer (i.e. with $p_a = 0$ and $p_w < 0$). For water contents below the optimum, the axis translation technique was used: a positive air pressure was applied to the sample until positive pore-water pressure was obtained; thus, cavitation problems in the measuring devices are avoided so long as air is prevented from passing through the porous stone, which was the case in all these tests.

Volume changes are derived from the water volume coming in or out of the cell, measured by the confining pressure controller.

Considering errors as a consequence of expansion of the Perspex envelope, the compression of the latex membrane or air bubbles during isotropic loading, this method was only used during application of the stress deviator, when the confining pressure is constant.

The consolidated undrained tests with measurement of pore-water pressure and control of air pressure were performed in two steps:

(i) isotropic consolidation under stresses of 7, 14, 28, 52 and 79 kPa;

(ii) compression under 79 kPa confining stress at a constant strain rate ($\approx 10^{-6}$ s^{-1}), up to failure.

During the consolidation phase, the changes in capillary pressure were recorded until equilibrium was reached.

6.3.3. *Normalization of data*

The results of SPTT and LPTT were all corrected to eliminate the effect of the different initial void ratios (e_0) of the specimens tested and a "normalized" value corresponding to $e_{ref} = 0.5$ was derived from the expression in Iwasaki *et al.* [IWA 78]:

$$E_{(e=0.5)} = E_{(e_0)} x \frac{f(0.5)}{f(e_0)} \qquad [6.22]$$

with $f(e_0) = \frac{(2.17 - e_0)^2}{1 + e_0}$ and $f(0.5) = 1.859$

6.4. Experimental results

6.4.1. *Isotropic compression paths*

When the specimen is submitted to an isotropic compression stress, its capillary pressure p_c tends to decrease, but the change depends to a large extent on the water content of the specimen (see Figure 6.13):

– when the soil is on the dry side of the optimum water content, the change is negligible;

– on the other hand, wetter specimens exhibit a suction decrease with the rise of stress, and significantly more so if their water content is high.

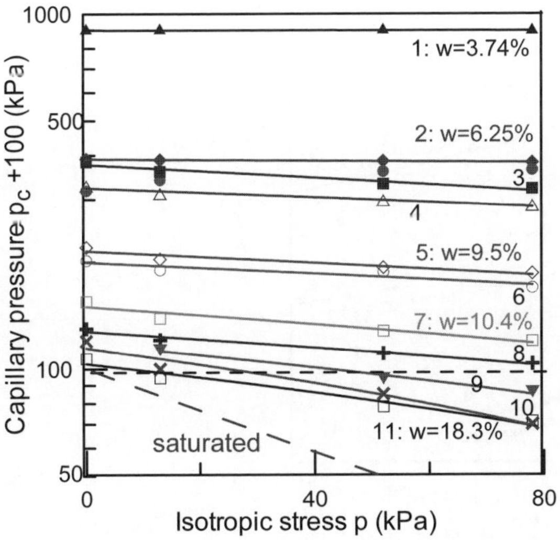

Figure 6.13. *Influence of isotropic stress and water content on capillary pressure [1: w=3.74%; 2: w=6.25%; 3: w=7.5%; 4: w=8%; 5: w=9.5%; 6: w=10%; 7: w=10.4%; 8: w=14.1%; 9: w=16.1%; 10: w= 17.9%; 11: w=18.3%]*

These results are consistent with volume change measurements showing that, even for a specimen compacted to the modified Proctor optimum water content and maximum density, the relative change in void ratio remains below 1% when the confining stress increases from 0 to 79 kPa [FLE 02a].

6.4.2. *Deviatoric compression paths*

At the end of the isotropic compressions, consolidated undrained triaxial tests, with measurement of the capillary pressure, were performed on the specimens under the same confining stress of 79 kPa. The test results are shown in Figure 6.14.

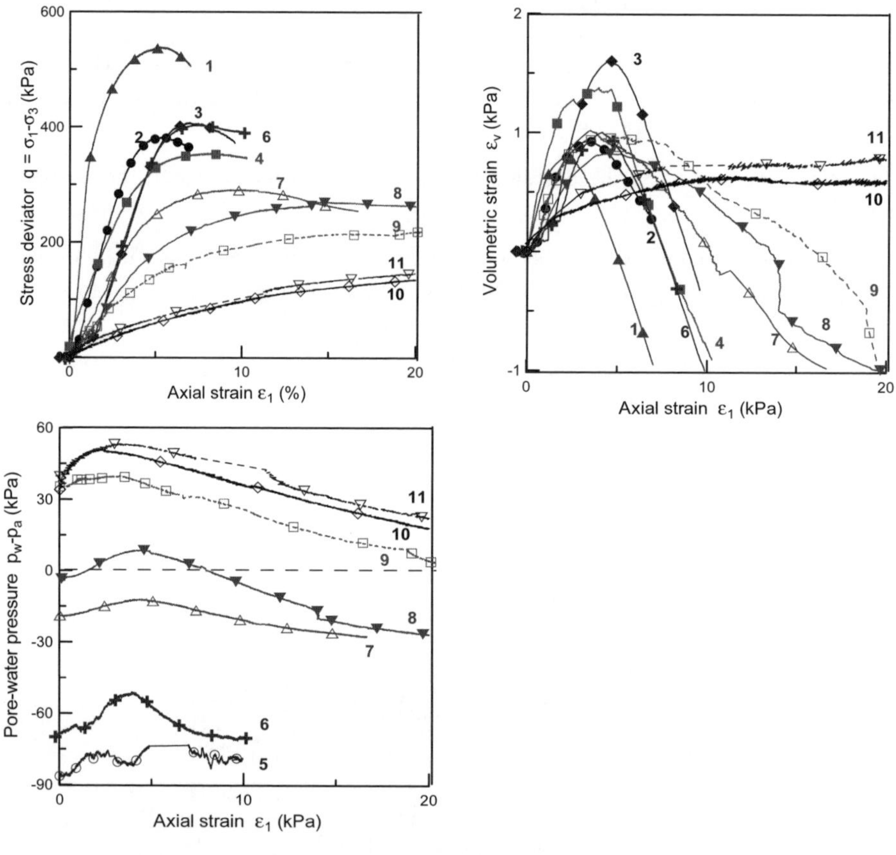

Figure 6.14. *Changes in stress deviator, volumetric strain and pore pressure versus axial strain for consolidated undrained tests on Perafita sand [1: w=3.74%; 2: w=6.25%; 3: w=7.5%; 4: w=8%; 5: w=9.5%; 6: w=10%; 7: w=10.4%; 8: w=14.1%; 9: w=16.1%; 10: w= 17.9%; 11: w=18.3%]*

When the water content increases, a progressive change from brittle to plastic behavior can be noted. For water contents below 8%, the stress–strain curves present a peak and the specimens seem strongly dilatant; on the other hand, for water contents higher than 16%, perfectly plastic and slightly contractant behaviors are observed. However, in all the tests, the volumetric deformations remain small (1.0 to 1.5 %). In the first case, the capillary pressure in the menisci creates strong bonds between the grains, which increase the strength of the material and prevent deformation. In fact, localization of strains are frequently observed in the samples under these conditions, so that the measured increase in volume is probably not wholly representative of the real behavior of the soil, but partly due to an apparent increase in the global volume after the formation of the discontinuity. A similar behavior was observed by Verbrugge in a quasi-dry loam [VER 02]. When the water content increases, the grains begin to be surrounded by water and the behavior of the soil tends towards that of the saturated soil at the same void ratio.

The changes in pore-water pressure $p_w - p_a$ during triaxial tests are shown only for water contents greater than 9.5 %. The curves for the driest samples (between 3.7 and 9.5 %) are not shown in this figure, as the pressure remains constant during the tests. They are at the same value as at the end of the isotropic compression tests (see Figure 6.13). The trend is the same for all the curves: (i) first, an increase in pressure corresponding to the contractant behavior already mentioned; and (ii) then a reduction in pressure, characteristic of dilatant behavior. This second phase is observed even in the nearly-saturated specimens that exhibit an overconsolidated behavior, corresponding to what is usually observed in saturated compacted specimens. In all cases, the changes in pressure are small (± 20 kPa or less), as a consequence of the low compressibility of sand.

6.4.3. *Small strain behavior*

The small strain properties of Perafita sand were measured in separate tests, not like those used to measure the large strain properties, but under approximately the same water content and void ratio conditions. Two tests (at $w = 6.5$ and 13.2%) were performed under several confining stresses and stress deviators, whereas the others were performed under isotropic stress states only. The influence of several factors was investigated. Figure 6.15 shows that, for unsaturated specimens, the changes in normalized modulus are approximately a power law of the total vertical stress, with an exponent $n = 0.35$ to 0.40. On the other hand, the curves for dry and quasi-saturated specimens are nearly superimposed, with a larger slope ($n = 0.68$). Under the same void ratio and vertical stress, the modulus increases when the water content decreases.

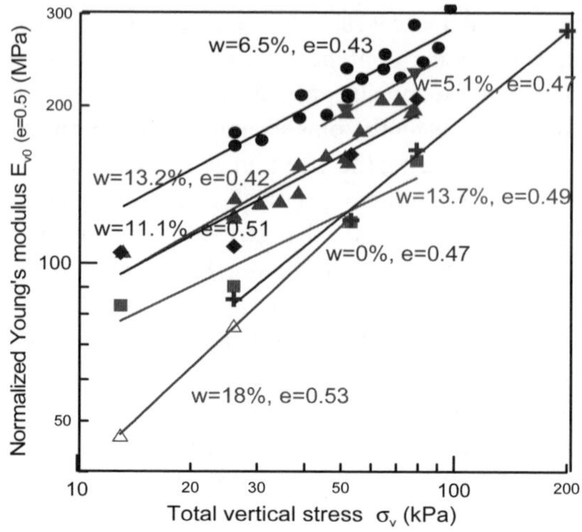

Figure 6.15. *Normalized Young's modulus versus total vertical stress for samples of Perafita sand at different water contents and densities*

6.5. Interpretation of results using the effective stress concept

The interpretation of the triaxial tests is based on the assumption that the effective stress concept is valid for determining both the failure criterion and the elastic moduli E_{v0}, and that a unique relationship exists between p' and q in the first case, $\log(\sigma'_v)$ and $\log(E_{v0})$ in the second, whether it is for saturated or unsaturated specimens.

When the soil is quasi-saturated, p' is given by Terzaghi's expression; in the case of unsaturated soils, the values of p' are derived from the micromechanical model described in section 6.2, provided that the "characteristic dimension" R has been determined. In the case of a real granular medium, R loses the physical meaning of a grain size, but becomes a fitting parameter of the model.

Considering first the tests on quasi-saturated specimens (i.e. with $S_w > 75\%$), it is possible to derive the maximum strength criterion from the large strain triaxial tests, with the equation:

$$q = \eta_{max}(p + \sigma'_{cap}), \quad \text{where} \quad \sigma'_{cap} = p_a - p_w \qquad [6.23]$$

Thereupon, the value of the "characteristic dimension" R of the micromechanical model is chosen so that the final points of the tests on the unsaturated samples (with $S_w < 50\%$) are located near the previously determined maximum strength criterion. In the case of the triaxial tests on Perafita sand, this leads to the function $\sigma'_{cap}(p_c)$, as shown in Figure 6.16. The "experimental" values of σ'_{cap} are derived from

$$\sigma'_{cap\_exp} = p' - (p - p_a) = (q/\eta_{max}) - (p - p_a) \quad [6.24]$$

The same approach is used to interpret the results of the small strain tests, with the same value of parameter R in the case of the unsaturated specimens.

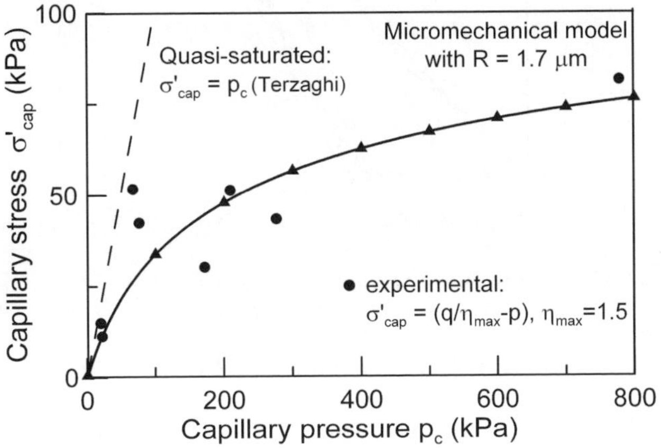

Figure 6.16. *Comparison between the values at failure derived from the triaxial tests and the results of the micromechanical model*

6.5.1. *Interpretation of large strain triaxial tests*

The stress paths for the different tests are shown in Figure 6.17, taking into account the effective vertical stress σ'_v. For the quasi-saturated specimens, the slope of the maximum strength criterion is $\eta_{max} \approx 1.5$. With a characteristic dimension R equal to 1.7 µm ($d = 3.4$ µm), the paths of the unsaturated specimens also finish near the same maximum strength criterion as the quasi-saturated tests. The shape of the stress paths is rather that of drained specimens, especially for the driest samples corresponding to the largest values of σ'_{cap}, and, therefore, of σ'_v. The reason is the fact that the capillary pressure remains constant under these conditions, since the sample strains are small. On the other hand, for the wettest specimens (under low suction values), the behavior is that of overconsolidated specimens. For some of the driest specimens, the localization of deformations, already mentioned, results in premature brittle failure of the specimens, which do not reach perfect plasticity.

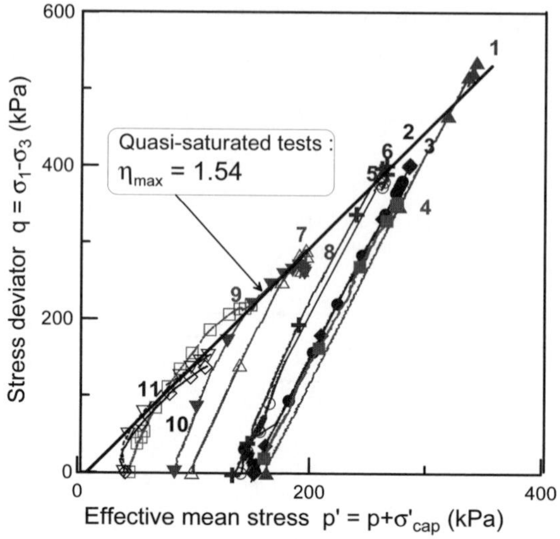

Figure 6.17. *Interpretation of results of large strain triaxial tests using the micromechanical model for determining effective stress [1: w=3.74%; 2: w=6.25%; 3: w=7.5%; 4: w=8%; 5: w=9.5%; 6: w=10% 7: w=10.4%; 8: w=14.1%; 9: w=16.1%; 10: w= 17.9%; 11: w=18.3%]*

6.5.2. *Interpretation of small strain modulus measurements*

The normalized values of the modulus for the different tests, shown in Figure 6.15, have been plotted in Figure 6.18 as a function of the effective vertical stress: $\sigma'_v = \sigma_v + \sigma'_{cap}$ ($p_a = 0$). To interpret the results, the values of the capillary pressure were derived from correlations based on the measurements made at different water contents, void ratios, and confining stresses. For quasi-saturated specimens, $\sigma'_{cap} = p_a - p_w$ and for dry specimens, $\sigma'_{cap} = 0$. For unsaturated specimens, the values of σ'_{cap} were derived from the micromechanical model, using the same "characteristic dimension" $R = 1.7$ µm as for the other triaxial tests and the function shown in Figure 6.16. Expressed against the effective stress, all the results (from $w = 0\%$ to $w = 18.3\%$) are correctly located near the regression line of the dry and quasi-saturated specimens, with equation

$$E_{v0} = 200\,(\sigma'_v/p_{atm})^{0.68} \qquad [6.25]$$

Considering that the results were obtained on samples of different sizes, under very different conditions, with different experimental devices, from values of pore-water pressure deduced from correlations and for deformations below 10^{-5}, the amount of scatter in the data is rather small.

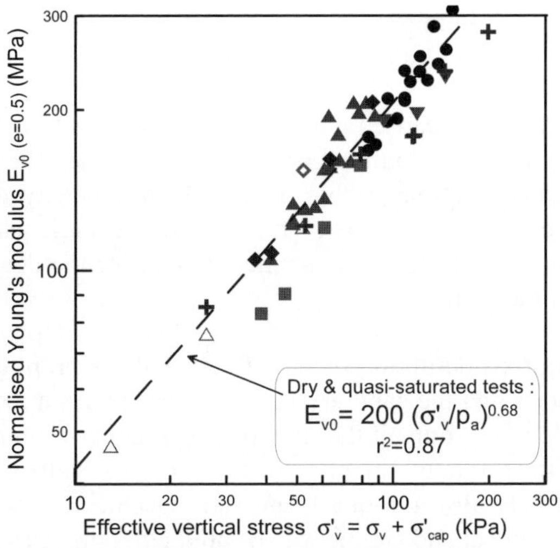

Figure 6.18. *Interpretation of the results of the small strain triaxial tests using the micromechanical model for determining effective stress*

6.6. Conclusions

A significant number of tests were performed on Perafita silty sand, under many different conditions of water content and density, isotropic and deviatoric stress, using non-standard devices for measuring very small strains or capillary pressures. The results from all these tests provide a fairly broad picture of the properties of the sand, from nearly dry to nearly saturated, from the elastic to the perfectly plastic domain. The residual material exhibits original properties with the grain size distribution of a fine sand, on the one hand, and noticeable plasticity, on the other hand that results in large changes in capillary pressure (from 3 to 2,000 kPa) in the domain of the water contents studied (7 to 21%).

Mechanically, the increase in capillary pressure results in an increase in maximal strength, stiffness and brittleness of soil. In fact, it is observed that, for soils compacted left of the modified Proctor optimum water content, volumetric changes due to alterations in isotropic or deviatoric stress become negligible. When the strains are very small, the curves relating the modulus values to the total stress do not cross the y-axis at 0, but present a positive ordinate for $\sigma_v = 0$. Moreover, when plotted in a bi-logarithmic coordinate system (see Figure 6.15), the paths seem fairly linear and parallel, but with a different slope from that of dry and quasi-saturated specimens. This type of behavior is systematically found in all partially saturated granular materials [TAI 08]. For the driest specimens, the increase in stiffness with capillary pressure may lead to precocious failure phenomena that need to be handled by introducing a damage criterion into the constitutive law.

Many studies have been carried out in the past 20 years to extend the use of the effective stress concept to unsaturated soils. The use of the effective stress concept allows a unified interpretation of the data, both in elasticity and on the failure criterion. The micromechanical model, which features a single parameter *R*, leads to a very good agreement with the experimental results for the determination of both the $q(p')$ and $E(\sigma'_v)$ functions.

With the generalized definition of the effective stress proposed in this chapter, sum of the total stress and the capillary pressure derived from the micromechanical model, it is possible to regroup all the final points of large strain triaxial tests on the failure criterion of the saturated samples and those of the small strain triaxial tests on the master curve of the dry and quasi-saturated specimens. Therefore, this model constitutes a very practical and simple way of predicting the strength and stiffness of a soil whenever its capillary pressure is known.

6.7. Acknowledgements

The authors would like to thank the ICCTI, the French Embassy in Portugal and the CNRS in France for their support in the organization of the exchanges between the IST and the ECP and for the realization of some of the tests in both institutions.

6.8. Bibliography

[BIA 89] BIAREZ J., FLEUREAU J.M., INDARTO TAIBI S., ZERHOUNI M.I., "Influence of water negative pore pressure on the flow of granular materials in silos", *Proceedings of the First International Conference on Micromechanics of Granular Media*, Clermont-Ferrand, 4-8 September 1989. BIAREZ J. & GOURVÈS R. (eds), A.A. Balkema, Rotterdam, The Netherlands, pp. 385-392, 1989.

[BIA 88] BIAREZ J., FLEUREAU J.M., ZERHOUNI M.I., SOEPANDJI B.S., "Variations de volume des sols argileux lors de cycles de draînage-humidification", *Rev. Franç. Géotechnique*, vol. 41, pp. 63-71, 1988.

[BIA 93] BIAREZ J., FLEUREAU J.M., TAIBI S., "Constitutive model for unsaturated granular media made up by spheres", *Proc. 2nd Int. Conf. on Micromechanics of Granular Media*, Birmingham, UK, Thornton (ed.), Balkema, Rotterdam, the Netherlands, pp.51-58, 1993.

[BIA 94] BIAREZ J., FLEUREAU J.M., TAIBI S., "Critère de résistance maximale des sols non saturés : Approche expérimentale et modélisation", *Proc. 13th Int. Conf. on Soil Mech. and Found. Eng.*, New Delhi, Oxford & I.B.H., pp. 385-388, 1994.

[BIO 41] BIOT M.A., "General theory of three-dimensional consolidation", *J. Appl. Phys.*, vol. 12, pp. 155-164, 1941.

[BIS 59] BISHOP A.W., "The principle of effective stress", *Teknisk Ukeblad*, vol. 106, no. 39, pp. 859-863, 1959.

[BIS 61] BISHOP A.W., DONALD I.B., "The experimental study of partly saturated soil in triaxial apparatus", *5th Int. Conf. on Soil Mechanics and Foundation Engineering*, vol. 1, pp. 13-21, Paris, France, 1961.

[BIS 63] BISHOP A.W., BLIGHT G.E., "Some aspects of effective stress in saturated and unsaturated soils", *Geotechnique*, no. 3, pp. 177-197, 1963.

[BOL 95] BOLZON G. SCHREFLER B.A., "State surfaces of partially saturated soils: an effective pressure approach", *Appl. Mech. Rev.*, vol. 48, no. 10, pp. 643-649, 1995.

[BOL 96] BOLZON G., SCHREFLER B.A, ZIENKIEWICZ O.C., "Elastoplastic soil constitutive laws generalised to partially saturated states", *Geotechnique*, vol. 46, pp. 279-89, 1996.

[BRO 64] BROOKS R.N., COREY A.T., "Hydraulic properties of porous media", *Colorado State Univ. Hydrol. Paper*, no. 3, 1964.

[CHA 95] CHATEAU X., DORMIEUX L., "Homogénéisation d'un milieu poreux non saturé: lemme de Hill et applications", *C.R. Acad. Sci.*, vol. 320, no. II b, pp. 627-634, 1995.

[COU 95] COUSSY O., *Mechanics of Porous Continua*, John Wiley and Sons, New York, USA, 1995.

[COU 02] COUSSY O., DANGLA P., "Approche énergétique du comportement des sols non saturés", in: COUSSY O. and FLEUREAU J.M. (eds.), *Mécanique des Sols non Saturés*, pp. 137-174, Hermès, Paris, France, 2002.

[DAN 02] DANGLA P., "Plasticité et hystérésis", in: COUSSY O. and FLEUREAU J.M. (eds), *Mécanique des sols non saturés*, pp. 231-271, Hermès, Paris, France, 2002.

[DEB 90] DE BOER R., EHLERS W., "The development of the concept of effective stresses", *Acta Mechanica*, vol. 83, pp. 77-92, 1990.

[DUF 02] DUFOUR LARIDAN E., Propriétés mécaniques des sols en petites déformations; étude expérimentale d'un sable silteux, Doctorate thesis, Ecole Centrale Paris, France, 2002.

[FLE 93] FLEUREAU J.M., KHEIRBEK-SAOUD S., SOEMITRO R., TAIBI S., "Behaviour of clayey soils on drying-wetting paths", *Canadian Geotechnical Journal*, vol. 30, no. 2, pp. 287-296, 1993.

[FLE 95] FLEUREAU J.M., KHEIRBEK-SAOUD S., TAIBI S., "Experimental behaviour of unsaturated soils or soils with a negative pressure", *Proc. 1st International Conference on Unsaturated Soils UNSAT'95*, Paris, France, Balkema, vol. 1, pp. 57-62, 1995.

[FLE 02a] FLEUREAU J.M., HADIWARDOYO S., DUFOUR-LARIDAN E., LANGLOIS V., GOMES CORREIA A., "Influence of suction and loading conditions on the dynamic properties of a silty sand", in: JUCA J., DE CAMPOS T. and MARINHO F. (eds), *Proc. 3rd Int. Conf. on Unsaturated Soils UNSAT'2002*, Recife, Brazil, Balkema, Lisse, vol. 2, pp. 463-472, 2002.

[FLE 02b] FLEUREAU J.M., VERBRUGGE J.C., HUERGO P.J., GOMES CORREIA A., KHEIRBEK-SAOUD S., "Description and modelling of the drying and wetting paths of compacted soils", *Canadian Geotechnical Journal*, vol. 39, pp. 1341-1357, 2002.

[GAL 03] GALLIPOLI D., GENS A., SHARMA R., VAUNAT J., "An elasto-plastic model for unsaturated soil incorporating the effects of suction and degree of saturation on mechanical behaviour", *Geotechnique*, vol. 53, no. 1, pp. 123-135, 2003.

[GRA 01] GRAY W.G., SCHREFLER B.A., "Thermodynamic approach to effective stress in partially saturated porous media", *Eur. J. Mech. A/Solids,* vol. 20. pp. 521-538, 2001.

[GUD 95] GUDEHUS G., "Constitutive relations for granulate-liquid mixtures with a pectic constituent", *Mechanics of Materials*, vol. 22, no. 2, pp. 93-103, 1995.

[HAD 02] HADIWARDOYO S. Caractérisation des matériaux routiers - Application à la modélisation du comportement des chaussées souples, Doctorate thesis, Ecole Centrale de Paris, France, 2002.

[HAS 80] HASSANIZADEH M., GRAY W.G., "General conservation equations for multiphase systems: 3. Constitutive theory for porous media", *Advances in Water Resources*, vol. 3, pp. 25-40, 1980.

[HUT 99] HUTTER K., LALOUI L., VULLIET L., "Thermo-dynamically based mixture models of saturated and unsaturated", *J. Coh. Frict. Mat.*, vol. 4, pp. 295-338, 1999.

[IWA 78] IWASAKI T., TATSUOKA F., TAKAGI Y., "Shear moduli of sands under cyclic torsional shear loading", *Soils and Foundations*, vol. 18, no. 1, pp. 39-50, 1978.

[JEN 62] JENNINGS J.E.B., BURLAND J.B., "Limitation to the use of effective stresses in partly saturated soils", *Géotechnique*, vol. 12, no. 2, pp. 125-144, 1962.

[KHA 98] KHALILI N., KHABBAZ M.H., "A unique relationship for the determination of the shear strengh of unsaturated soils", *Geotechnique*, vol.48, no. 5, pp. 681-687, 1998.

[KHA 01] KHALILI N., LORET B., "An elasto-plastic model for non-isothermal analysis of flow and deformation in unsaturated porous media: formulation", *Int. J. Solids and Structures*, vol. 38, pp. 8305-8330, 2001.

[KHA 04] KHALILI N., GEISER F., BLIGHT G.E., "Effective stress in unsaturated soils: A critical review with new evidence", *International Journal of Geomechanics. ASCE*, vol. 4, no. 2, pp. 115-126, 2004.

[KOH 93] KOHGO Y., NAKANO M., MIYAZAKI T., "Theoretical aspects of constitutive modelling for unsaturated soils", *Soil and Foundations*, vol. 33, no. 4, pp. 49-63, 1993.

[LAN 01] LANGLOIS V., DEA Report, Ecole Centrale Paris, France, 2001.

[LEW 87] LEWIS R.W., SCHREFLER B.A., *The Finite Element Method in the Deformation and Consolidation of Porous Media* (1st edition), John Wiley & Sons, New York, USA, 1987.

[LOR 00] LORET B., KHALILI N., "A three phase model for unsaturated soils", *International Journal for Numerical Analytical methods in Geomechanics*, vol. 24, pp. 893-927, 2000.

[LOR 02] LORET B., KHALILI N., "An effective stress elastic-plastic model for unsaturated porous media", *Mechanics of Materials*, vol. 34, pp. 97-116, 2002.

[MOD 94] MODARESSI A., ABOUBEKR N., "Constitutive model for unsaturated soils; validation on a silty material", *Proc. of Numerical Methods in Geotechnical Engineering*, Manchester, vol. 1, pp. 91-96, Balkema, Rotterdam, The Netherlands, 1994.

[NUR 71] NUR A., BYERLEE J.D., "Exact effective stress law for elastic deformation of rock with fluids", *Journal of Geophysical Research*, vol. 76, no. 26, pp. 6414-6419, 1971.

[RIC 76] RICE J.R., CLEARY M. P., "Some basic stress-diffusion solutions for fluid-saturated elastic porous media with compressible constituents", *Reviews of Geophysics and Space Physics*, vol. 14, pp. 227-241, 1976.

[RUS 06] RUSSELL A.R., KHALILI N., "A unified bounding surface plasticity model for unsaturated soils", *Int. J. Num. Analy. Methods in Geomechanics*, vol. 30, no. 3, pp. 181-212, 2006.

[SHE 03a] SHENG D., SLOAN S.W., GENS A., SMITH D.W., "Finite-element formulation and algorithms for unsaturated soils. Part I: Theory", *International Journal for Numerical and Analytical Methods in Geomechanics*, vol. 27, no. 9, pp. 745-765, 2003.

[SHE 03b] SHENG D., SMITH D.W, SLOAN S.W., GENS A., "Finite element formulation and algorithms for unsaturated soils. Part II: Verification and application", *International Journal for Numerical and Analytical Methods in Geomechanics*, vol. 27, no. 9, pp. 767-790, 2003.

[SKE 61] SKEMPTON A.W., "Horizontal stresses in an overconsolidated Eocene clay", *Proceedings, 5th International Conference on Soil Mechanics and Fondation Engineering*, vol. 1, pp. 352-357, 1961.

[SOU 06] SOULIÉ F., CHERBLANC F., EL YOUSSOUFI M.S., SAIX C. "Influence of liquid bridges on the mechanical behaviour of polydisperse granular materials", *International Journal for Numerical Methods in Geomechanics*, vol. 30, pp. 213-228, 2006.

[SUN 03] SUN D.A., MATSUOKA H., CUI H.B., XU Y.F., "Three-dimensional elasto-plastie model for unsaturated compacted soils with different initial densities", *International Journal for Numerical and Analytical Methods in Geomechanics*, vol. 27, no. 12, pp. 1079-1098, 2003.

[TAÏ 94] TAÏBI S., Comportement mécanique et hydraulique des sols soumis à une pression interstitielle négative- Etude expérimentale et modélisation, Doctorate thesis, Ecole Centrale Paris, France, 1994.

[TAÏ 08] TAÏBI S., FLEUREAU J.-M., HADIWARDOYO S., KHEIRBEK-SAOUD S. "Small and large strain behaviour of an unsaturated compacted silt", *European Journal of Environmental and Civil Engineering*, vol. 12, no. 3, pp. 203-228, 2008.

[TER 36] TERZAGHI K., "The shearing resistance of saturated soils and the angle between the planes of shear", *First Int. Conf. Soil Mech. Found. Eng.*, pp. 54-56, 1936.

[TER 43] TERZAGHI K., *Theoretical Soil Mechanics*, John Wiley & Sons, New York, USA, 1943.

[VER 02] VERBRUGGE J.C., FLEUREAU J.M., "Bases expérimentales du comportement des sols non saturés", in: COUSSY O., FLEUREAU J.M. (eds), *Mécanique des sols non saturés*, pp. 69-112, Hermès, Paris, France, 2002.

[WHE 03] WHEELER S.L., SHARMA R.L., BUISSON M.S.R., "Coupling of hydraulic hysteresis and stress-strain behaviour in unsaturated soils", *Géotechnique*, vol. 53, no. l, pp. 41-54, 2003.

Chapter 7

A Microstructural Model for Soils and Granular Materials

7.1. Introduction

The mechanical behavior of granular materials is dependent on the properties of the grains that constitute the material. It therefore appears to be logical to establish a reasoning framework that allows us to pass from a discontinuous medium to an equivalent continuous medium. To achieve this goal, a methodology was created by Jean Biarez and his collaborators [BIA 62, BIA 77, BIA 89, BIA 94]. Subsequently, a method of classifying the representative parameters of granular materials was developed. Two classes of parameters can be distinguished:

– the parameters representing the mechanical properties of the grains and of the local contact law;

– the parameters describing the geometrical conditions of the grain assembly. The geometry of the grain themselves can be defined by parameters representing the size, shape and grain size distribution, which can generally be considered invariable. The geometry of the arrangement, which can vary, includes a parameter expressing the compacity (a scalar) and a parameter describing its anisotropy (a tensor).

Based on this approach linking discontinuous and equivalent continuous media, in this chapter we present a constitutive model for the continuous medium constructed from the inter-granular properties and from the geometry of the granular assembly. Homogenization techniques for granular materials have been developed in

---

Chapter written by Pierre-Yves HICHER.

the first place to predict their elastic properties [CHA 88, CHA 89, EME 96, LIA 00, WAL 87]. More recently, the same techniques have been applied with more or less success for modeling non reversible behavior [JEN 93, NEM 02]. The basic idea for most of these models is to consider the granular medium as a collection of tangent planes at the contact points between grains interacting with each other. This approach requires us to write the contact law along these tangent planes and develop homogenization operators in order to integrate all these contacts at the scale of the representative elementary volume.

Two techniques have been considered (see Figure 7.1). The kinematic approach is based on a localization operator, which provides a relationship between intergranular contact displacements and the assembly strain, and a homogenization operator, which relates the average stress of the granular assembly to a mean field of particle contact forces (see, for example, [NIC 05]). The static approach is based on an inverse technique consisting of a localization operator relating the macro stresses to the local contact forces and a homogenization operator relating the grain displacements to the macro strains. Both approaches relate local forces and displacements through a constitutive law at the contact level. The micro-structural model presented in this chapter is based on a static approach presented in the following section [CHA 05], which appears to be more suitable for describing the behavior of granular materials. The kinematically constrained models are more popularly used in concrete models. The overriding reason for using this approach is that, in the case of strain softening, it is easier to construct a stable model using a kinematic rather than a static constraint. However, the kinematic constraint obviously gives a more restrictive deformation pattern and thus provides a stiffer result, especially when granular materials are subjected to high deviatoric stresses [CHA 90].

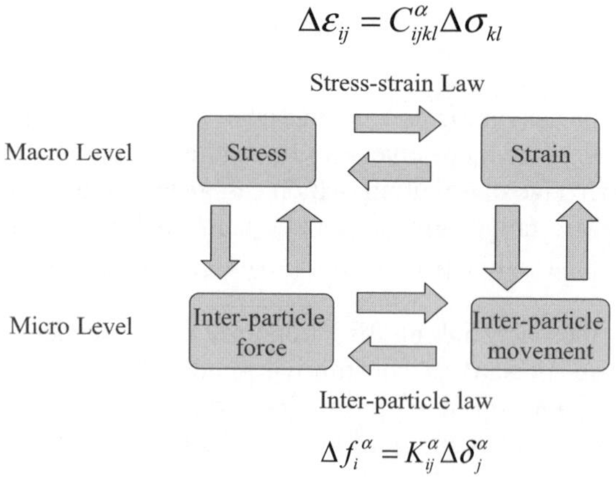

Figure 7.1. *Homogenization techniques for granular materials*

7.2. The micro-structural model

In this model, we conceive a granular material as a collection of particles. The deformation of a representative volume of the material is generated by mobilizing contact particles in all orientations. Thus, the stress–strain relationship can be derived as an average of the mobilization behavior of local contact planes in all orientations. For a contact plane in the αth orientation, the local forces f_i^α and the local movements δ_i^α can be denoted:

$$f_i^\alpha = \left\{ f_n^\alpha, f_s^\alpha, f_t^\alpha \right\} \text{ and } \delta_i^\alpha = \left\{ \delta_n^\alpha, \delta_s^\alpha, \delta_t^\alpha \right\}$$

where the subscripts *n*, s, and *t* represent the components in the three directions of the local coordinate system. The direction normal to the plane is denoted *n*; the other two orthogonal directions, *s* and *t*, are tangential to the plane (see Figure 7.2).

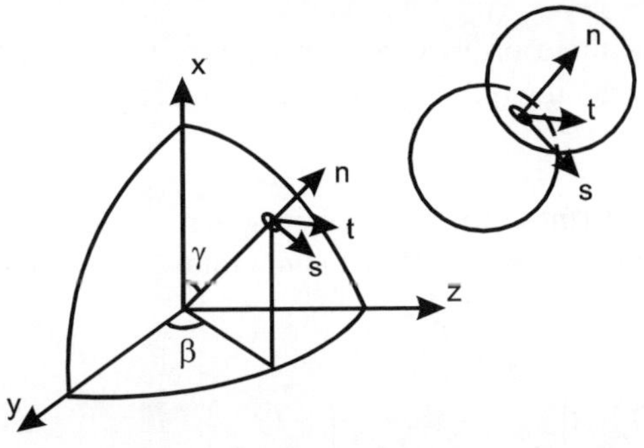

Figure 7.2. *Local coordinates*

The forces and movements at the contact planes of all orientations are suitably superimposed to obtain the macroscopic stress–strain tensors. The macroscopic stiffness tensor is obtained on the condition that the rate of energy dissipation expressed in terms of the macro stress and strain must be equivalent to that expressed in terms of micro forces and movements. Under such a formulation, it has usually been assumed that the microstructure is statically constrained, which means that the forces on each contact plane are assumed to be equal to the resolve components of the macroscopic stress tensor.

7.2.1. *Inter-particle behavior*

7.2.1.1. *Elastic part*

The orientation of a contact plane between two particles is defined by the vector perpendicular to this plane. On each contact plane, a local auxiliary coordinate can be established, as shown in Figure 7.1. The contact stiffness of a contact plane includes normal stiffness, k_n^α, and shear stiffness, k_r^α. The elastic stiffness tensor is defined by

$$f_i^\alpha = k_{ij}^{\alpha e} \delta_j^{\alpha e} \quad [7.1]$$

which can be related to the contact normal and shear stiffness:

$$k_{ij}^{\alpha e} = k_n^\alpha n_i^\alpha n_j^\alpha + k_r^\alpha \left(s_i^\alpha s_j^\alpha + t_i^\alpha t_j^\alpha \right) \quad [7.2]$$

where *n*, *s* and *t* are three orthogonal unit vectors that form the local coordinate system. The vector *n* is outward normal to the contact plane. Vectors *s* and *t* are on the contact plane.

The value of the stiffness for two elastic spheres can be estimated from Hertz-Mindlin's formulation [MIN 69]. For sand grains, a revised form was adopted [CHA 89b], as follows:

$$k_n = k_{n0} \left(\frac{f_n}{G_g l^2} \right)^n ; \quad k_t = k_{t0} \left(\frac{f_n}{G_g l^2} \right)^n \quad [7.3]$$

where G_g is the elastic modulus for the grains, f_n is the contact force in normal direction, l is the branch length between the two particles, and k_{no}, k_{ro} and *n* are material constants. For two spherical particles, the branch length is the same as the particle size $l = d$. Let $n = 1/3$, and

$$k_{n0} = G_g \frac{d}{2} \left(\frac{\sqrt{12}}{1 - \nu_g} \right)^{2/3} \quad [7.4]$$

Equation [7.4] is equivalent to the Hertz-Mindlin's contact formulation [MIN 69].

7.2.1.2. *Plastic part*

The shear force T and the rate of plastic sliding $d\Delta^p$ are defined as

$$T = \sqrt{f_s^2 + f_t^2} \text{ and } d\Delta^p = \sqrt{\left(d\delta_s^p\right)^2 + \left(d\delta_t^p\right)^2} \qquad [7.5]$$

The yield function is assumed to be of Mohr-Coulomb type:

$$F(f_i, \kappa) = T - f_n \kappa\left(\Delta^p\right) = 0 \qquad [7.6]$$

where $\kappa(\Delta^P)$ is an isotropic hardening/softening parameter. The hardening function is defined by a hyperbolic curve in the $\kappa - \Delta^p$ plane, which involves two material constants: ϕ_p and k_{p0}.

$$\kappa = \frac{k_{p0} \tan\phi_p \Delta^p}{\left|f_n\right| \tan\phi_p + k_{p0}\Delta^p} \qquad [7.7]$$

The value of κ asymptotically approaches $\tan\phi_p$. The initial slope of the hyperbolic curve is k_{p0}.

The elastic sliding behavior between two particles does not have a coupling effect (i.e. there are no shear-induced normal movements). However, plastic sliding often occurs along the tangential direction of the contact plane with an upward or downward movement; thus, shear dilation/contraction work, due to plastic movements, to the friction loss of contact planes. The relationship between normal and tangential displacements is governed by the dilatancy equation:

$$\frac{d\delta_n^p}{d\Delta^p} = \frac{T}{f_n} - \tan\phi_0 \qquad [7.8]$$

where ϕ_0 is a material constant which, in most cases, can be considered equal to the internal friction angle ϕ_μ. This equation can be derived by equating the dissipation work due to plastic movements to the friction loss of contact planes in the same orientation.

On the yield surface, under a loading condition, the shear plastic flow is determined by a normality rule applied to the yield function. However, the plastic flow in the direction perpendicular to the contact plane is governed by the stress–dilatancy equation in equation [7.8]. Thus, the flow rule is non-associated.

7.2.1.3. *Interlocking influence*

The internal friction angle ϕ_μ is a constant for the material. However, the peak friction angle, ϕ_p, on a contact plane is dependent on the degree of interlocking by neighboring particles, which can be related to the state of the packing void ratio e by:

$$\tan\phi_p = \left(\frac{e_c}{e}\right)^m \tan\phi_\mu \qquad [7.9]$$

where m is a material constant [BIA 94].

For dense packing, the peak frictional angle ϕ_p is greater than ϕ_μ. When the packing structure dilates, the degree of interlocking and the peak frictional angle are reduced, which results in a strain-softening phenomenon.

The dilative behavior of a granular material does not depend on the absolute value of the assembly void ratio, but rather on its relative value compared to the critical void ratio. Under critical state, the granular material will remain at a constant volume, while it is subjected to a continuous distortion. The void ratio corresponding to this state is e_c.

The critical void ratio e_c is a function of the mean stress. The relationship has traditionally been written as:

$$e_c = \Gamma - \lambda\log(p') \quad \text{or} \quad e_c = e_{ref} - \lambda\log\left(\frac{p'}{p_{ref}}\right) \qquad [7.10]$$

where Γ and λ are two material constants and p' is the mean stress of the packing, and (e_{ref}, p_{ref}) is a reference point on the critical state line.

7.2.1.4. *Local elasto-plastic relationship*

With the elements discussed above, the final incremental stress–strain relation of the material can be derived that includes both elastic and plastic behavior, given by

$$\dot{f}_i^\alpha = k_{ij}^{\alpha p} \dot{\delta}_j^\alpha \quad [7.11]$$

Detailed expression of the elasto-plastic stiffness tensor is given in [CHA 05].

7.2.2. *Stress–strain relationship*

7.2.2.1. *Macro–micro relationship*

The stress–strain relationship for an assembly can be determined from integrating the behavior of interparticle contacts in all orientations. During the integration process, a relationship is required to link the macro and micro variables. The stress increment $\dot{\sigma}_{ij}$ can be related to the contact forces and branch vectors for contacts in all orientations [CHR 81, ROT 81]:

$$\dot{\sigma}_{ij} = \frac{1}{V} \sum_{\alpha=1}^{N} \dot{f}_j^\alpha l_i^\alpha \quad [7.12]$$

Using the principle of energy balance, the mean force on the contact plane of each orientation can be written

$$\dot{f}_j^\alpha = \dot{\sigma}_{ij} A_{ik}^{-1} l_k^\alpha V \quad [7.13]$$

The fabric tensor in equation [7.13] is defined by:

$$A_{ik} = \sum_{\alpha=1}^{N} l_i^\alpha l_k^\alpha \quad [7.14]$$

Using the static hypotheses proposed by Liao *et al.* [LIA 97], we obtain the relation between the macro strain and interparticle displacement (here, we do not consider the finite strain condition):

$$\dot{u}_{j,i} = A_{ik}^{-1} \sum_{\alpha=1}^{N} \dot{\delta}_j^\alpha l_k^\alpha \quad [7.15]$$

where $\dot{\delta}_j$ is the relative displacement between two contact particles and the branch vector l_k is the vector joining the centers of two contact particles. It is noted that contact particles include both direct and indirect contact of neighboring particles associated with a Voronoi polyhedron, as discussed by Cambou *et al.* [CAM 00]. N is the total number of contact orientations. The variables $\dot{\delta}_j^{\alpha}$ and l_k^{α} are defined as the averaged values of $\dot{\delta}_j$ and l_k, respectively, for all contacts belonging to the αth orientation.

7.2.3. *Model parameters*

We can summarize the material parameters as:

– normalized contact number per unit volume: Nl^3 / V;

– mean particle size, d;

– interparticle elastic constants: k_{n0}, k_{t0} and n;

– interparticle friction angle: ϕ_μ and m;

– interparticle hardening rule: k_{p0} and ϕ_0;

– critical state for the packing: λ and Γ or e_{ref} and p_{ref}.

We obtain two classes as discussed in the introduction:

– those that are independent of the arrangement and are connected to the contact law: k_{n0}, k_{t0} , n, ϕ_μ , m k_{p0} and ϕ_0;

– those that describe the geometry of the arrangement and its mechanical state: Nl^3 / V , d, λ and Γ or e_{ref} and p_{ref}, and the fabric tensor $A_{ik} = \sum_{\alpha=1}^{N} l_i^{\alpha} l_k^{\alpha}$ in the case of an anisotropic assembly. The latter case will not be discussed in this study.

Standard values for k_{p0} and ϕ_0 are the following: $k_{p0} = k_n$ and $\phi_0 = \phi_\mu$ and a typical ratio $k_{t0} / k_{n0} = 0.5$ can be generally assumed. Therefore, only six parameters have to be derived from experimental results and they can all be determined from the experimental data obtained from the material studied. The following section shows some examples on Hostun sand.

7.3. Results of numerical simulation on Hostun sand

7.3.1. *Drained triaxial tests*

A series of drained triaxial tests on coarse Hostun sand performed by Bouvard [BOU 82] were reported and analyzed by Biarez and Hicher [BIA 94]. The tests were performed at different confining pressures on samples prepared at different initial void ratios. Typical results are presented in Figure 7.3, which shows the triaxial testing results for both dense and loose specimens made of Hostun sand. The stress–strain curves are plotted for three different confining stresses in Figure 7.3a. The void ratio change with respect to strain is plotted in Figure 7.3b.

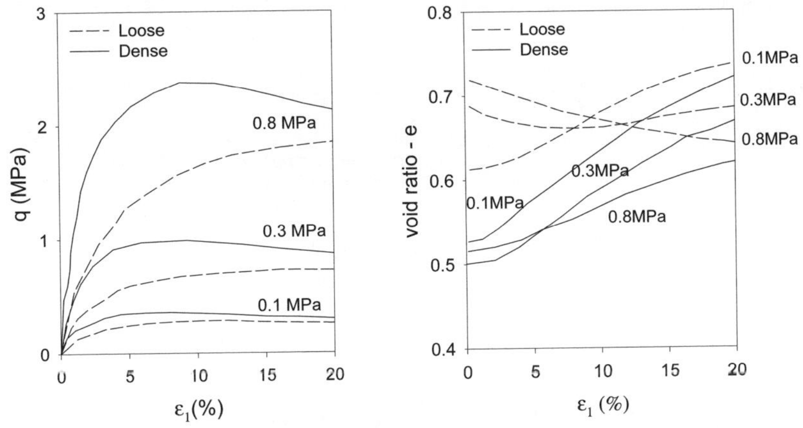

Figure 7.3. *Drained triaxial tests on Hostun sand [BOU 92]*

We can see the combined influence of the initial void ratio and the mean effective stress on the stress–strain curves and the volumetric change. The stress–strain curve presents a peak corresponding to the dilatancy of the sample, which increases with its initial relative density. For dilative materials, the deviatoric stress reduces after the peak and moves towards a constant state of stress for a constant void ratio corresponding to the critical state. In practice, it is difficult to reach this state because of strain localization, especially in dense materials. It is possible, however, to limit the effect of the localization by reducing the thickness of the specimen and taking measures to minimize the friction between the rigid end plates and the specimen. This was done in the tests presented here, which allowed the strain localization to be delayed at large deformations. The critical state could, therefore, be determined with sufficient accuracy [BIA 94]. If this were not the case, Biarez showed that the position of the critical state line could be obtained by using the following correlations:

-$e_c = e_{max}$ ($Dr = 0$) with $p' = 0.1$ MPa;

-$e_c = e_{min}$ ($Dr = 1$) with $p' = 5$ MPa.

The mean size of the particles for Hostun sand is d = 1.3 mm. The normalized contact number per volume Nl^3 / V can be obtained by matching the predicted and the experimentally measured elastic modulus for specimens with different void ratios [HIC 06]. The relationship calculated for Nl^3 / V versus the void ratio for Hostun sand is shown in Figure 7.4. The relationship for the regular packing of equally sized spheres is also plotted in Figure 7.4. for the sake of comparison. The evolution of Nl^3 / V due to new contact generation with neighboring particles was not considered.

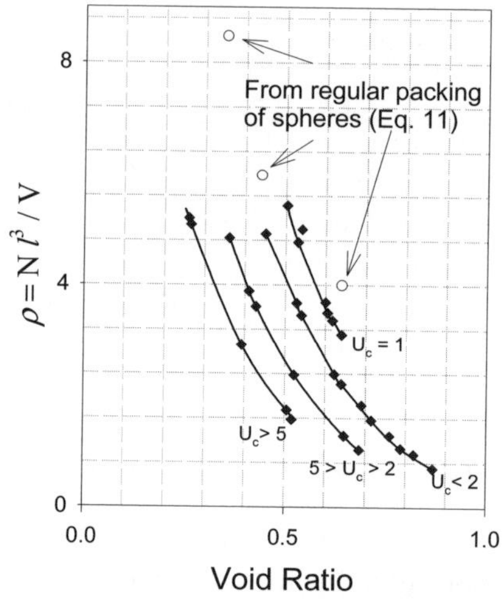

Figure 7.4. *Empirical relationship for packing density as a function of void ratio and of grain size distribution*

The interparticle elastic constant k_{n0} is assumed to be 61,000 N/mm. The value of k_{t0}/k_{n0} is commonly around 0.4, corresponding to a Poisson's ratio for Hostun sand ν = 0.2 and the exponent n = 0.5 [BIA 94, HIC 96]. From the results in Figure 7.3, we were able to derive the values of the two parameters corresponding to the position of the critical state in the $e - p'$ plane: $\lambda = 0.16$ and $p_{ref} = 0.1$ MPa for $e_{ref} = e_{max}$ = 0.81. The friction angle ϕ_μ was also determined from the stress state corresponding to the critical state: ϕ_μ = 33°. The equation governing the dilatancy rate requires the ϕ_0 parameter to be determined. A value of $\phi_0 = \phi_\mu$ was retained in accordance with the fact that, for contractive samples, the critical state was reached for the maximum value of the deviatoric stress, i.e. the sample was contractant all the way up to the critical state.

The peak friction angle is not an intrinsic parameter, but varies with the void ratio according to equation [7.9]. A value of $m = 0.6$ was determined from the test results. The values of k_{p0} are directly connected to the elastic properties by considering the relation: $k_{p0} = ak_n$. Here we simply considered that $k_{p0} = k_n$.

The set of parameters for Hostun sand is presented in Table 7.1.

e_{ref}	p_{ref}(Mpa)	λ	ϕ_μ(°)	ϕ_0(°)	m
0.81	0.1	0.16	33	33	0.6

Table 7.1. *Model parameters for Hostun sand*

The numerical simulations are presented in Figure 7.5. A reasonable simulation of the behavior of Hostun sand at various mean effective stresses and various initial void ratios can be obtained with a single set of model parameters, capturing both the contractive and dilative behavior of the sand, as well as the influence of this contractive or dilative behavior on the stress–strain curves and on the maximum strength. For large deformations, the curves corresponding to the same confining stress and different initial void ratios converge towards an identical stress state and void ratio, in accordance with the definition of critical state.

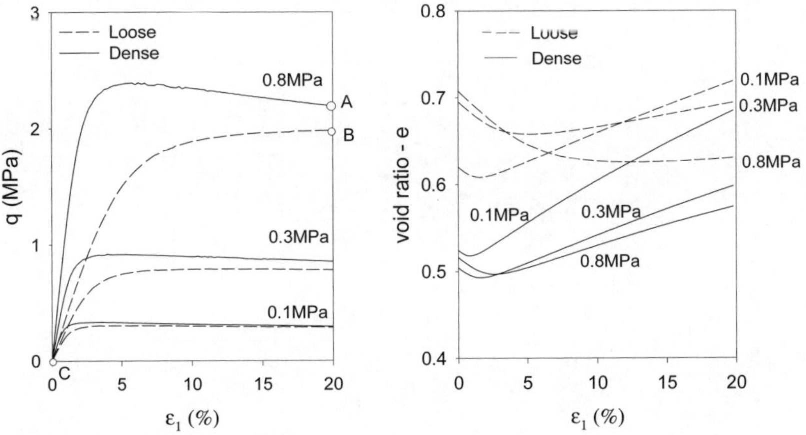

Figure 7.5. *Numerical simulations of drained triaxial tests on Hostun sand*

The set of parameters in Table 7.1, which can be directly determined from the experimental curves, has the advantage of being commonly used in the soil mechanics community: elastic stiffness, friction angle, and critical state parameters.

Figure 7.6 presents a set of measured stress–strain behaviors showing the influence of the initial void ratio. The representation of the curves obtained by Biarez and Hicher [BIA 08] from a comparison of results on different sands at different initial void ratios shows the correspondence between the stress–strain curve and the volumetric change for contractive and dilative materials. Figure 7.7 presents a set of numerical results obtained with the set of parameters from Table 7.1 with different initial void ratios, corresponding to relative densities between 0.05 and 1.00. Both stress–strain relations and volume changes are in agreement with the experimentally measured curves in Figure 7.6, showing the capability of the model to account for the influence of the initial density of the granular assembly.

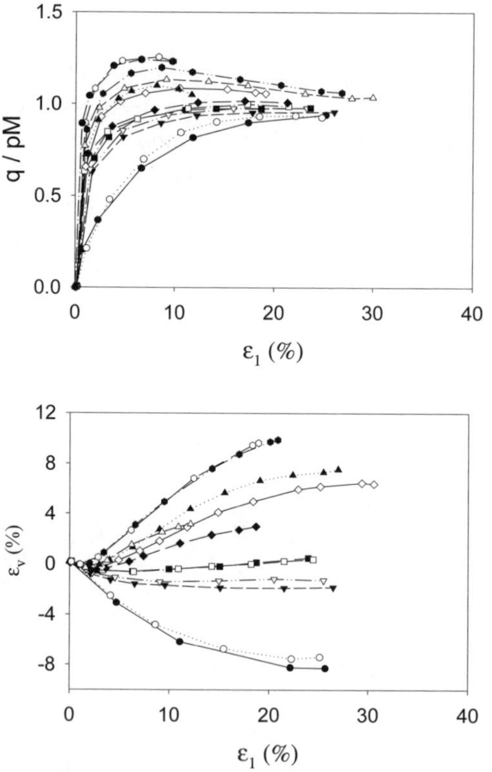

Figure 7.6. *Influence of initial density on the behavior of sand along drained triaxial loading*

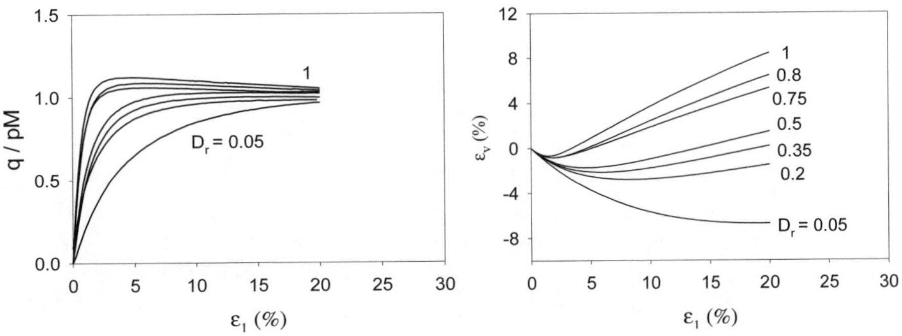

Figure 7.7. *Numerical simulations for various initial relative densities*

7.3.2. *Undrained triaxial tests*

Undrained triaxial tests have also been simulated, under the same set of parameters. The experimental results are plotted in Figure 7.8 and are compared with the predicted results in Figure 7.9. We can see that for strongly contractive materials, i.e. small values of the relative density, a maximum strength is obtained in the $q - \varepsilon_l$ plane followed by a decrease in the deviatoric stress down to a minimum strength that is almost zero for a relative density close to zero. This represents the phenomenon called static liquefaction. For slightly higher relative densities, the same pattern is first observed, but it is then followed by an increase in the deviatoric stress up to the ultimate strength at large deformations, corresponding to the critical state. For dilative materials, the deviatoric stress increases monotonically up to an ultimate stress achieved at large deformations when the critical state is reached.

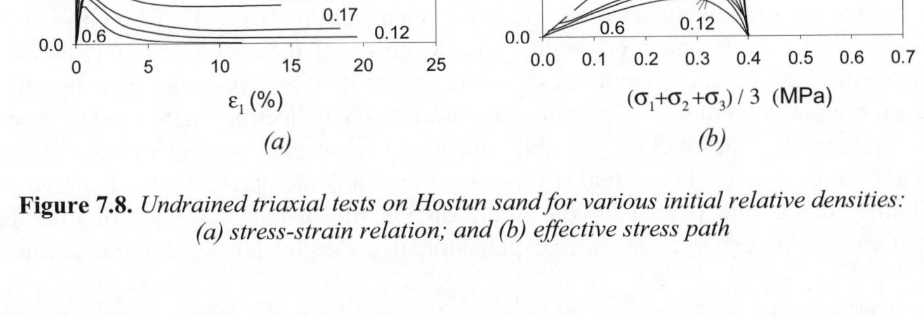

Figure 7.8. *Undrained triaxial tests on Hostun sand for various initial relative densities: (a) stress-strain relation; and (b) effective stress path*

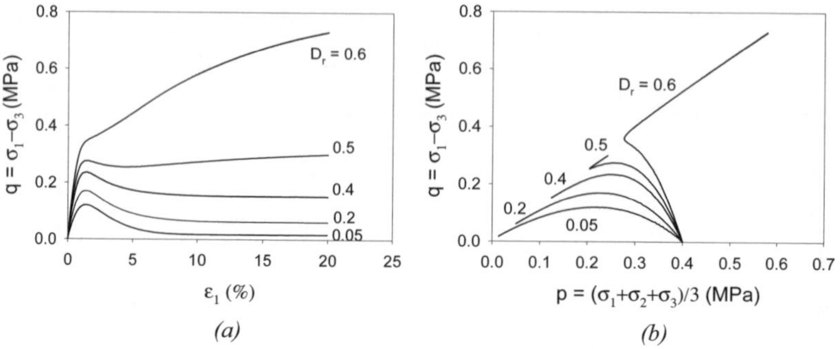

Figure 7.9. *Numerical simulations of undrained triaxial tests: (a) stress–strain relation; and (b) effective stress path*

These different evolutions can be related to the stress paths followed in a p', q plane: continuous decrease of the mean effective stress for strongly contractive materials, with a decrease followed by an increase up to the critical state for dilative materials. These results indicate that, in undrained condition, the model is capable of capturing the general trend observed for contractive and dilative sands [HIC 98].

7.4. Model extension to clayey materials

The representative scale of the constitutive elements that affect the mechanical behavior of the equivalent continuous medium is less evident in this case. Several scales coexist in a clayey material:

– the elementary clay layer, consisting of the basic element of a clay particle;

– the clay particle made up of an assembly of layers, usually organized in stacks;

– the aggregate made up of particles whose size and internal arrangement depend on the history of the material. The size of the aggregates can be measured by sedimentation.

A research study has been undertaken to relate the mechanical behavior of saturated remolded clays to their structural organization [HIC 00]. Two clays were studied: a kaolinite and a bentonite. The evolution of the shape, the size, and the concentration and the orientation of the elements that constitute the clay structure were examined by means of scanning and transmission electron microscopes, thanks to techniques that obviate possible disturbances of the micro-structure during observation. The results of this study show the main role played by the aggregates, similar to the role played by the grains in the mechanical behavior of granular materials. This explains the similar behavior of the equivalent continuous medium for the different nature of the constituents, as seen in the previous section. The difference in nature between grains and aggregates is, however, insufficient to explain the quantitative differences in the stress–strain relationship. In particular, the deformability of the aggregates can play an important role in the relationship. This deformability depends on the way in which the clay particles are assembled, which is a function of the mineralogy, adsorbed ions, etc. The photographs in Figures 7.10 and 7.11 show two different examples of clayey materials:

– the kaolinite is made of small, rigid particles assembled in compact stacks;

– the bentonite is made of larger, more flexible particles creating less compact and, therefore, forming more deformable aggregates.

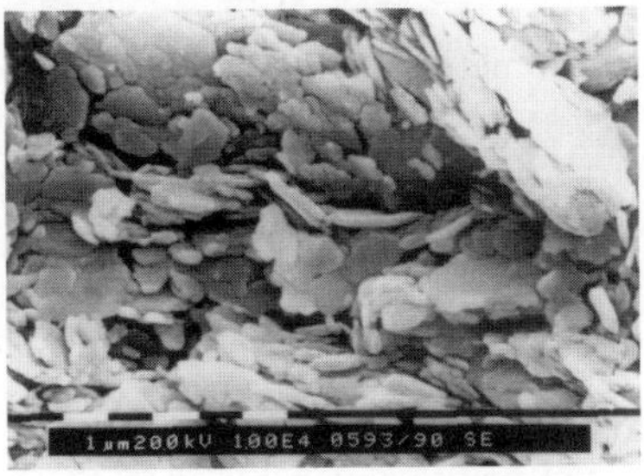

Figure 7.10. *Scanning electron microscope photo of kaolinite P300*

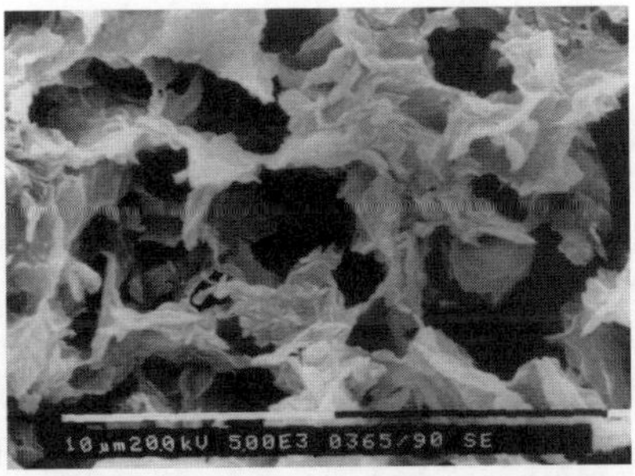

Figure 7.11. *Scanning electron microscope photo of bentonite*

Analysis of the pore size distribution in the kaolinite by means of mercury intrusion porosimetry confirmed the existence of two major groups of pore sizes: one centered around 1 μm and the other around 10 nm. The first group corresponds to the inter-cluster pores, whereas the second is related to the interparticle pores. The number and size of the big pores progressively decreased when consolidation stress increased, while the small pore size remained relatively unchanged under moderate loading stresses. This result confirmed that volume change during loading is mainly due to the rearrangement of the clay clusters.

Therefore, we can extend the micro-structural model presented above by considering the clayey material as a collection of clusters. The clusters interact with each other mainly on mechanical bases and play the same role as the grains in a granular material. In order to account for the compressible behavior between clay clusters during consolidation, a second yield surface is added. The second yield function is assumed to be [CHA 09]:

$$F_2(\sigma, \kappa_2) = \sigma - \kappa_2(\varepsilon^p) \quad for\ \sigma > p_p \qquad [7.16]$$

k_2 is another hardening parameter, which can be written as

$$\kappa_2 = p_p 10^{\varepsilon^p / c_p} \qquad [7.17]$$

where c_p is the compression coefficient in the $\varepsilon^p - \log\sigma'$ plane. When the compression σ is less than p_p, the plastic strain produced by the second yield function is null. Thus, p_p in equation [7.11] corresponds to the pre-consolidation stress in soil mechanics.

7.4.1. *Remolded clays*

The black kaolinite clay is a remolded clay with plastic limit (*wp*) of 30%, and liquid limit (*wl*) of 70%. The macro parameter λ was determined experimentally from the measured slope of the critical state line in the $e - \log p'$ curve: $\lambda = 0.53$.

From the photograph in Figure 7.10, we assume that the mean particle size d is 0.001 mm. The influence of the aggregate deformability is particularly pronounced on the elastic properties of clayey materials, since the elastic domain is restricted to very small strains ($< 10^{-5}$), for which the relative displacements of the constituents are negligible. Under these conditions, the elastic moduli measured in sands and gravels are much higher than in clays. From test results at very small strains, we estimated the value of the inter-aggregate stiffness to be k_{n0} = 35 N/mm, and the value of $k_{t0}/k_{n0} = 0.4$.

The position of the critical state line is given by e_{ref} = 1.39, corresponding to reference pressure p_{ref} = 0.01 MPa. The isotropic compression line gave a value of

c_e equal to 0.0367. Figure 7.12c shows the experimental and calibrated isotropic compression and critical state lines in the e – log$p$ plane. The other parameters, presented in Table 7.2, were calibrated from a triaxial test on a normally consolidated specimen.

e_{ref}	p_{ref}(MPa)	λ	φ_μ (°)	φ_0 (°)	m	c_e	k_{p0}/k_{n0}
1.39	0.01	0.53	20.70	20.70	2.00	0.03	0.20

Table 7.2. *Model parameters for the black kaolinite clay*

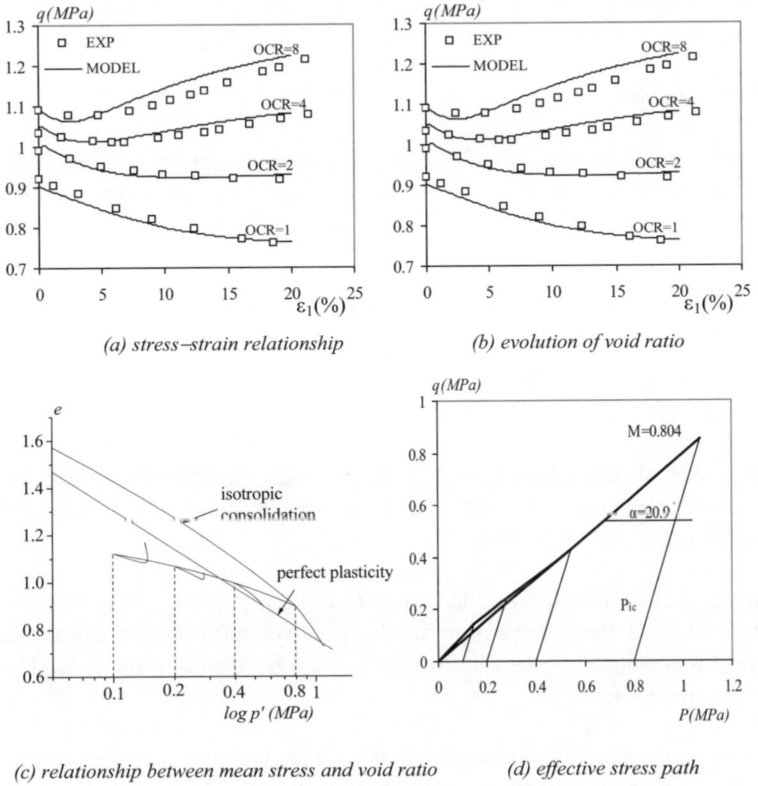

(a) stress–strain relationship *(b) evolution of void ratio*

(c) relationship between mean stress and void ratio *(d) effective stress path*

Figure 7.12. *Comparison between experimental results and numerical simulations of drained triaxial tests on black clay for various overconsolidation ratios*

Tests on black kaolinite clay samples were performed by Zervoyanis and Bard [ZER 82] and analyzed by Biarez and Hicher [BIA 94]. The tests began with an isotropic consolidation of up to 0.8 MPa, then unloading down to 0.4 MPa, 0.2 MPa, and 0.1 MPa, respectively. The corresponding overconsolidation ratios (OCR) are 1, 2, 4, and 8, respectively. The results are presented in Figure 7.12.

The predicted void ratio changes in Figure 7.12b, showing contractive behavior for OCR = 1 and 2, and shows a dilative behavior for OCR = 4 and 8. The contractive and dilative behavior can also be seen in the predicted paths on $e - \log p$ curves in Figure 7.12c. The corresponding stress–strain curves in Figure 7.12a are hardening for OCR = 1 and 2 and softening for OCR = 4 and 8. The stress paths in Figure 7.12d show stress states above the critical state line for OCR = 4 and 8 and on the critical state line for OCR = 1 and 2. Figure 7.12 shows overall good agreement between the experimental and predicted results for different overconsolidation ratios.

7.4.2. *Natural clays*

The geological history of sedimentary depositions has an important impact on mechanical properties, in particular the maximum stress state borne by the material during its history, called the consolidation stress. We can roughly distinguish two large families:

– the recent soils, which have been deposited since the last Ice Age that are normally or slightly overconsolidated;

– the older soils, which are often strongly overconsolidated, at least in the superficial layers.

Their mode of deposition as well as their evolution with time produces specific characteristics that cannot be found in laboratory-prepared soils, such as the ones examined in the previous section.

First of all, their structure or the geometry of their particle arrangement is usually anisotropic due to their deposition in the gravity field. This creates mechanical properties higher in the vertical rather than in the horizontal directions. This aspect will not be studied here.

Another important characteristic is the existence of an intergranular “glue”, mainly due to physicochemical evolutions at the level of particle contact. The influence of this glue can be found in the mechanical properties of the equivalent continuous medium. It can significantly increase the initial stiffness as well as its maximum strength.

7.4.2.1. *Saint-Herblain clay*

Saint-Herblain clay is a clayey river alluvial deposit from the Loire Valley's Paleolithic period. It is slightly or moderately organic and characterized by a high plasticity index (I_p = 42) with plastic limit wp = 48% and liquid limit wl = 102%. Its compression index is equal to 0.96. The samples were taken from a depth of 5.5 to 7.5 m. They correspond to a very soft clay with an *in situ* void ratio of between 2.25 and 2.6, and a mean undrained shear strength c_u equal to 26 kPa. The mean cluster size d was taken to be equal to 0.001 mm, the initial stiffness k_{n0} = 60 N/mm^2, k_{t0}/k_{n0} = 0.3. The other parameters were determined either directly from experimental curves or are presented in Table 7.3.

e_{ref}	p_{ref}(MPa)	λ	φ_μ (°)	φ_0 (°)	c_c	β	k_{p0}/k_{n0}
2.60	0.01	0.96	31.00	31.00	0. 20	1.00	0.50

Table 7.3. *Model parameters for Saint-Herblain clay*

Results of tests reported by Zentar [ZEN 99] were used here for the evaluation of the model. Four K consolidation tests were performed with K= 0.9, 0.92, 0.77 and 0.51 ($K = \sigma'_3/\sigma'_1$). They were consolidated to different points in the $p' - q$ plane, as shown in Figure 7.13. The samples were then sheared in an undrained condition up to failure.

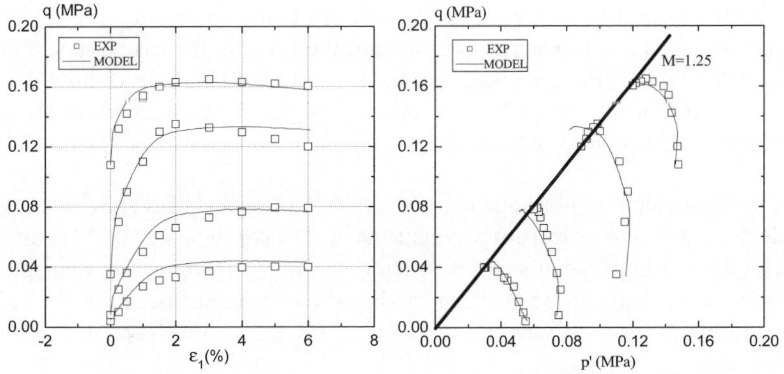

(a) stress-strain relationship *(b) evolution of void ratio*

Figure 7.13. *Experimental results and numerical simulations of undrained triaxial tests on anisotropically consolidated Saint-Herblain clay*

The influence of the K-consolidation can be seen at the beginning of the stress–strain curves. When the specimens are consolidated under a constant stress ratio K, anisotropy develops. Thus the initial stiffness of the specimens is influenced by the different values of K during consolidation. The model can take into account the influence of a stress-induced anisotropy by means of the fabric tensor A, defined in equation [7.14], as well as by the dependency of the elastic modulus with the stress state (equation [7.3]). The model is capable of taking into account the existence of an initial anisotropic stress state and a good agreement between experimental results and numerical simulations can be achieved for the stress–strain relationship (see Figure 7.13a), as well as the effective stress path (see Figure 7.13b) during undrained triaxial tests.

7.4.2.2. *Guiche clay*

The previous results correspond to materials for which the contact law between two grains is reduced to a solid friction. If two grains in contact are also "glued", the change in the contact law also produces a change in the mechanical properties of the equivalent continuous medium. Its overall behavior depends on the behavior of the glue. The main visible change in the material behavior is the fact that the maximum strength envelope does not intercept the axes origin and that the material exhibits a non null strength in traction: this is usually expressed by the term "cohesion". However, the role of the intergranular glue cannot be reduced to a scalar representing this cohesion and experimental results show that the global stress–strain relationship is affected by the glue's properties. If the glue has a stiffness similar to that of the grains, a significant decrease in the role of mean stress on the elastic properties follows. At the same time, the force necessary to displace two grains in contact is proportional to the normal force in the case of a simple solid friction. If the two grains are glued, we have to add the strength of the glue; if this strength is constant, the role of the friction is negligible provided that the normal force is small and becomes preponderant at elevated normal forces.

For the equivalent continuous material, this implies that its behavior is strongly dependent on the glue properties at small mean stresses, when $c >> Mp'$, and much less dependent at high mean stresses when $c << Mp'$. A progressive damage of the glue during a mechanical loading can modify the relative influence of the cohesion and the friction. Generally, natural soils present a higher mechanical strength than remolded soils.

Sometimes, the intergranular glue is ductile enough to sustain macroscopic deformations of several percentage points. Under these conditions, the maximum strength can remain constant after reaching the peak, which creates the equivalent of a perfectly plastic state for a cohesive material. This is the case for the Guiche clay that was extensively tested by Khemissa [KHE 92]. It corresponds to a silty clay from the Adour Valley. The Atterberg's limits are wl = 75% and wp = 25%. The undrained test results on samples around 13.5 m deep are presented in Figure 7.14 [FAY 00]. The maximum strength envelope in the p',q plane shows a macroscopic cohesion of the clay around 10 kPa.

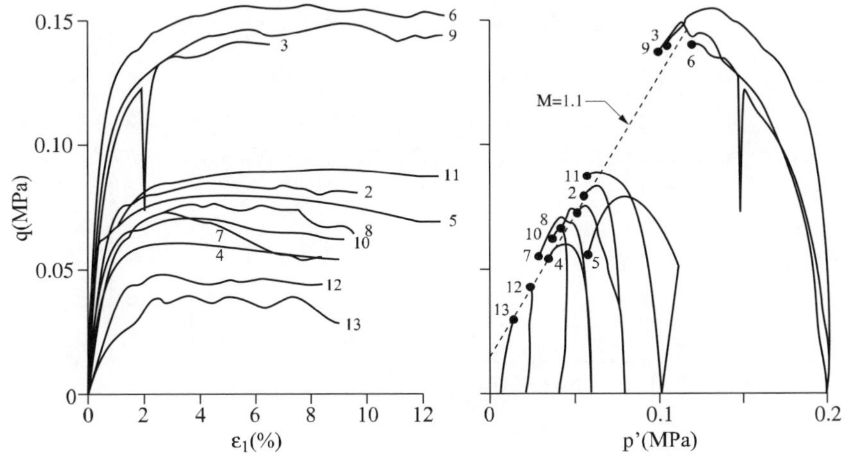

Figure 7.14. *Triaxial tests on Guiche clay [FAY 00, KHE 92]*

The model parameters were determined from a series of drained and undrained triaxial tests, as well as from the physical characteristics of the material. They are summarized in Table 7.4. In order to account for macroscopic cohesion, the contact law was modified by introducing an additional adhesive force perpendicular to the contact plane, which plays the role of the intergranular glue, as discussed above. In this example, the glue is ductile and keeps the same value throughout the mechanical loading. Its value was calibrated in order to obtain the correct value of the measured macroscopic cohesion: $f = 0.022$N.

e_{ref}	p_{ref}(MPa)	λ	φ_μ (°)	φ_0 (°)	C_e	β	m	e_0
1.80	0.01	0.80	27.00	27.00	0. 10	1.00	0.50	1.62

Table 7.4. *Model parameters for Guiche clay*

The results of the numerical simulations of three undrained triaxial tests are presented in Figure 7.15. They demonstrate that the model can reproduce both the stress–strain relationships and the effective stress paths for three different consolidation stresses with good accuracy. The maximum strength envelope was similar to the experimental one, which indicated that the model could simulate the effect of the cohesion of this natural clay.

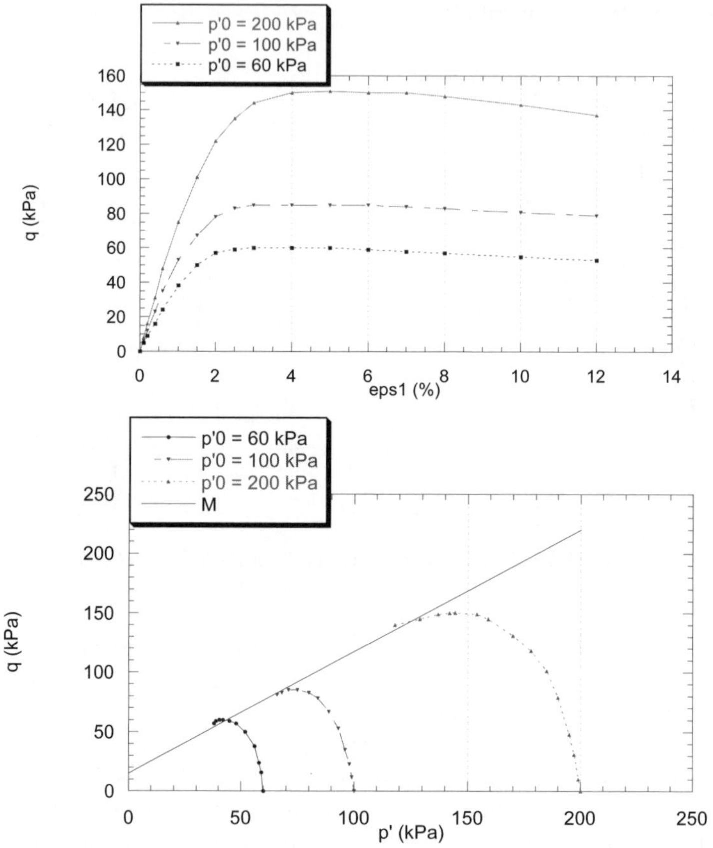

Figure 7.15. *Numerical simulations of undrained triaxial tests on Guiche clay*

7.5. Unsaturated granular materials

One of the main advantages of the micromechanical modeling approach is that it allows us to introduce various physical mechanisms in a natural way at the grain scale. One example concerns the unsaturated granular materials in which capillary forces develop between grains, depending on the degree of saturation and the morphology of the granular assembly. For dry samples, the contact forces are directly determined from the external stresses σ applied on the granular assembly (equation [7.13]). In the case of wetted samples, different stages of saturation can be identified. The fully saturated regime corresponds to a two-phase material with water completely filling the voids between grains. The water pressure p_w can be positive or negative (suction) but, in both cases, the effective stress concept [TER 25] can be applied and the contact forces determined by considering the effective stresses σ' as the external stresses [DEB 96, HIC 98]:

$$\boldsymbol{\sigma}' = \boldsymbol{\sigma} - p_w \mathbf{I} \qquad [7.18]$$

In the case of partially saturated samples, the liquid phase is distributed in menisci located between close grains and, as a consequence, capillary forces are applied on the grains and are added to the contact forces defined above. When the water content decreases inside a saturated granular sample, the air breaks through at a given point. The negative water pressure or suction corresponding to that point is called the air-entry pressure. Its value depends on the pore sizes. Afterwards, the sample becomes unsaturated, with the water phase being connected inside the pores.

This state is called the funicular regime, in which a constant decrease in the degree of saturation corresponds to a gentle increase in pore water pressure, which is homogenized along the continuous water phase. The menisci start to form between two grains, not necessarily in contact, and are connected to each other. The pendular regime starts when the water menisci become disconnected. The water is no longer continuous and equilibrium in the water pressure is obtained by the vapor pressure. At this stage, the capillary forces start to increase drastically according to the pore size. If the drying process continues, the water bridges begin to fail, starting with grains that are not in contact, until a completely dry state is achieved and no capillary forces are left inside the granular assembly.

Several studies have shown that the attractive capillary forces between two grains connected by a water bridge are a decreasing function of the distance between the grains until the bridge fails (see, for example, [BIA 93, CHA 02]). This function depends on the volume of liquid found between the grains. Different mathematical expressions have been proposed for these capillary forces, which are the sum of the pressure forces exerted by the liquid on the wetted area and the surface tension forces acting along the wetted perimeter. In this study, we retained the following expression:

$$f_n^{cap} = f_{\max} e^{-c\left(\frac{d}{R}\right)} \qquad [7.19]$$

where

– f_{cap} is the capillary force between two neighboring grains, not necessarily in contact;

– f_{max} is the value of f_{cap} for two grains in contact;

– R is the mean grain radius;

– d represents the distance between two grains and is equal to $l - 2R$, l being the branch length given as a distribution function of the grain size and the void ratio;

– c is a material parameter, dependent on the grain morphology and on the water content;

– f_{max} depends on the capillary pressure, defined as the pressure jump across the liquid–air interface, depending on the liquid–air interface surface tension, as well as on the geometry of the menisci governed by the solid–liquid contact angle and the filling angle.

In this study, a simplified approach was taken considering an empirical relation between f_{max} and the degree of saturation S_r, without taking into account the hysteresis along drying and wetting paths:

$$\begin{aligned} f_{\max} &= f_0 \frac{S_r}{S_0} && \textit{for } 0 < S_r < S_0 \\ f_{\max} &= f_0 \frac{S_0 (1 - S_r)}{S_r (1 - S_0)} && \textit{for } S_0 < S_r < 1 \end{aligned} \quad [7.20]$$

where f_0 and S_0 are material parameters. f_0 depends on the grain size distribution and S_0 represents the degree of saturation at which any further drying of the specimen will cause substantial breaking of the menisci in the pendular domain. S_0 depends on the nature of the granular material. The following empirical expression was proposed by Wu *et al.* [WU 84] for compacted granular materials:

$$(S_r)_{opt} = (-0.65 \log(d_{10}) + 1.5) / 100 \quad [7.21]$$

in which $(S_r)_{opt}$ is equivalent to S_0 in equation [7.5] and d_{10} is the effective grain size in mm.

Since the menisci are not necessarily all formed in the funicular regime, equation [7.20] may not be applicable for high degrees of saturation. In this first approach, however, we decided to extend it to the whole range of saturation, considering that the amplitudes of capillary forces were small for degrees of saturation higher than 80% and could, therefore, be approached with sufficient accuracy by using the same equation.

The first example concerns the results obtained by Wu *et al.* [WU 84] on a silty material called Glacier Way silt compacted in dry condition in a mould and then saturated. The samples are then dried to various degrees of saturation for the different specimens tested in the resonant column in order to obtain their elastic properties as a function of the degree of saturation. Figure 7.16 presents the results obtained for the Glacier Way silt for two different confining pressures. The value of G was the same for S_r = 0% and S_r = 100%. A well-marked peak for the two confining pressures could be observed, which occurred for a value of S_r around 17%. The maximum value of G was 1.8 times its value for dry and saturated states, at a confining pressure of 50 kPa and 1.6 times at a confining pressure of 100 kPa. The capillary effect appeared to be more pronounced at lower confining pressures.

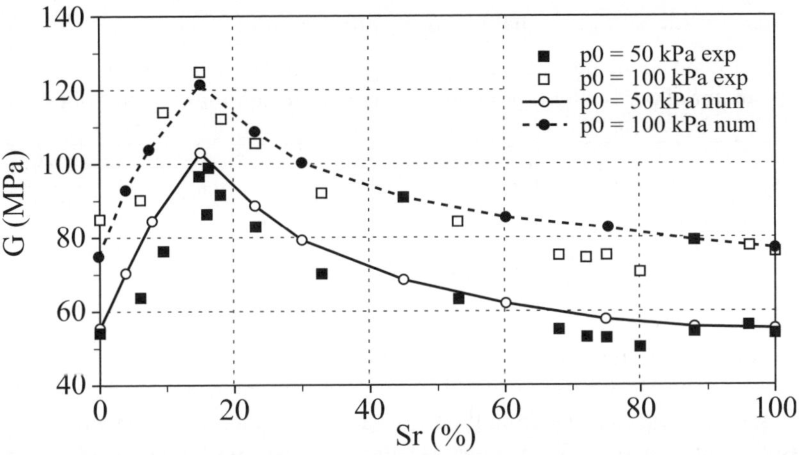

Figure 7.16. *Shear modulus function of the degree of saturation for Glacier Way silt*

The general trend showing the evolution of the shear modulus with the degree of saturation is in agreement with the model's concepts. This implies an identical value of G at dry and fully saturated states in the absence of capillary forces and a maximum influence of these forces at an optimum degree of saturation, as given by equation [7.20], in which S_0 corresponds to $\left(S_r\right)_{opt}$ in equation [7.21].

To determine the capillary forces, equation [7.19] includes two material parameters, c and d. Very little information could be gathered on these two parameters by simply using the test results. We therefore decided to take a constant value $c = 8$ and a value d function of the mean grain size d_{50}: $d = 1.1\ d_{50}$. These values gave a standard evolution for the capillary forces, function of the distance between particles, as well as an initial isotropic distribution of these forces.

For the numerical simulations, the following parameters had to be determined:

– the parameters of the elastic non-linear law: k_{n0}, k_{t0} and n;

– the parameters of the capillary forces equation: f_0 and S_0.

k_{n0} and n were determined from the values of G obtained at dry (or saturated states). A ratio of $k_{t0}/k_{n0} = 0.35$ was assumed, leading to a Poisson's ratio $\nu = 0.25$, which corresponds to a standard value for this type of material [HIC 96].

The value of N/V was determined by using the chart in Figure 7.4. For a material with U_c>5, a void ratio $e = 0.58$ corresponded to $N/V = 0.31$.

S_0 was considered to be equal to $(S_r)_{opt}$ in equation [7.21] for $d_{10} = 0.0024$ mm. f_0 was computed to fit the maximum value of G for the confining pressure of 100 kPa. The material parameters for Glacier Way silt are summarized in Table 7.5.

k_{n0}	α	n	N/V	S_0	f_0 (N)
2,200	0.35	0.50	0.31	0.175	0.02

Table 7.5. *Model parameters for Glacier Way silt*

The numerical results are presented in Figure 7.16, together with the experimental ones [HIC 08]. A good overall agreement was obtained for the two confining pressures. Starting from the fully saturated state, the shear modulus values increased slowly, at first, for degrees of saturation varying between 100% and 70% and then faster up to the maximum values obtained for $S_r = S_0$. The computed peak value of G was close to the experimental one for the confining pressure of 50 kPa, showing a more significant effect of the capillary forces at low confining pressure. For degrees of saturation below S_0, the shear modulus values decreased toward the values obtained for dry states.

Well-documented experimental results on a loose decomposed granitic soil have been presented by Ng and Chiu [NG 03]. The soil can be described as a silty, clayey, gravelly sand. Triaxial samples were prepared by the wet tamping method, compacted with a water content of 19% and dry densities ranging from 74 to 84% of the Proctor maximum, which corresponded to a void ratio ranging from 0.85 to 1.10 and degrees of saturation ranging from 45 to 58%. Undrained triaxial tests on saturated samples were performed at different confining pressures and initial void ratios. They allowed us to determine the model parameters. Table 7.6 summarizes the set of parameters used for the modeling of the saturated sample behavior.

e_{ref}	p_{ref} (kPa)	λ	ϕ_μ (°)	ϕ_0 (°)	m
1.08	10.00	0.11	38.00	38.00	1.00

Table 7.6. *Model parameters for the decomposed granitic soil*

After applying an initial suction of 40 kPa, unsaturated samples were isotropically compressed and then sheared at constant water content. Figure 7.17 presents the results obtained for three different confining pressures, 50 kPa, 100 kPa and 150 kPa, on specimens with the same initial void ratio e_0 = 1.09. The results show a constant increase in deviatoric stress up to a constant value for axial strain over 25%. The deviatoric stress increased with the applied mean stress. The three specimens exhibited a contractive volumetric behavior, with higher contraction for the specimen subjected to higher confining stress.

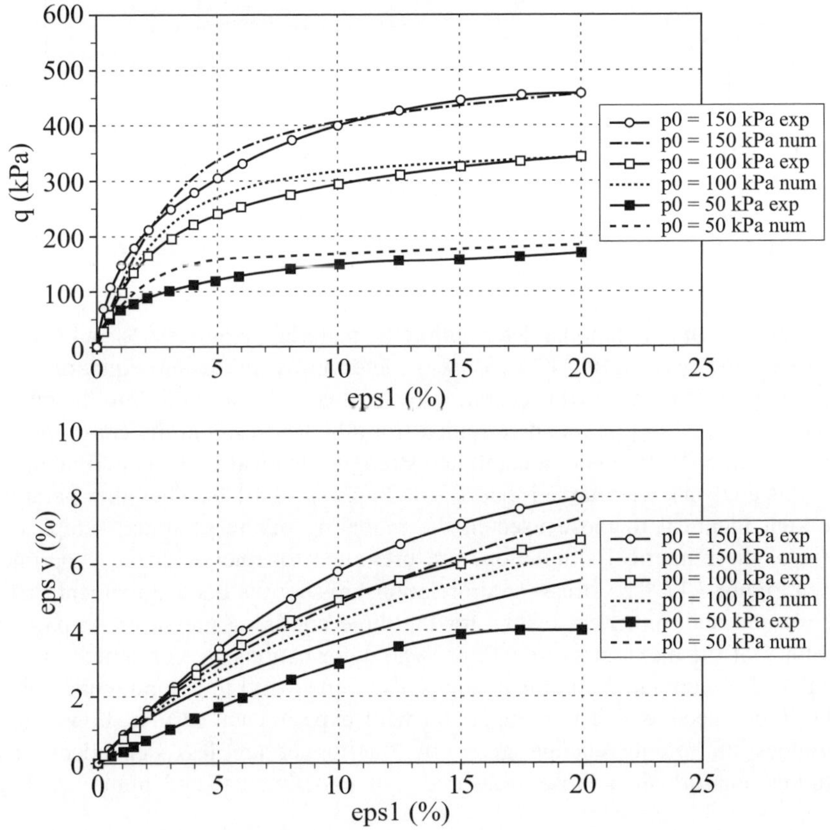

Figure 7.17. *Triaxial tests at constant water content on unsaturated silty sand*

A stabilization of the volume change was observed at large deformations. This contractive behavior led to an increase in the degree of saturation for all samples, since the water content was kept constant. The initial degree of saturation was equal to 43% for the three specimens and the final degree of saturation reached 48 to 61%, according to the amount of volume change during the tests. Measured suctions showed a decrease of 8 to 15 kPa for the most contractive specimen (Figure 7.18).

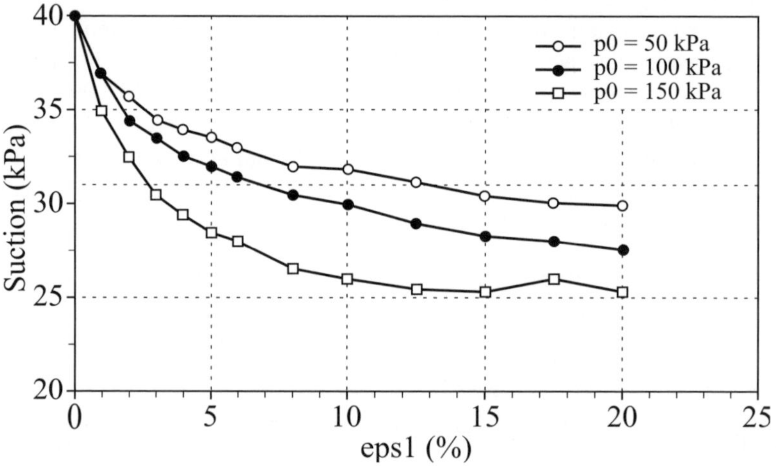

Figure 7.18. *Evolution of suction during triaxial tests at constant water content*

For numerical simulations, determining the material parameters S_0 and f_0 was the necessary first step [HIC 07]. S_0 was determined by using equation [7.21], considering that the mode of preparation by compaction was similar to the one used by Wu *et al.* [WU 84]. f_0 was determined from the test result at the confining stress of 100 kPa in order to obtain a computed strength identical to the experimental one. From this analysis, we retained that $S_0 = 0.2$ and $f_0 = 0.01$ N. The other parameters were kept identical to those used in the modeling of the saturated samples. The results of the numerical simulations are presented together with the experimental ones in Figure 7.17. The stress–strain relationship shows a good agreement between experimental and numerical results for the three confining pressures. In particular, the values of the maximum strength at large deformations are computed with good accuracy. The computed volume changes also correspond to a contractive behavior of the three specimens. The comparison with experimental results shows that the amplitudes of volume change given by the model are less dependent on the confining pressure than those measured, but experimental and numerical results remain reasonably close.

The final states of the three specimens were plotted in the total stress plane p,q together with the critical state line obtained from the tests on saturated samples. The state at the end of each test on unsaturated specimens was located above the critical state line. This demonstrated the increase in strength due to the capillary forces. It can be noted that the final points were closer to the critical state line for higher applied net mean stress. This was due to the volumetric contraction that increases with the confining pressure. As a consequence, the degree of saturation increased during shearing, causing a decrease in the capillary forces, which were more pronounced at a higher confining pressure (Figure 7.19).

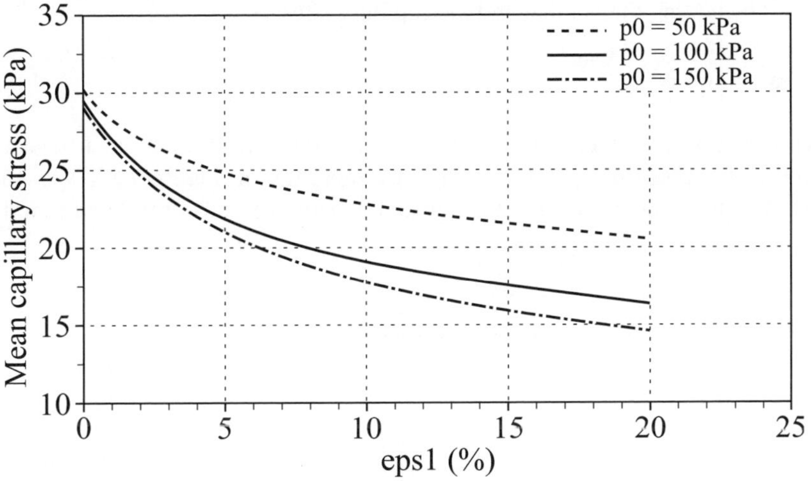

Figure 7.19. *Evolution of the mean capillary stress during triaxial tests at constant water content*

If we were to extend the maximum strength locus for the unsaturated samples toward the origin of the axes, we would obtain an intercept $q_{unsat} = 10$ kPa for $p = 0$. This would reflect the contribution of the capillary forces on the maximum strength. If we transfered these results to a Mohr-Coulomb plane, we would obtain an apparent cohesion for the unsaturated samples $c_{ap} = 5.3$ kPa and a friction angle similar to the one obtained for saturated samples: $\phi = 37.7°$.

Using the definition of equation [7.12], the stress induced by capillary forces can be computed and is termed capillary stress, which is given by

$$\left(\dot{\sigma}_{ij}\right)^{Cap} = \frac{1}{V}\sum_{\alpha=1}^{N} \dot{f}_j^{\alpha(cap)} l_i^{\alpha} \qquad [7.22]$$

It is worth noting that this term is not analogous with the usual concept of capillary pressure or suction, which represents the negative pore water pressure inside the unsaturated material. In this model, the capillary stress depends on the geometry of the pores and is a tensor rather than a scalar. The capillary stress can only can be reduced to an isotropic tensor for an isotropic distribution of the branch lengths l^{α}. This can be the case for an initially isotropic structure during isotropic loading, but during deviatoric loading an induced anisotropy is created and the capillary tensor is no longer isotropic.

From equations [7.12] and [7.22] we can define a generalized effective stress tensor σ^* by adding the applied and the capillary stresses:

$$\boldsymbol{\sigma}^* = \boldsymbol{\sigma} + \boldsymbol{\sigma}^{cap} \quad [7.23]$$

We can now plot the final states of the unsaturated samples in a generalized effective stress plane p^*, q^*. The results are presented in Figure 7.20. We can see that, in this plane, the critical state is unique for saturated and unsaturated samples.

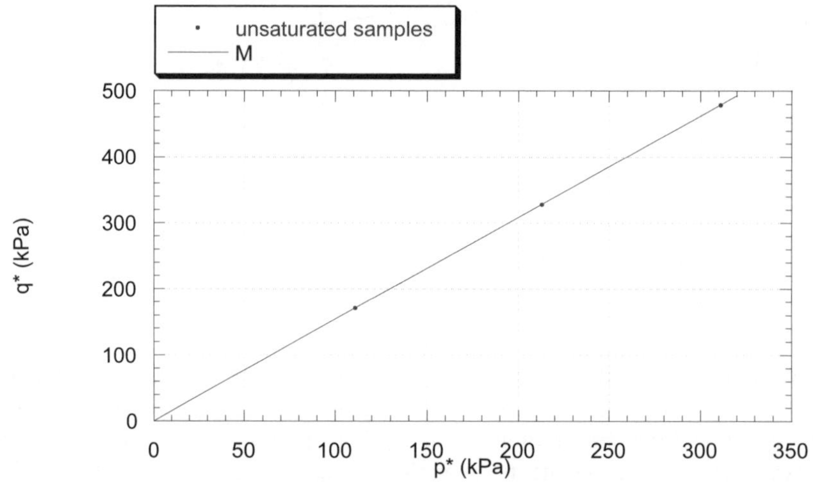

Figure 7.20. *Maximum strength locus in the generalized effective stress plane*

There is no direct relationship between the measured suction and the capillary stress tensor, but their evolution during the shearing phase is similar and controlled by the evolution of the degree of saturation due to the volume change. Figures 7.18 and 7.19 show the evolution of the suction and the mean capillary stress, respectively, with the axial strain for the three unsaturated specimens. We can see a very similar evolution, even if the absolute values are different.

Ng and Chiu [NG 03] also performed several tests, consisting of wetting specimens of the same silty sand at a constant stress state. After an isotropic consolidation up to 100 kPa, a specimen was sheared at constant net mean stress and constant suction, s = 40 kPa, until it reached a net stress ratio $\eta = q/p = 1.4$. Then, the specimen was wetted by reducing the suction while the deviatoric stress and net mean stress were kept constant. Figure 7.21 shows the evolution of the axial strain and the volumetric strain during the wetting phase as a function of suction. A continuous increase in the axial strain can be noted throughout the process. At the beginning of the wetting, this increase remains limited; toward the end of the test, the rate of increase accelerates. The same kind of evolution can be observed for the volumetric strain. At first, a very limited volumetric contraction takes place, but, as the suction continues to decrease, toward the end of the test the rate of increase in volumetric contraction also increases. The authors attribute this increase in the rate of volumetric strain to the collapse of the soil skeleton.

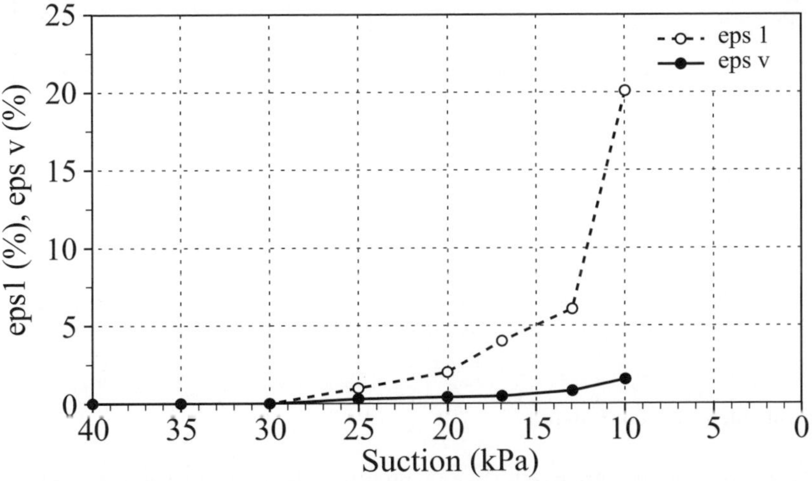

Figure 7.21. *Evolution of axial and volumetric deformations during a wetting test (p_0 = 100 kPa) (experimental results from [NG 03])*

A numerical simulation of the same wetting test was undertaken. The shearing phase was conducted at a constant capillary stress in order to reproduce the condition of constant suction as closely as possible. The numerical results are presented in Figure 7.22. An increase in both the axial and the volumetric strain was obtained, as in the experiment, with a continuous increase in the strain rate when the capillary stress decreased. However, the evolution was more progressive than in the experimental results, the specimen starting to strain as early as the beginning of the wetting phase. This could be due to the choice made for the relationship between capillary forces and the degree of saturation (equation [7.19]), as well as the way in which parameters S_0 and f_0 were determined. The model therefore appears capable of reproducing the phenomenon called the collapse of unsaturated soils during a wetting phase under a deviatoric stress. During the wetting phase, the reduction in the capillary forces with the increase in the degree of saturation weakens the mechanical strength of the unsaturated specimen, which, at a certain point, can no longer sustain the applied deviatoric stress.

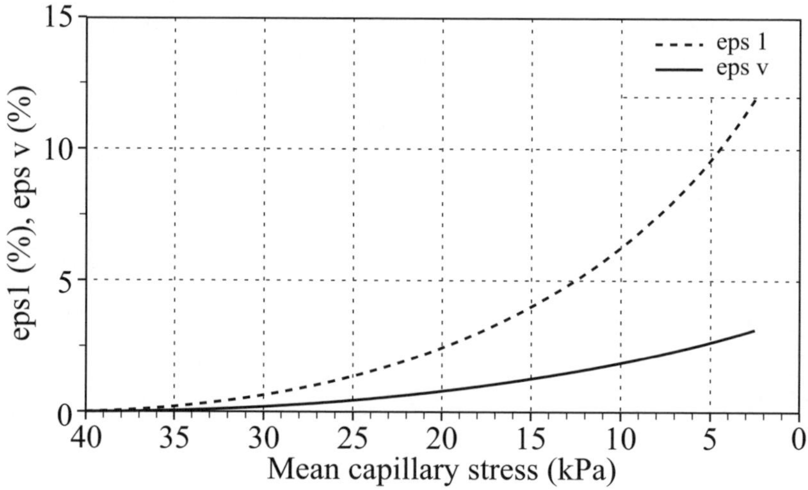

Figure 7.22. *Evolution of axial and volumetric deformations during a wetting test (p_0 = 100 kPa). Numerical simulations with the micromechanical model*

7.6. Summary and conclusion

The microstructural approach is conceptually attractive because it dispenses with having to deal with relations between the stress and strain tensors. We only need to describe a simple relation between the vectors of forces and relative displacements on a contact plane, which thus requires fewer material parameters. The stress and strain tensors are automatically obtained by integration over all spatial orientations. The local contact law can be written based on the conceptual framework that allows us to pass from a discontinuous medium to an equivalent continuous medium, as developed by Biarez and his collaborators. It is then possible to introduce various physical mechanisms at the grain scale

In the model, a simple elastic-plastic behavior was assumed on each contact plane. The elastic part is based on Hertz-Mindlin's contact formulation, while the plastic part is based on a Mohr-Coulomb friction law with an isotropic hardening assumption and a non-associated flow rule that permits the developments of contractive or dilative volume changes at the scale of the representative elementary volume. For the whole packing, a critical state behavior is assumed at large deformations and the friction angle on each plane is related to the actual void ratio compared to the critical void ratio at the same state of stress. A strain softening behavior can, therefore, be obtained for dense materials. On the whole, the model merely requires a limited number of parameters, which can easily be determined from conventional triaxial testing.

The ability of the model to reproduce the main features of sand behavior has been demonstrated. Model simulations were compared with drained triaxial test results at different initial void ratios and different confining stresses, leading to contractive or dilative behavior of the sand specimens. A good accuracy was obtained between numerical simulations and experimental results by use of a unique set of parameters. Simulations of undrained triaxial tests with the same set of parameters demonstrated that the model was capable of reproducing the general trend for both loose and dense sands.

The micro-structural model has been extended to clayed materials by considering clay as a collection of clusters. The clusters interact with each other, mainly on mechanical bases, and play the same role as the grains in a granular material. In order to account for the compressible plastic behavior between clay clusters during consolidation, a second yield surface was added. The numerical simulations of the behavior of remolded and intact clays along drained and undrained triaxial loading paths demonstrated that this extended version of the model was able to reproduce the main features of clayey materials to a suitable level of satisfaction. Natural clays often exhibit an intrinsic cohesion whose amplitude depends on their geological history. In order to account for this cohesion, we added an adhesive force at the level of the cluster contacts. The numerical results obtained in the simulation of natural soft clay behavior showed that this approach could reproduce the stress–strain relationship, as well as the effective stress paths during undrained triaxial tests.

An example of the introduction of an additional physical mechanism at the level of the grain contact was presented for unsaturated granular materials by considering capillary forces. The capillary forces were computed as a function of the pore morphology and of the water content, using a simple empirical relationship. Several examples of simulation demonstrated the ability of the model to reproduce the main features of an unsaturated granular material's behavior. The numerical simulations included resonant column tests at various degrees of saturation, constant water content triaxial tests at a given degree of saturation and various net confining pressures, and a wetting test at constant deviatoric stress and constant net mean stress. For constant water content triaxial tests, the results showed an increase in the mechanical properties whenever the degree of saturation decreased. The results also demonstrated that the collapse of an unsaturated specimen could be obtained during wetting, provided that a sufficiently high stress ratio was applied throughout the process.

The integration of the capillary forces over all spatial orientations permits us to define a capillary stress tensor. This tensor is isotropic, i.e. it can be reduced to a capillary pressure, but only if the particulate structure is itself isotropic. A generalized effective stress tensor can, therefore, be determined by adding the total stress and the capillary stress tensors. Numerical simulations demonstrate that the maximum strength criterion, expressed in terms of generalized effective stresses, is unique for dry, partially and fully saturated samples.

7.7. Bibliography

[BIA 62] BIAREZ J., Contribution à l'étude des propriétés mécaniques des sols et des matériaux pulvérulents, PhD thesis in engineering sciences, University of Grenoble, France, 1962.

[BIA 77] BIAREZ J., FAVRE J.-L., "Statistical estimation and extrapolation from observations", *Reports of Organisers, Special Session 6*, IX ICSMFE, vol. 3, pp. 205-209, Tokyo, Japan, 1977.

[BIA 89] BIAREZ J., HICHER P.-Y., "An introduction to the study of the relation between the mechanics of discontinuous granular media and the rheological behaviours of continuous equivalent media – application to compaction", *Powders and Grains*, Balkema, London, 1989.

[BIA 93] BIAREZ J., FLEUREAU J-M., TAIBI S., "Mechanical constitutive model for unsaturated granular media", *Second International Conference on the Micromechanics of Granular Media*, pp. 51-58, Thorton, Balkema, London, 1993.

[BIA 94] BIAREZ J., HICHER P.Y., *Elementary Mechanics of Soil Behaviour*, Balkema, London, p. 208, 1994.

[BIA 02] BIAREZ J., HICHER P.Y., "Mechanisms of soil deformation", in: HICHER P.-Y. and SHAO J.F. (eds), *Constitutive Modeling of Soils and Rocks*, ISTE, London, John Wiley & Sons, New York, 2008.

[BOU 82] BOUVARD D., Rhéologie des milieux pulvérulents: étude expérimentale et identification d'une loi de comportement, PhD thesis, engineering, University of Grenoble, France, 1982.

[CAM 00] CAMBOU B., DEDECKER, F., CHAZE M., "Relevant local variables for the change of scale in granular materials", in: DIMITRIOS K. (ed.), *Constitutive Modelling of Granular* Materials Springer, Berlin, Germany, pp. 275-290, 2000.

[CHA 88] CHANG C.S., "Micromechanical modeling of constructive relations for granular material", in: SATAKE M. and JENKINS J.T. (eds), *Micromechanics of Granular Materials*, pp. 271-279, 1988.

[CHA 89a] CHANG C.S., SUNDARAM S.S., MISRA A., "Initial moduli of particulate mass with frictional contacts", *International Journal of Numerical Analytical Methods in Geomechanics*, vol. 13, no. 6, pp. 626-641, 1989.

[CHA 89b] CHANG C.S., MISRA A., WEERARATNE S.P., "A slip mechanism based constitutive model for granular soils", *Journal of Engineering Mechanics*, vol. 115, no. 4, pp. 790-807, 1989.

[CHA 92] CHANG C.S., MISRA A., ACHEAMPON K., "Elastoplastic deformation of granulates with frictional contacts", *Journal of Engineering Mechanics*, vol. 118, no. 8, pp. 1692-1708, 1992.

[CHA 05] CHANG C., HICHER P.-Y., "An elastoplastic Model for granular materials with microstructural consideration", *International Journal of Solids and Structures*, vol. 42, no. 14, pp. 4258-4277, 2005.

[CHA 09] CHANG C., HICHER P.Y., YIN Z., KONG L., "An elasto-plastic model for clay with microstructural consideration", *Journal of Engineering Mechanics*, vol. 135, no. 9, pp. 917-931, 2009.

[CHA 02] CHATEAU X., MOUCHERONT P., PITOIS O., "Micromechanics of unsaturated granular media", *Journal of Engineering Mechanics*, vol. 128, no. 8, pp. 856-863, 2002.

[CHR 81] CHRISTOFFERSON J., MEHRABADI M.M., NEMAT-NASSAR S., "A micromechanical description on granular material behavior", *ASME Journal of Applied Mechanics*, vol. 48, pp. 339-344, 1981.

[DEB 96] DE BUHAN P., DORMIEUX L., "On the validity of the effective stress concept for assessing the strength of saturated porous materials: a homogenization approach", *J. Mech. Phys. Solids*, vol. 44, no. 10, pp. 1649-16667, 1996.

[EMR 96] EMERIAULT F., CAMBOU B., "Micromechanical modelling of anisotropic non-linear elasticity of granular medium", *International Journal of Solids and Structures*, vol. 33, no. 18, pp. 2591-2607, 1996.

[FAY 00] FAYAD T., Mécanique des grains avec et sans colle pour une comparaison avec des sols naturels et remaniés, PhD thesis, Ecole Centrale de Paris, 2000.

[HIC 96] HICHER P.-Y., "Elastic properties of soils", *Journal of Geotechnical Engineering*, vol. 122, no. 8, pp. 641-648, 1996.

[HIC 98] HICHER P-Y., "Experimental behavior of granular materials" in: CAMBOU B. (ed.), *Behavior of Granular Materials*, pp. 1-97, Springer, Vienna, Austria & New York, USA, 1998.

[HIC 00] HICHER P.Y., WAYUDI H., TESSIER D., "Microstructural analysis of inherent and induced anisotropy in clay", *Mechanics of Cohesive-frictional Materials*, vol. 5, pp. 341-371, 2000.

[HIC 06] HICHER P.Y., CHANG C.S., "Anisotropic non linear elastic model for particulate materials", *Journal of Geotechnical and Environmental Engineering*, vol. 132, no. 8, pp. 1052-1061, 2006.

[HIC 07] HICHER P.Y., CHANG C.S., "A microstructural elastoplastic model for unsaturated granular materials", *International Journal of Solids and* Structures, vol. 44, no. 7-8, pp. 2304-2323, 2007.

[HIC 08] HICHER P.Y., CHANG C.S., "An elastic model for partially saturated granular materials", *J. Eng. Mechanics*, vol. 134, no. 6, pp. 505-513, 2008.

[JEN 93] JENKINS J.T., STRACK O.D.L., "Mean-field inelastic behavior of random arrays of identical spheres", *Mechanics of Material*, vol. 16, pp. 25-33, 1993.

[KHE 92] KHEMISSA M., Recherches expérimentales sur les propriétés mécaniques d'une argile molle naturelle, PhD thesis, Ecole Nationale des Ponts et Chaussées, Paris, France, 1992.

[LIA 97] LIAO C.L., CHANG T.P., YOUNG D., CHANG C.S., "Stress-strain relationship for granular materials bases on hypothesis of best fit", *International Journal of Solids and Structures*, vol. 34 , no. 31-32, pp. 4087-4100, 1997.

[LIA 00] LIAO C.L., CHAN T.C., SUIKER A.S.J., CHANG C.S., "Pressure-dependent elastic moduli of granular assemblies", *International Journal for Analytical and Numerical Methods in Geomechanics*, vol. 24, pp.265-279, 2000.

[LUO 80] LUONG M. P., "Stress-strain aspects of cohesionless soils under cyclic and transient loading", *International Symposium on Soils under Cyclic and Transient Loading*, Swansea, pp. 353-376, 1980.

[MIN 69] MINDLIN R.D., "Microstructure in linear elasticity", *Arch. Rational Mech. Anal.*, vol. 16, pp. 51-78,1969.

[NEM 02] NEMAT-NASSER S., ZHANG J., "Constitutive relations for cohesionless frictional granular material", *International Journal of Plasticity*, vol. 18, pp. 531-547, 2002.

[NG 03] NG C., CHIU A., "Laboratory study of loose saturated and unsaturated decomposed granitic soil", *Journal of Geotechnical Eng.*, vol. 129, no. 6, pp. 550-559, 2003.

[NIC 05] NICOT F., DARVE F., "A multi-scale approach to granular materials", *Mechanics of Materials*, vol. 37, pp. 980-1006, 2005.

[ODA 93] ODA M., "Inherent and induced anisotropy in plasticity theory of granular soils", *Mechanics of Materials*, vol. 16, no. 1-2, pp. 35-45, 1993.

[RAN 02] RANGEARD D., Identification des caractéristiques hydro-mécaniques d'une argile par analyse inverse d'essais pressiométriques, PhD thesis, University of Nantes and Ecole Centrale de Nantes, France, 2002.

[ROT 81] ROTHENBURG L., SELVADURAI A.P.S., "Micromechanical definitions of the Cauchy stress tensor for particular media", in: SELVADURAI A.P.S. (ed.), *Mechanics of Structured Media*, pp. 469-486, Elsevier, Amsterdam, The Netherlands, 1981.

[TAI 94] TAIBI S., Comportement mécanique et hydraulique des sols partiellement saturés, PhD thesis, Ecole Centrale de Paris, France, 1994.

[WAL 87] WALLTON K., "The effective elastic moduli of a random packing of spheres", *J. Mech. Phys. Solids*, vol. 35, pp. 213-226, 1987.

[WU 84] WU S., GRAY D.H., RICHART F.E. JR, "Capillary effects on dynamic modulus of sands and silts", *Journal of Geotechnical Eng.*, vol. 110, no. 9, pp. 1188-1203, 1984.

[ZEN 99] ZENTAR R., Analyse inverse des essais pressiométriques, application à l'argile de Saint Herblain, PhD thesis, University of Nantes and Ecole Centrale de Nantes, France, 1999.

[ZER 92] ZERVOYANIS C., Étude synthétique des propriétés mécaniques des argiles et des sables sur chemin oedométrique et triaxial de révolution, PhD thesis in engineering, Ecole Centrale Paris, France, 1992.

Chapter 8

Modeling Landslides with a Material Instability Criterion

8.1. Introduction

We propose to introduce an approach for detecting the phenomenon of instabilities within structures built from elastoplastic granular materials. This phenomenon has been discussed in other chapters using approaches which are based on either micromechanical descriptions of the media [CAM 00, CHR 81, NIC 05] or direct numerical simulations by use of the discrete element method [CUN 79, MAG 98, SIB 07]. In the present study, we consider the material as a continuous medium represented by an elastoplastic macroscopic model and we solve boundary problems by the finite elements method.

One of the principal characteristics of a geomaterial is its non-associatedness: the normality rule is not satisfied and, consequently, the elastoplastic tensor loses its major symmetry [HIL 67, MAN 66, MRÓ 63, MRÓ 66]. From a geotechnical point of view, this non-associated state is generally described by a dilatancy angle different from the friction angle [BIA 94]. Because of this feature, it is now widely known that failures can occur before the plasticity limit is reached [DAR 87, DAR 04, LAD 92, NOV 91]. Two failure modes have been experimentally observed before this plasticity limit [BIA 08]: plastic strain localization in shear bands can develop before the peak stress deviator during a drained triaxial path [DES 04, VAR 95] and liquefaction has also been observed at this peak during an undrained triaxial test controlled with axial force [DAR 07, KHO 06]. Failure modes which do not display a localization pattern, such as liquefaction, are, called diffuse modes.

Chapter written by Florent PRUNIER, Sylvain LIGNON, Farid LAOUAFA and Félix DARVE.

Failure in materials can be translated as the loss of stability in an equilibrium state, which is manifested by the fact that, by slightly disturbing the current condition (position or rate) of the equilibrium position, an unbounded response of the material will be induced. Besides, we consider only loadings for which mass and inertial effects can be omitted. Hence, flutter instabilities will be not studied. Under these conditions, the Hill's criterion [HIL 58] permits the detection of all kinds of material instabilities, which are localized or diffused. In the spirit of Hill, a stress strain state is unstable, if material deformation can be pursued in a loading direction without any energy input. Stability is translated by the following relationship:

$$\int_V \delta s_{ij}\, d\left(\frac{\partial u_j}{\partial x_i}\right) dV > 0 \quad \forall\, d\left(\frac{\partial u_j}{\partial x_i}\right) \neq 0 \qquad [8.1]$$

where V denotes the body volume at time t, s_{ij} the components of the stress nominal tensor (the transposed of the Piola-Kirchoff tensor) and $d(\frac{\partial u_j}{\partial x_i})$ the gradient of the deformation rate. Whenever effects of configuration change can be disregarded, this relation boils down to the following simplification:

$$\int_V d\sigma_{ij}\, d\varepsilon_{ij}\, dV > 0 \quad \forall\ \|d\varepsilon\| \neq 0 \qquad [8.2]$$

with $d\sigma$ being the incremental Cauchy stress tensor and $d\varepsilon$ the infinitesimal strain rate tensor. In the first part of this chapter, this condition is analyzed at the material point level, hence the relation effectively studied is the following,

$$d^2W = d\sigma_{ij}\, d\varepsilon_{ij} > 0 \quad \forall\ \|d\varepsilon\| \neq 0 \qquad [8.3]$$

where d^2W is called second-order work.

This chapter consists of two sections. Section 8.2 is devoted to an analysis of relation [8.3] in three-dimensional conditions: the contour of the domain in which instabilities can arise, called the bifurcation domain, is given for Darve's models [DAR 95] in the three-dimensional stress space. On the other hand, cones of unstable loading directions for a stress state located inside this domain are also presented in the stress rate space for the models mentioned above. Section 8.3 presents an application of this criterion to a slope stability case. The Petacciato landslide, which occurred in March 1996 in Italy, under a slope of only 6°, is modeled. Simulation is done with the finite elements Lagamine code [LAG 07] developed at Liège University. In addition to a non-associated elastoplastic model for soils, this code benefits from a coupled hydromechanical model, making a simplified description of unsaturated soils possible. With the aid of this model, we can shed useful light on certain triggering mechanisms due to the heavy rainfall which preceded the landslide.

8.2. Study of the second-order work criterion

8.2.1. *Analytical study*

In this section we present a detailed analysis of condition [8.3] given in the introduction. In the following we denote by $\underline{\underline{N}}$ the matrix linking strain rate to stress rate, such as:

$$\underline{d\varepsilon} = \underline{\underline{N}}\,\underline{d\sigma} \qquad [8.4]$$

and by $\underline{\underline{M}}$ the inverse matrix as $\underline{\underline{M}} = \underline{\underline{N}}^{-1}$. In principal axes and for a given tensorial zone[1], $\underline{\underline{N}}$ takes the following form:

$$\underline{\underline{N}} = \begin{bmatrix} \frac{1}{E_1} & -\frac{\nu_{21}}{E_2} & -\frac{\nu_{31}}{E_3} \\ -\frac{\nu_{12}}{E_1} & \frac{1}{E_2} & -\frac{\nu_{32}}{E_3} \\ -\frac{\nu_{13}}{E_1} & -\frac{\nu_{23}}{E_2} & \frac{1}{E_3} \end{bmatrix} \qquad [8.5]$$

With these notations, the shallow link between the second-order work criterion and the condition given by:

$$det\left(\underline{\underline{N_s}}\right) = 0 \qquad [8.6]$$

is also described. $\underline{\underline{N_s}}$ is the symmetric part of $\underline{\underline{N}}$. Using the constitutive relation, we obtain:

$$d^2W = 0 \Leftrightarrow {}^t\underline{d\sigma}\,\underline{\underline{N_s}}\,\underline{d\sigma} = 0 \qquad [8.7]$$

For all incrementally piece-wise linear models, equation [8.7] can be developed in the following manner in the principal stress rate space:

$$\begin{aligned} &\frac{d\sigma_1^2}{E_1} + \frac{d\sigma_2^2}{E_2} + \frac{d\sigma_3^2}{E_3} - \left(\frac{\nu_{12}}{E_1} + \frac{\nu_{21}}{E_2}\right) d\sigma_1\, d\sigma_2 \\ &- \left(\frac{\nu_{32}}{E_3} + \frac{\nu_{23}}{E_3}\right) d\sigma_3\, d\sigma_2 - \left(\frac{\nu_{13}}{E_1} + \frac{\nu_{31}}{E_2}\right) d\sigma_1\, d\sigma_3 = 0 \end{aligned} \qquad [8.8]$$

This expression describes an elliptical cone, but real solutions depend on the sign of the eigenvalues of the quadratic form: the cone can be degenerated. After quadric reduction, equation [8.8] takes the following form in the diagonal base:

$$\lambda_1 X^2 + \lambda_2 Y^2 + \lambda_3 Z^2 = 0 \qquad [8.9]$$

---

1 A tensorial zone is a part of the loading space for which the constitutive relation is incrementally linear.

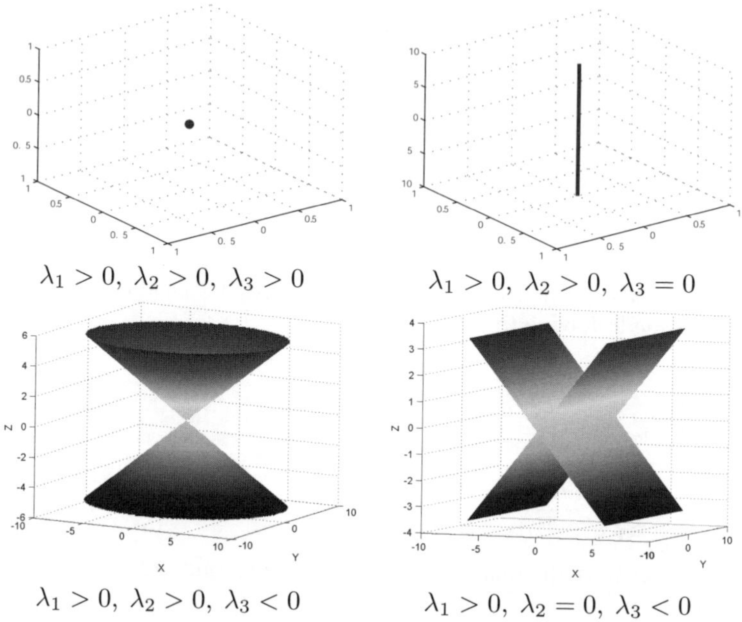

Figure 8.1. *Solutions of equation: $\lambda_1 X^2 + \lambda_2 Y^2 + \lambda_3 Z^2 = 0$. The four complementary solutions with opposite signs of eigenvalues are the same, but the sign of second-order work is geometrically also complementary. $\lambda_1 > 0,\ \lambda_2 > 0,\ \lambda_3 < 0$ give positive second-order work outside the cone and negative inside while $\lambda_1 < 0,\ \lambda_2 < 0,\ \lambda_3 > 0$ give positive second-order work outside the cone and positive inside*

with $\lambda_1,\ \lambda_2,\ \lambda_3$ being the eigenvalues of a quadratic form matrix. Figure 8.1 illustrates the different mathematical solutions. It should be noted that the solutions displayed in the right column (line and planes intersection) are never reached numerically. In fact, chances are poor for obtaining a numerically exactly null value of the eigenvalues. Nevertheless, a solution given by a straight line is almost reached, if λ_3 is very low, the elliptical cone is almost closed. On the other hand, for the solution given by the two planes, even if λ_2 is very low, we retrieve instantly an elliptical cone.

Given these results, we understand that the second-order work condition refers to the direction of the loading rate. A material point which admits none of the solutions of equation [8.8] is stable without other conditions. Conversely, the material point is potentially unstable: stability is conditioned by the loading path direction. So, it is possible to define the bifurcation domain limit as the set of stress states which admit a unique unstable loading direction. Assuming that all eigenvalues are strictly positive at the initial virgin state and that they evolve in a continuous manner with the loading parameter, then the bifurcation domain limit is given by the first vanishing of $det(\underline{\underline{N_s}})$:

$$\min_{i=1\ldots,n}\left(det\left(\underline{\underline{N_s}}\right)_i\right)=0 \quad \text{with} \quad \underline{u_i}\subset Z_i \qquad [8.10]$$

with n the number of tensorial zones of the constitutive model, u_i the eigenvector associated with the null eigenvalue and Z_i the tensorial zone examined.

It is of interest to note that a similar analysis can be made in the strain space and that the condition $det\left(\underline{\underline{M_s}}\right)=0$ is equivalent to the condition $det\left(\underline{\underline{N_s}}\right)=0$ for non-associated materials. In fact, it is possible to prove the following relation [PRU 09a, PRU 09b]:

$$det\left(\underline{\underline{M_s}}\right)=\frac{det\left(\underline{\underline{N_s}}\right)}{\left(det\left(\underline{\underline{N}}\right)\right)^2} \qquad [8.11]$$

For associated materials, the bifurcation domain contour coincides with the plasticity limit surface and relation [8.11] remains valid because $det\left(\underline{\underline{M}}\right)=1/det\left(\underline{\underline{N}}\right)$, but the plasticity limit cannot be expressed in a similar way with the matrix $\underline{\underline{N}}$.

To illustrate the results presented above, Figures 8.2 and 8.3 display limits of the bifurcation domain as well as instability cones obtained along a drained triaxial path

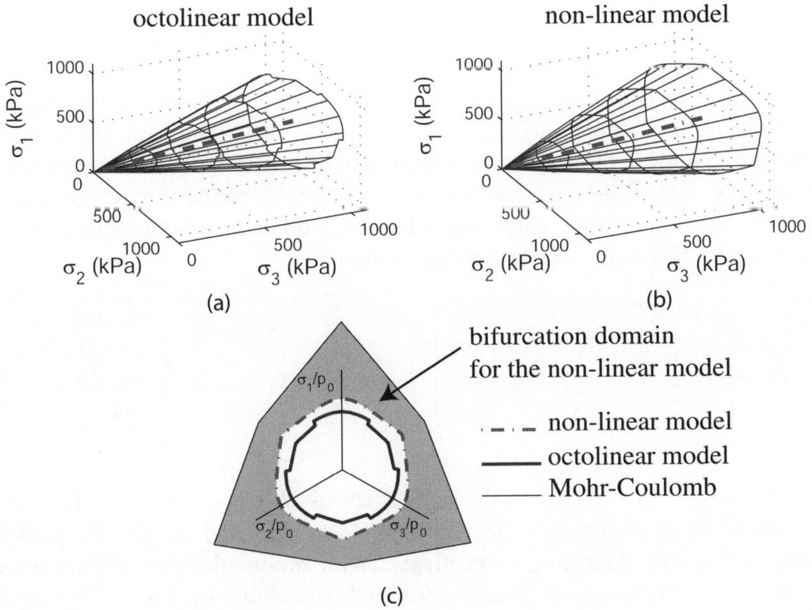

Figure 8.2. *Bifurcation domain limit in the 3D stress space for the constitutive Darve's models*

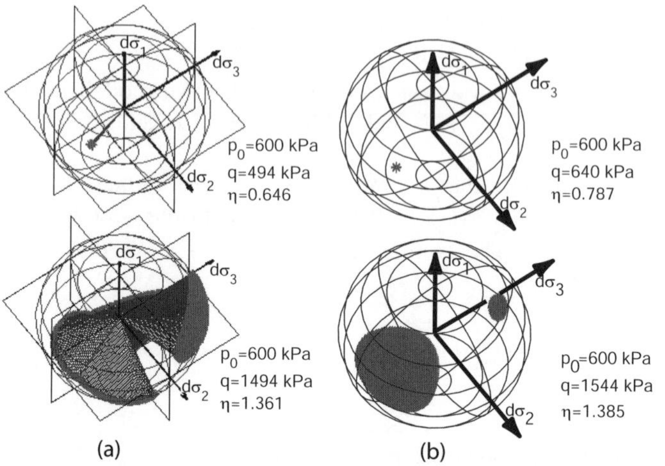

Figure 8.3. *3D instability cones for a dense Hostun sand. The Figure 8.3(a) displays cones obtained with the octolinear model. Planes delineate the eight tensorial zones, meshes correspond to the analytical solution of equation [8.8] and point clouds to solution obtained with a numerical method. The Figure 8.3(b) shows results obtained with the non-linear model with the numerical method.* p_o *is the initial confing pressure,* $q = \sigma_1 - \sigma_3$, $\eta = \frac{q}{p}$

according to Darve's models [DAR 95]. Without entering into the details involving these models, let us just remark that they are not based on the classical concept of elastoplasticity: no hypothesis has been made on strain decomposition into elastic and plastic part and no loading surface and flow rule have been defined. The irreversible non-linear behavior is directly described by a phenomenological incrementally non-linear relation. Relationships have been developed to describe triaxial loading paths on granular materials, after which the behavior in other loading directions has been extrapolated by one of the two following relations:

$$\begin{bmatrix} d\varepsilon_1 \\ d\varepsilon_2 \\ d\varepsilon_3 \end{bmatrix} = \frac{1}{2}\left[N^+ + N^-\right]\begin{bmatrix} d\sigma_1 \\ d\sigma_2 \\ d\sigma_3 \end{bmatrix} + \frac{1}{2\,\|\underline{d\sigma}\|}\left[N^+ - N^-\right]\begin{bmatrix} d\sigma_1^2 \\ d\sigma_2^2 \\ d\sigma_3^2 \end{bmatrix} \qquad [8.12]$$

where matrices N^+ and N^- have the same shape as N in equation [8.5]. Exponents + and − allow us to distinguish tangent Young's moduli and tangent Poisson ratios in compression and extension, respectively. With this mathematical formulation, a different material behavior is obtained for each stress rate direction. This model is called INL for incrementally non-linear. The second model is based on the same formulation as the INL, but with a simpler interpolation relation, which corresponds to the incrementally piece-wise linear model:

$$\begin{bmatrix} d\varepsilon_1 \\ d\varepsilon_2 \\ d\varepsilon_3 \end{bmatrix} = \frac{1}{2}\left[N^+ + N^-\right] \begin{bmatrix} d\sigma_1 \\ d\sigma_2 \\ d\sigma_3 \end{bmatrix} + \frac{1}{2}\left[N^+ - N^-\right] \begin{bmatrix} |d\sigma_1| \\ |d\sigma_2| \\ |d\sigma_3| \end{bmatrix} \quad [8.13]$$

Hence, this second model has eight tensorial zones and is called Model $8L$ for being octolinear. To conclude this analysis of the second-order work criterion, we now propose a physical interpretation of the results presented above.

8.2.2. *Physical interpretation*

As we saw in section 8.2.1, the second-order work criterion induces a directional dependence of the stability; regarding a certain tensorial zone, elliptical cones of unstable loading directions can open up (Figure 8.1) strictly inside the zone. Hence, the real cone, according to a given stress strain state, is the compilation of several elliptical cones truncated in their tensorial zone (Figure 8.3).

Because of this directional pattern, it is worthwhile considering proportional loading paths, making it possible to follow any loading rate direction. In what follows, the selected loading variables are the stress increments, but the same mathematical developments can be made by using the strain increments instead (as loading variables). We propose to study the following loading paths:

$$\begin{cases} d\sigma_1 = constant & constant \in \mathbb{R} \\ d\sigma_1 + R\,d\sigma_3 = 0 & R \in \mathbb{R} - \{0\} \\ d\sigma_2 - R'\,d\sigma_3 = 0 & R' \in \mathbb{R} \end{cases} \quad [8.14]$$

Given these paths, the second-order work can be re-written with the following conjugated variables:

$$\begin{aligned} d^2W = {} & d\sigma_1 \left(d\varepsilon_1 - \frac{d\varepsilon_3}{R} - \frac{R'}{R}\,d\varepsilon_2\right) \\ & + (d\sigma_1 + R\,d\sigma_3)\left(\frac{d\varepsilon_3}{R} + \frac{R'}{R}\,d\varepsilon_2\right) \\ & + (d\sigma_2 - R'\,d\sigma_3)\,d\varepsilon_2 \end{aligned} \quad [8.15]$$

After a drained triaxial compression until the deviatoric ratio $q/p = 0.787$, some loading paths for fixed R and R' are simulated. Figure 8.4 displays the responses of these paths. Classical notations are used: $p = \frac{1}{3}tr(\underline{\sigma})$ and $q = \sqrt{\frac{3}{2}}\|dev(\underline{\sigma})\|$ with $dev\left(\underline{\sigma}\right) = \underline{\sigma} - p\underline{\underline{I}}$ in three-dimensions. In Figure 8.4b, the response to the triaxial compression is not shown in order not to clutter the graph. Then, if we choose to control the following variables:

$$d\varepsilon_1 - d\varepsilon_3/R - R'/R\,d\varepsilon_2,\ d\sigma_1 + R\,d\sigma_3 \text{ and } d\sigma_2 - R'\,d\sigma_3$$

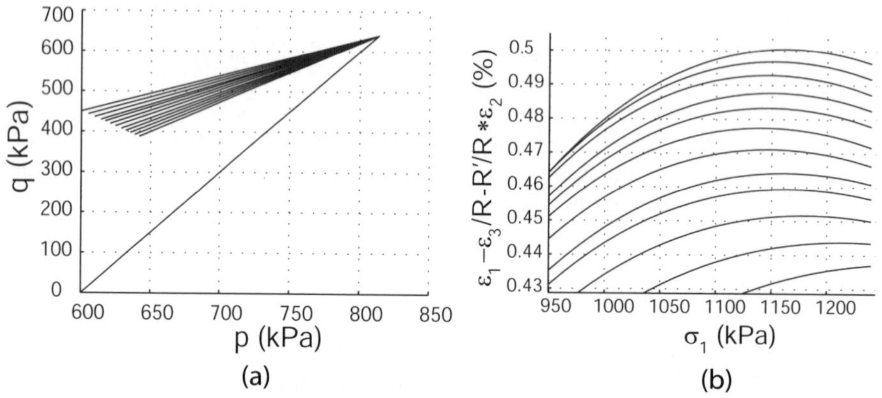

Figure 8.4. *Stress proportional loading path responses obtained with the octolinear model [DAR 82] after a triaxial compression at $p_o = 600\ kPa$ and $q = 640\ kPa$. $R \in [-3.5; -2.5]$ and with $R' \in [0.8; 1.2]$*

it is possible to build the constitutive matrix $\underline{\underline{S}}$ such as

$$\begin{bmatrix} d\varepsilon_1 - \frac{d\varepsilon_3}{R} - \frac{R'}{R}\, d\varepsilon_2 \\ d\sigma_1 + R\, d\sigma_3 \\ d\sigma_2 - R'\, d\sigma_3 \end{bmatrix} = \underline{\underline{S}} \begin{bmatrix} d\sigma_1 \\ \frac{d\varepsilon_3}{R} + \frac{R'}{R}\, d\varepsilon_2 \\ d\varepsilon_2 \end{bmatrix} \qquad [8.16]$$

Given the static condition [8.14], second-order work vanishes at the $\varepsilon_1 - \varepsilon_3/R - R'/R\varepsilon_2$ peak. Furthermore the $\underline{\underline{S}}$ determinant also vanishes because

$$\underline{\underline{S}} \begin{bmatrix} d\sigma_1 \\ \frac{d\varepsilon_3}{R} + \frac{R'}{R}\, d\varepsilon_2 \\ d\varepsilon_2 \end{bmatrix} = \begin{bmatrix} 0 \\ 0 \\ 0 \end{bmatrix} \qquad [8.17]$$

with a non-zero vector $(d\sigma_1,\ d\varepsilon_3/R + R'/R\, d\varepsilon_2,\ d\varepsilon_2)$. Hence, it is possible to define a generalized flow rule (called failure rule) according to these mixed conjugated variables (see [8.17]). Finally, it should be noted that when R and R' are considered to be variables, $det\left(\underline{\underline{S}}\right) = 0$ is equivalent to equation [8.8] [PRU 09c].

To conclude with these aspects, it must be stated that every stress-strain state, where $\underline{\underline{N_s}}$ or $\underline{\underline{M_s}}$ is not positive definite, is included inside the bifurcation domain and, consequently, unstable loading directions exist. Furthermore, given a particular loading path, when the second-order work vanishes, it means that a peak does exist for properly conjugated variables and that a failure rule can be defined for these variables. Eventually, effective failure takes place only when the variable showing a peak is controlled. If, on the contrary, the conjugated variable is controlled, it is possible to pass through this peak without obvious failure. Nevertheless, the loading path after the peak remains unstable.

8.3. Petacciato landslide modeling

8.3.1. *Site presentation*

Here we present modeling work done within the European project Lessloss [LES 04]. The Petacciato landslide is located on the Italian Adriatic coast with 7 km along the sea and 2 km inland (Figure 8.5). The difference in altitude between the highest point and the sea level is about 200 meters. This site experienced a series of small movements during the 20th Century until a major event occurred on March 14, 1996 after a period

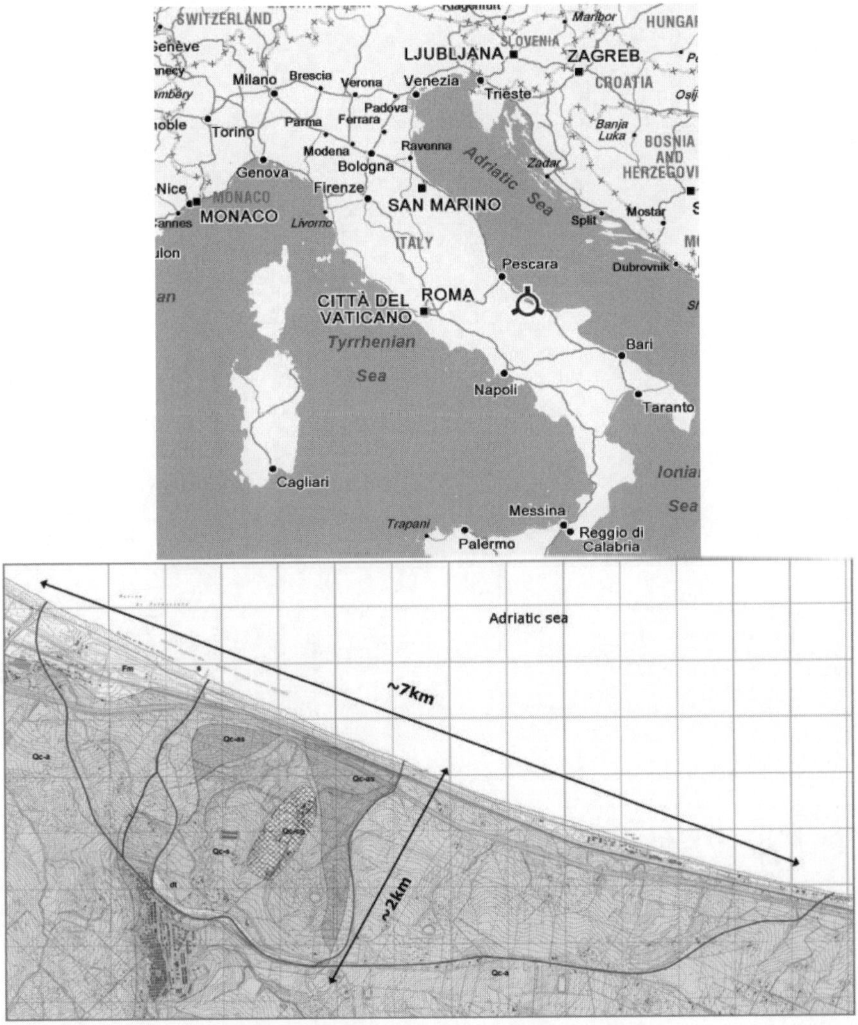

Figure 8.5. *Petacciato geographical situation*

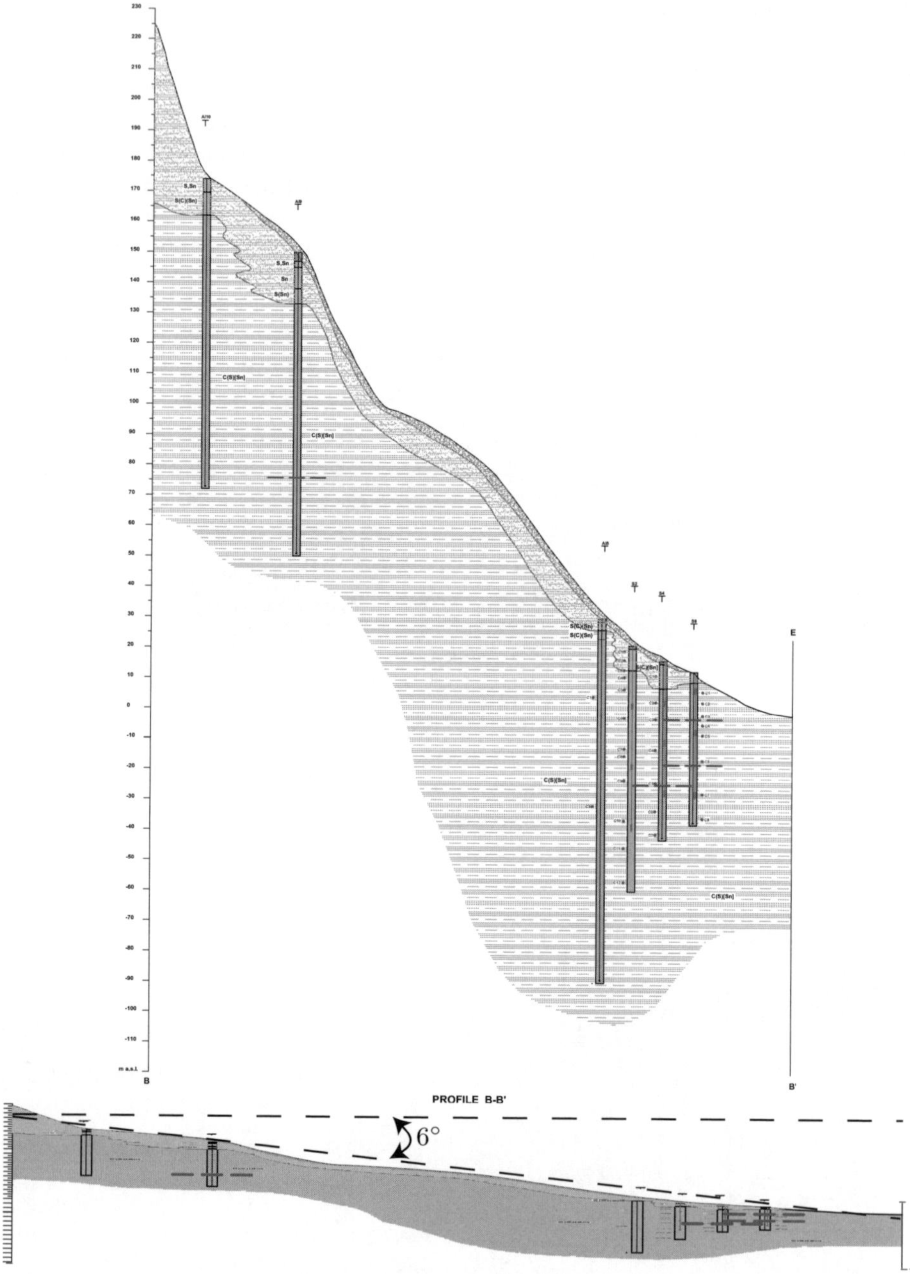

Figure 8.6. *Morphological cut to model*

of heavy rainfall. The infrastructure of the commune: roads, railway, buildings, etc. was heavily damaged. A morphological cut of the site was reconstituted for the purposes of the project and is reproduced in Figure 8.6. In this image, the dashed lines cutting boreholes depict the zones where the *in situ* slip surface was found; that is, between 70 and 90 meters deep. The grayscale distinguishes the different materials. Aside the superficial layers, the soil is largely homogeneous, with a grain size distribution situated between clays and silts. Table 8.1 gives the principal morphological features of the soil.

Features of particle size	
large particles ($\phi > 2$ mm)	0.0%
sands ($0.06 < \phi < 2$ mm)	6.1%
silts ($0.002 < \phi < 0.06$ mm)	55.7%
clays ($\phi < 0.002$ mm)	38.2%
General features	
grain density (γ_s)	2.74 g/cm^3
initial void ratio (e)	0.517
initial saturation degree (S_r)	99.90%
liquid limit (LL)	58.41%
plasticity limit (LP)	21.44%

Table 8.1. *General features of the Petacciato soil*

8.3.2. *Description of the model used*

To perform this modeling, we used the Lagamine code [LAG 07] developed at the Liège University. This code has at its disposal a non-associated elastoplastic law for the description of the soil's mechanical behavior [BAR 98]. This model is based on the Van-Eekelen loading surface [VAN 80a] where the conic shape depends on the current friction angle φ and the current cohesion C. Figure 8.7 depicts a comparison between the Mohr-Coulomb, Drücker-Prager and Van-Eekelen surfaces in the deviatoric plane. Hardening is taken into account by a hyperbolic evolution of the cohesion C and the friction angle φ with the Von Mises equivalent plastic strain. The initial values of C and φ provide the elastic limit and their maximum values provide the plastic limit:

$$\begin{cases} \varphi_c = \varphi_{c0} + \frac{(\varphi_{cf} - \varphi_{c0})\varepsilon^p_{eq}}{B_p + \varepsilon^p_{eq}} \\ \varphi_e = \varphi_{e0} + \frac{(\varphi_{ef} - \varphi_{e0})\varepsilon^p_{eq}}{B_p + \varepsilon^p_{eq}} \\ c = c_0 + \frac{(c_f - c_0)\varepsilon^p_{eq}}{B_c + \varepsilon^p_{eq}} \end{cases} \quad [8.18]$$

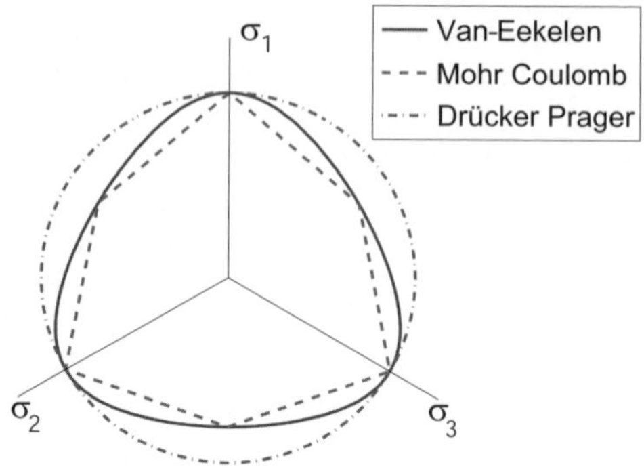

Figure 8.7. *Mohr-Coulomb, Drücker-Prager and Van-Eekelen limit surfaces in the deviatoric plane with* $\varphi_c = \varphi_e = 19°$

φ_c and φ_e are the friction angles in compression and extension, respectively. B_p and B_c are two hardening constants permitting the evolution of this hardening to be calibrated. Finally, ε^p_{eq} is the total Von Mises plastic strain obtained during the loading. The non-associated feature is described with a plastic potential defined in the same manner as the loading surface, but using the dilatancy angle instead of the friction angle. Evolution of ψ is given, assuming that

$$\varphi - \psi = cte \quad cte \in \mathbb{R} \tag{8.19}$$

all along the loading.

The Petacciato landslide was probably triggered by the heavy rainfalls which preceded the event. For this additional reason, we used a model which allowed the main characteristics of unsaturated soils to be described. This behavior is governed by the equation of the transfer of flow water within unsaturated soils. This equation is formulated with the generalized Darcy law and the mass balance equation:

$$\underline{v} = -k(S_r)\ \underline{\nabla(h)} \tag{8.20}$$

with k the permeability depending on the saturation degree S_r and h the hydraulic head defined as follows:

$$h = \frac{p}{\rho\, g} + z \tag{8.21}$$

Furthermore, if we hypothesize that the water is incompressible, the balance equation for the water mass could be written as

$$\text{div}\,(\underline{v}) = -\frac{\partial \theta}{\partial t} \tag{8.22}$$

with θ the volumetric water content:

$$\theta = n\, S_r \tag{8.23}$$

The variable n designates the porosity defined as the ratio between the voids and the total volume (voids plus solid particles). The right member of this equation [8.22] is called the storage term. Assuming that the solid particles are not deformable, the following relation:

$$\frac{\partial n}{\partial t} = (1-n)\ \frac{\partial \varepsilon_v}{\partial t} \tag{8.24}$$

can be obtained where ε_v is the relative total volume change. Hence the storage term can be formulated as

$$\frac{\partial \theta}{\partial t} = (1-n)\ S_r \frac{\partial \varepsilon_v}{\partial t} + n\ \frac{\partial S_r}{\partial p_c}\ \frac{\partial p_c}{\partial t} \tag{8.25}$$

Then, by substituting the expressions of $\underline{v}$ and $\partial\theta/\partial t$ in [8.22], we obtain the equation of the water transfer in unsaturated soils:

$$\text{div}\left(k(S_r)\ \underline{\nabla(h)}\right) = (1-n)\ S_r \frac{\partial \varepsilon_v}{\partial t} + n\ \frac{\partial S_r}{\partial p_c}\ \frac{\partial p_c}{\partial t} \tag{8.26}$$

This equation involves two characteristics of the water behavior in soil: the permeability as a function of the degree of saturation and the water retention curve. This curve gives the degree of saturation according to the capillary pressure $p_c = u_a - u_w$, with u_a the air pressure, and u_w the water pressure.

As far as the permeability is concerned, we assume simply that it varies in a linear manner with the degree of saturation and the water isotropic permeability k_0:

$$k = S_r\, k_0 \tag{8.27}$$

As for the water retention curve, we use the Van Genuchten model [VAN 80b]:

$$S_r = S_{r0} + \frac{1 - S_{r0}}{\left(1 + (\alpha\, p_c)^{\beta}\right)^{1-1/\beta}} \tag{8.28}$$

with S_{r0} the residual saturation degree, and α and β the calibration parameters. Even if this model is commonly used, it does not take into account the hysteresis effect experimentally observed between drying and wetting phases.

Finally, the hydromechanical coupling is done with the help of Bishop's effective stress concept [BIS 63]:

$$\sigma'_{ij} = \sigma_{ij} - u_a\, \delta_{ij} + \chi\ (u_a - u_w)\ \delta_{ij} \tag{8.29}$$

Moreover, it is assumed that the air pressure remains constant and equal to atmospheric pressure and that χ parameter is equal to the degree of saturation. Hence, we use the following relation:

$$\sigma'_{ij} = \sigma_{ij} - S_r\, u_w\, \delta_{ij} \qquad [8.30]$$

with σ'_{ij} as the stress linked to strains by the elastoplastic constitutive model.

8.3.3. *Landslide computation*

For computational purposes, the elastoplastic model was calibrated on the experimental tests performed by other people working on the LessLoss project. Figure 8.8 presents calibration curves.

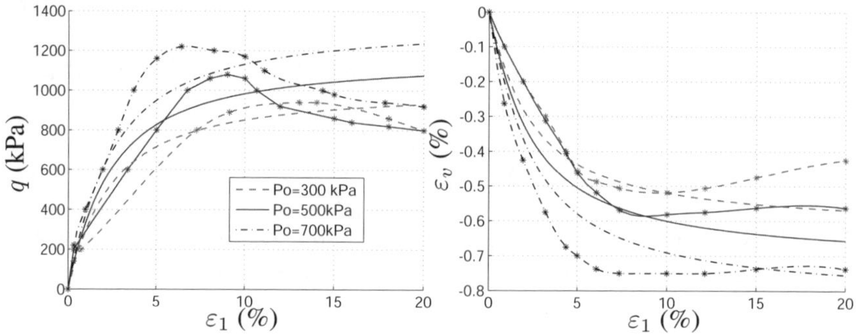

Figure 8.8. *Calibration of Plasol model for the Petacciato soil*

It should be noted that the model employed does not permit us to describe softening. Hence, only initial tangent mechanical properties and peak values are reproduced on average. Table 8.2 gives the input values of the model.

As for the water retention curve calibration, no experimental results from the project were provided. We, therefore, used experimental data from [FLE 02] soils having a gradation more or less similar to that of the site. Figure 8.9 shows these results and Table 8.3 provides the hydraulic parameters of the model.

Figure 8.10 displays the adopted mesh as well as the kinematic boundary conditions considered. Concerning the imposed loading, our aim was to reproduce the influence of the water table level on slope stability. For this reason, water pressure was imposed on the left and right boundaries of the mesh. A hydrostatic pressure distribution, as defined in Figure 8.11, was assumed. Below the point $u_w = 0$, the soil is fully saturated with positive water pressures and above the soil is partially saturated with negative water pressure, which furnishes a supplementary cohesion to the soil (suction).

Soil parameters	Symbols	Units	Blue clayey silt
grain specific weight	ρ_s	kN/m^3	27.40
initial tangent Young modulus	E	MPa	95.0
initial tangent Poisson ratio	ν	–	0.21
porosity	n	–	0.3
initial friction angle	$\varphi_{c0} = \varphi_{e0}$	°	1.0
friction angle at failure	$\varphi_{cf} = \varphi_{ef}$	°	19.0
hardening parameter 1	B_p	–	0.01
initial cohesion	c_0	kPa	10^{-3}
cohesion at failure	c_f	kPa	171
hardening parameter 2	B_c	–	0.02
dilatancy angle at failure	ψ_f	°	0

Table 8.2. *Calibration parameters*

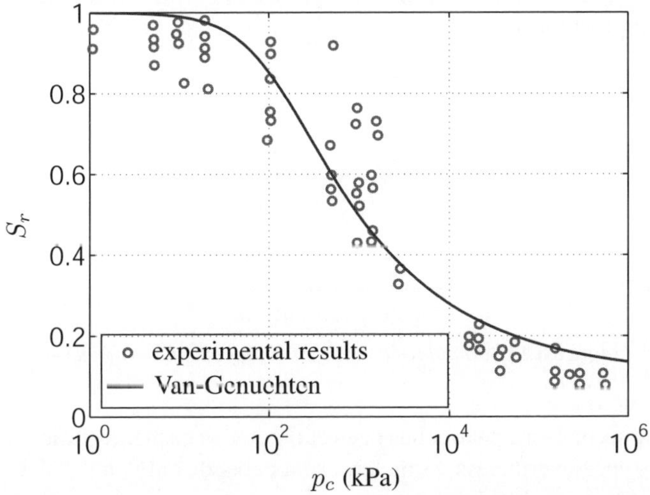

Figure 8.9. *Typical water retention curve of a clay [FLE 02] used for the Petacciato soil*

Hydraulic parameters	Symbols	Units	Clay
intrinsic permeability	k_0	m^2	10^{-17}
residual saturation degree	S_{r0}	–	0.1
first retention parameter	α	Pa^{-1}	1.0 10^{-5}
second retention parameter	β	–	1.35

Table 8.3. *Water retention parameters*

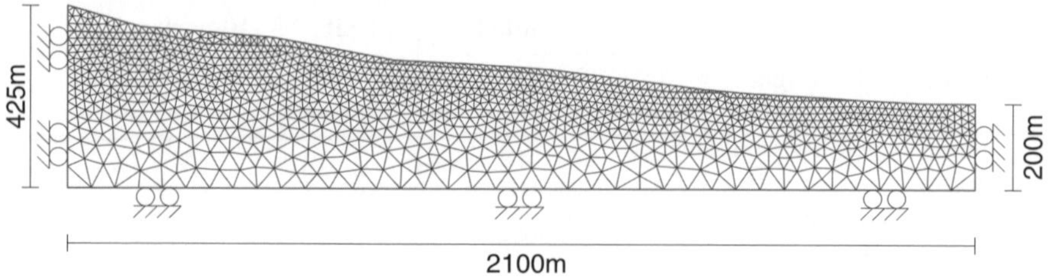

Figure 8.10. *Mesh of the Petacciato cut*

On the other hand, since we do not know the degree of water transfer during the regular rainy seasons, we chose to start with null values of water pressure at the bottom of the site. This pressure is augmented step by step at the boundaries, as shown in Figure 8.11. The physical time of flow between two steps is high enough for there to be, at the end of each step, a permanent water flow between the left and right banks of the soil. The water table slope was chosen arbitrarily in order to be more or less parallel to the mean slope of the site.

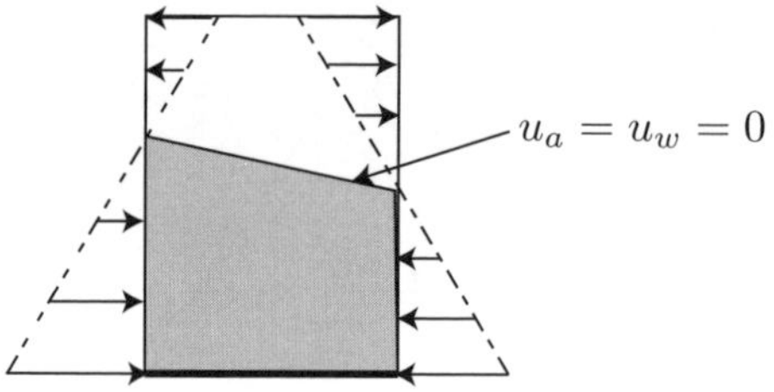

Figure 8.11. *Boundary conditions to impose the water level*

Finally, the second-order work has been computed at each integration point for each computation step. Nevertheless, as the criterion depends only on the sign of d^2W, the effective quantity computed is:

$$d^2W_n = \frac{d^2W}{\|\underline{d\sigma}\| \; \|\underline{d\varepsilon}\|} \tag{8.31}$$

d^2W_n is called normalized second-order work and represents the cosine of the angle between stress rate and strain rate vectors. This quantity decreases continuously when the second-order work tends to zero. As a consequence, changes of the normalized second-order work's isovalues with the loading make it possible to follow the evolution of the weakened zones. Figure 8.12 displays these results for typical water levels. It can be noted that the first instabilities appear at the surface when the water table level is at about 75% from the maximum level. We should note that potentially unstable zones develop not around this water table but at the surface. This means that the gravitational

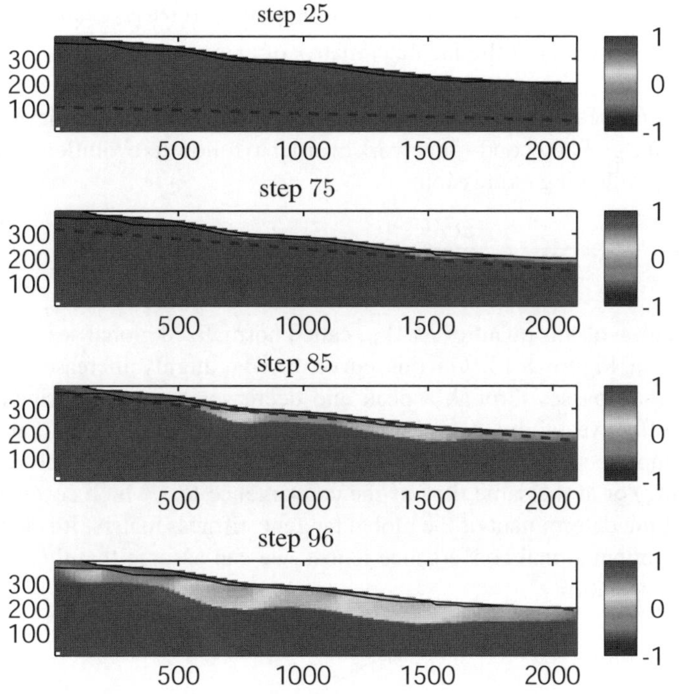

Figure 8.12. *d^2W_n isovalues for distinct water level (in dashed line)*

effect, although small at the surface under a low slope, manages to override the level of the suction imposed. After this loading step, affected zones develop in depth because water pressure increases beyond the confinement imposed by gravity. When the water table level reaches 96% of the maximum height, convergence of the computation is lost. It can be assumed that, at this step, effective failure occurs because the plasticity limit has been reached at several integration points. Furthermore, isovalues of the normalized second-order work permit us to correctly describe the slip surfaces observed *in situ*.

The criterion for being able to pronounce the global stability of the body is none other than the second-order work integrated over the total volume as in equation [8.2]. From a numerical point of view, this integration is done as follows:

$$D^2W = \int_V d\sigma_{ij}\, d\varepsilon_{ij}\, dV = \sum_{e=1}^{nel} \sum_{k=1}^{npi} {}^t\underline{d\sigma}_k . \underline{d\varepsilon}_k . \omega_k . j_e \qquad [8.32]$$

with:

– nel the number of elements;

– npi the integration point number on each element;

– ω_k the weight affected at each integration point k (Gauss quadrature);

– j_e the determinant of the Jacobian matrix of element e.

For the same reasons as the ones discussed at the material point level, it is worth normalizing the global second-order work in order to follow its evolution. Khoa [KHO 05] suggested the following expression:

$$D^2W_n = \frac{D^2W}{\sum_e \sum_k \omega_k \, j_e \sum_e \sum_k {}^t\underline{d\sigma}_k \, \underline{d\varepsilon}_k} \qquad [8.33]$$

The variation of this quantity D^2W_n, called normalized global second-order work is displayed in Figure 8.13. On this curve, D^2W_n highly increases at the outset of loading, then passes through a peak and decreases to tend in principle towards zero at the effective failure state. From a theoretical point of view, if we overcome numerical imperfections, global second-order work should vanish before (in case of diffuse failure) or at the same time as the convergence loss which corresponds to the vanishing of the determinant of the global tangent stiffness matrix. It is for this reason that, when computational convergence is lost, we can assume that the body loses its ultimate global stability.

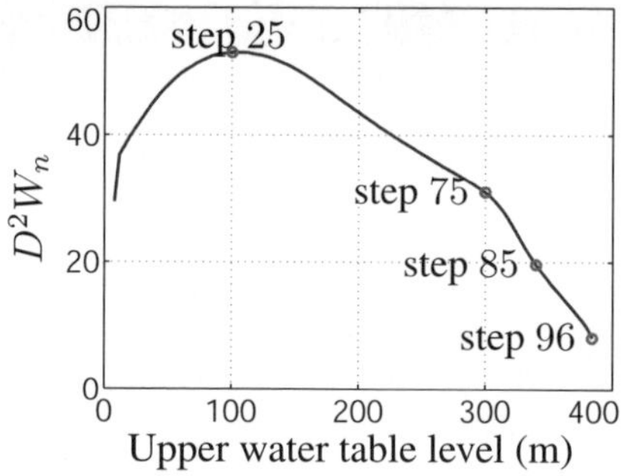

Figure 8.13. D^2W_n *change during the wetting*

8.4. Conclusion

The subject of this chapter has been the local second-order work criterion, a condition which allows the material instability of elastoplastic materials to be detected. For associated materials (which validate normality rule), we have found that this condition is attained at the same time as the plasticity limit. For non-associated materials, however, we see that it can be reached not only before the plasticity limit, but also before a Rice strain localization criterion.

From a general point of view, the sign of second-order work depends essentially on loading directions: after any given loading path, a material point is stable in the sense of second-order work criterion if no direction of the incremental stress space or of the incremental strain space causes this quantity to be cancelled. Otherwise, the material point is potentially unstable. Instability is effective only when the loading path follows an unstable direction with proper loading variables. All the unstable directions at a material point constitute a cone called the instability cone. All the stress-strain states, where the cone is reduced to a single unstable direction, define the bifurcation domain limit.

When a piece-wise incrementally linear model is under consideration (this being the overwhelming majority of cases), instability cones are elliptical in the incremental stress space or in the incremental strain space with the origin of this space as a summit (see equation [8.8]). The bifurcation domain limit, as well as the instability cones, were shown for Darve's models.

Indeed, it is important to note that instability is not synonymous with failure. When a material is subjected to a loading path in an unstable direction, the effective failure is dependent on the variables which control the loading conditions. In order to reach effective failure, the controlled loading variables are necessarily mixed. We have seen that when second-order work becomes annulled, it means that a peak arises for one of these mixed loading variables. At this peak, a generalized flow rule can be defined and an exponential increase of strains comes into effect (effective failure). Otherwise, if the conjugated variable is controlled, the peak can be passed without collapse. Nevertheless, after the peak, the loading path remains unstable: a little perturbation on controlled variables will lead instantaneously to failure.

Section 8.3 of the chapter was devoted to modeling the Petacciato landslide by using a coupled hydromechanical model for unsaturated soils and an analysis based on the second-order work criterion. Concerning this simulation, isovalues of the local normalized second-order work provided a good description of instability zones in the soil slope without postulating in principle any particular slip surface. In this case study, unstable zones would not have been detected with only a plasticity limit criterion, because the slope is quite gentle. The natural loading path imposed upon this slope by boundary conditions, water flow conditions and gravity, is then for some points included inside instability cones.

Hence, with the help of these tools, we were able to show that an unusual saturation of the site could lead to failures even at great depths. But these numerical simulations of large natural sites can only give qualitative information. In fact, more realistic physical models which could more quantitatively describe hydraulic behavior could be used and a three-dimensional simulation could also enable the model to take into account the finer geometric effects. However, the main difficulties we face are probably in the limitations of our current capacities to characterize the media constituting the site, particularly from the hydraulic viewpoint, for the data from the hydraulic field are still difficult to obtain.

8.5. Bibliography

[BAR 98] BARNICHON J.D., Finite element modeling in structural and petroleum geology, Doctorate thesis, University of Liège, 1998.

[BIA 94] BIAREZ J., HICHER P.Y. (eds), *Elementary Mechanics of Soil Behavior: Saturated Remoulded Soils*, Balkema, 1994.

[BIA 08] BIAREZ J., HICHER P.Y., "Mechanisms of soil deformation", *Constitutive Modeling of Soils and Rocks*, ISTE, London, John Wiley & sons, New York, 2008.

[BIS 63] BISHOP A.M., BLIGHT G.E., "Some aspects of the effective stress in saturated and unsaturated soils", *Géotechnique*, vol. 13, no. 3, p. 177-197, 1963.

[CAM 00] CAMBOU B., CHAZE M., DEDECKER F., "Change of scale in granular materials", *European Journal of Mechanics-A/Solids*, vol. 19, p. 999-1014, 2000.

[CHR 81] CHRISTOFFERSEN J., MEHRABADI M.M., NEMAT-NASSER S., "A micro-mechanical description of granular material behavior", *Journal of Applied Mechanics*, vol. 48, p. 339-344, 1981.

[COL 03] COLLIN F., Couplage thermo-tydro-mécanique dans les sols et les roches tendres, PhD thesis, University of Liège, 2003.

[CUN 79] CUNDALL P.A., STRACK O.D.L., "A discrete numerical model for granular assemblies", *Géotechnique*, vol. 29, no. 1, p. 47-65, 1979.

[DAR 82] DARVE F., LABANIEH S., "Incremental constitutive law for sands and clays. Simulations of monotonic and cyclic tests", *International Journal for Numerical and Analytical Methods in Geomechanics*, vol. 6, p. 243-275, 1982.

[DAR 87] DARVE F., CHAU B., "Constitutive instabilities in incrementally non-linear modeling", in DESAI C.S., KREMPL E., KIOUSIS P.D., KUNDU T. (eds), *Constitutive Laws for Engineering Materials, Theory and Application*, p. 301-310, Elsevier Science, New York, 1987.

[DAR 95] DARVE F., FLAVIGNY E., MEGHACHOU M., "Yield surfaces and principle of superposition revisited by incrementally non-linear constitutive relations", *International Journal of Plasticity*, vol. 11, no. 8, p. 927-948, 1995.

[DAR 04] DARVE F., VARDOULAKIS I. (eds), *Degradations and Instabilities in Geomaterials*, CISM courses, vol. 461, Springer, New York, 2004.

[DAR 07] DARVE F., SIBILLE L., DAOUADJI A., NICOT F., "Bifurcations in granular media: macro- and micro-mechanics approaches", *Comptes Rendus Académie des Sciences Mécaniques*, vol. 335, p. 496-515, 2007.

[DES 04] DESRUES J., VIGGIANI J., "Strain localization in sand: on overview of the experimental results obtained in Grenoble using stereophotogrammetry", *International Journal for Numerical and Analytical Methods in Geomechanics*, vol. 28, no. 4, p. 279-321, 2004.

[FLE 02] FLEUREAU J.M., VERBRUGGE J.C., HUERGO P.J., GOMEZ CORREIA A., KHEIRBEK-SAOUD S., "Aspects of the behavior of compacted clayey soils on drying and wetting paths", *Canadian Geotechnical Journal*, vol. 39, p. 1341-1357, 2002.

[HIL 58] HILL R., "A general theory of uniqueness and stability in elasto-plastic solids", *Journal of the Mechanics and Physics of Solids*, vol. 6, p. 236-249, 1958.

[HIL 67] HILL R., "Eigenmodal deformations in elastic/plastic continua", *Journal of the Mechanics and Physics of Solids*, vol. 15, p. 371-386, 1967.

[KHO 05] KHOA H.D.V., Modélisation des glissements de terrains comme un problème de bifurcation, doctorate thesis, INPG, Grenoble, 2005.

[KHO 06] KHOA H.D.V., GEORGOPOULOS I.O., DARVE F., LAOUAFA F., "Diffuse failure in geomaterials: Experiments and modeling", *Computer and Geotechnics*, vol. 33, p. 1-14, 2006.

[LAD 92] LADE P.V., "Static instability and liquefaction of loose fine sandy slopes", *Journal of Geotechnical Engineering*, vol. 118, no. 1, p. 51-71, 1992.

[LAG 07] LAGAMINE, *User's Manual, Département GEOMAC*, Université de Liège, 2007.

[LES 04] LESSLOSS REPORT, The Petacciato Landslide: Geological and Geotechnical Data, Report, SGI: Studio Geotechnico Italiano, Campobasso, Italy, 2004.

[MAG 98] MAGNIER S.A., DONZÉ F.V., "Numerical simulations of impacts using a discrete element method", *Mechanics of Cohesive-Frictional Materials*, vol. 3, p. 257-276, 1998.

[MAN 66] MANDEL J., "Conditions de stabilité et postulat de Drücker", in KRAVTCHENKO J., SIRIEYS P.M. (eds), *Rheology and Soils Mechanics*, p. 58-68, Springer, Berlin, 1966.

[MRÓ 63] MRÓZ Z., "Non-associated flow laws in plasticity", *Journal de Mécanique*, vol. 2, p. 21-42, 1963.

[MRÓ 66] MRÓZ Z., "On forms of constitutive laws for elastic-plastic solids", *Archive of Mechanics Stosovanej*, vol. 18, p. 1-34, 1966.

[NIC 05] NICOT F., DARVE F., "A multiscale approach to granular materials", *Mechanics of Materials*, vol. 37, no. 9, p. 980-1006, 2005.

[NOV 91] NOVA R., "A note on sand liquefaction and soil stability", *3rd International Conference on Constitutive Laws for Engineering Materials: Theory and Applications*, p. 153-156, Tuscon, USA, 1991.

[PRU 09a] PRUNIER F., LAOUAFA F., DARVE F., "3D bifurcation analysis in geomaterials, Investigation of the second-order work criterion", *European Journal of Environmental and Civil Engineering*, vol. 13, no. 2, p. 135-147, 2009.

[PRU 09b] PRUNIER F., LAOUAFA F., LIGNON S., DARVE F., "Bifurcation modeling in geomaterials: from the second-order work criterion to spectral analyses", *International Journal for Numerical and Analytical Methods in Geomechanics*, vol. 33, no. 9, p. 1169-1202, 2009.

[PRU 09c] PRUNIER F., NICOT F., DARVE F., LAOUAFA F., LIGNON S., "3D multiscale bifurcation analysis of granular media", *Journal of Engineering Mechanics (ASCE)*, vol. 135, no. 6, p. 493-509, 2009.

[SIB 07] SIBILLE L., DONZÉ F.V., NICOT F., CHAREYRE B., DARVE F., "From bifurcation to failure in a granular material, a DEM analysis", *Acta Geotechnica*, vol. 3, no. 1, p. 15-24, 2007.

[VAN 80a] VAN EEKELEN H.A.M., "Isotropic yield surfaces in three-dimensions for use in soil mechanics", *International Journal for Numerical and Analytical Methods in Geomechanics*, vol. 4, p. 89-101, 1980.

[VAN 80b] VAN GENUCHTEN M.T., "A closed form for predicting the hydraulic conductivity of unsaturated soils", *Soil Science Society of America Journal*, vol. 4, p. 892-898, 1980.

[VAR 95] VARDOULAKIS I., SULEM J. (eds), *Bifurcation Analysis in Geomechanics*, Blackie Academic and Professional, Glasgow, 1995.

Chapter 9

Numerical Modeling: An Efficient Tool for Analyzing the Behavior of Constructions

9.1. Notations

In some sections the symbols listed below can have a different meaning. In this case, it is clearly explained.

Latin alphabet

a, a_1, a_2	Deviatoric mechanism's hardening parameter
b	Plasticity yield function form parameter varying between 0: *Coulomb* and 1: *Cam Clay*
$\underline{\underline{\underline{\underline{C}}}}$	Elasticity tensor
$\underline{\underline{\underline{\underline{C}}}}^{ep}$	Elastoplasticity tensor
c, c_{cyc}	Hardening parameter of isotropic mechanism
C_c	Compression index in the $(e - \log p')$ plane
C_z	Coefficient of curvature $d_{30}^2/(d_{60} \cdot d_{10})$
D	Damping ratio
d	Distance between critical state and isotropic consolidation lines in $(\varepsilon_v^p - \ln p')$ plane
d_k	Diameter of the sieve through which $k\%$ of the grains pass

Chapter written by Arezou MODARESSI-FARAHMAND-RAZAVI.

D_R	Relative mass density
I_D	Density index
E	Young's modulus
E_p	Shear hardening modulus
e	Void ratio
$\underline{e}_i$	Basic vector
e_{w_L}	Void ratio for a water content of w_L
e_{w_P}	Void ratio for a water content of w_P
f	Stress yield function
$\mathcal{F}$	Strain yield function
$\underline{\mathrm{g}}$	Gravity acceleration
G	Shear modulus
G_{max}	Maximum shear modulus
G_{ref}	Shear modulus at reference pressure
G_s	Specific density of soil grains
$\underline{\underline{I}}, \underline{\underline{\underline{\underline{I}}}}$	Unit tensor of second order and fourth order
I_p	Plasticity index
K	Elastic bulk modulus
K_w	Water compressibility
M	Slope of the perfect plasticity line in $(q - p')$ plane
M^p_{act}	Set of active plastic mechanisms
M^p_{pact}	Set of potential active plastic mechanisms
$\underline{n}$	Normal vector
n_e	Elasticity law exponent
n_p	Deviatoric hardening rule's exponent
OCR	Overconsolidation ratio
p_a	Atmospheric pressure
p_{desat}	Pressure at which the liquid phase becomes discontinuous
p_c	Critical pressure
P_{c0}	Initial critical pressure
P^{ns}_c	Unsaturated critical pressure
p'	Mean effective stress $(\sigma'_1 + \sigma'_2 + \sigma'_3)/3$
p_c	Capillary pressure (or suction in the soil)
p_i	Pressure in phase i
p'_{c0}	Initial critical pressure in $(\varepsilon^p_v - \ln p'_c)$ plane
p_{ref}	Reference pressure
p_{sat}	Pressure at which saturation is reached
p_w	Pore pressure
S_i	Degree of saturation of phase i
R	Radius of the spheres
S_w	Degree of saturation in water
S_a	Degree of saturation in air

r	Degree of friction mobilization
$r_{ela}, r_{hys}, r_{mob}$	Parameters of the elastoplastic model
$\underline{\underline{s}}$	Stress deviator tensor
S_u	Undrained shear strength $(\sigma'_1 - \sigma'_3)/2$ or τ_{max}
T	Superficial tension
U_c	Uniformity coefficient d_{60}/d_{10}
W	Work
w_L	Liquid limit
w_P	Plastic limit

Greek alphabet

α	Biot's parameter
α	Hardening variable (generic), specific to each yield function
β	Plastic compressibility modulus
γ	Distortion
$\gamma_{ela}, \gamma_{hys}, \gamma_{mob}$	Different plastic behavior domains
γ_{tv}	Volume variation threshold
$\underline{\underline{\varepsilon}}$	Strain tensor
$\underline{\underline{\varepsilon}}$	Deviatoric strain tensor
$\underline{\underline{\dot{\varepsilon}}}^e$	Reversible strain rate tensor
$\underline{\underline{\dot{\varepsilon}}}^p$	Plastic strain rate tensor
$\underline{\underline{\dot{\varepsilon}}}^{vp}$	Viscoplastic strain rate tensor
ε_v	Volumetric strain
ε_v^p	Plastic volumetric strain
η	Hardening variables common to several mechanisms
λ	Compression index in $(e - \ln p')$ plane
λ	Lamé's coefficient
$\dot{\lambda}^p$	Plastic multiplyer
μ, G_{max}	Lamé's coefficient, elastic shear modulus
π_c^{max}	Maximum capillary stress
$\underline{\underline{\sigma}}$	Cauchy total stress tensor
$\underline{\underline{\sigma}}'$	Cauchy effective stress tensor
$\underline{\underline{\sigma}}^*$	Cauchy net stress tensor
σ_{c0}	Initial critical stress
σ_c	Critical stress
σ_I	Major principal stress
$\sigma_1, \sigma_2, \sigma_3,$	Principal stresses
σ_{nn}	Normal stress to plane with normal $\underline{n}$

τ	Shear stress
ϕ'_{pp}	Friction angle at perfect plasticity
ψ	Dilatancy angle
$\underline{\underline{\Psi}}$	Flow direction
ψ_{BJ}	State index
χ	Bishop's coefficient

Subscripts and superscripts

e	Elastic
i, j, k	Plasticity mechanism indicator
p	Plastic
vp	Viscoplastic

Symbols

$[.]$	Discontinuity in.
$\cdot$	Scalar product
$\otimes$	Tensorial product
$:$	Contracted product
$\underline{a}$	Vector a
$\underline{\underline{a}}$	Second order tensor
$\underline{\underline{\underline{\underline{a}}}}$	Fourth order tensor
$\lvert . \rvert$	Absolute value of a scalar.
$\lVert \underline{.} \rVert$	Norm of a vector $\underline{.}$
a_n	Normal component of vector $\underline{a}$
$\underline{a}_t$	Tangential projection of vector $\underline{a}$
$\mathrm{tr}\underline{\underline{.}}$	Trace of tensor
log	Decimal logarithm
ln	Natural logarithm

9.2. Introduction

Macroscopic models used within the field of continuum mechanics are still the most rational tools for modeling the behavior of constructions, though with the steady development of computational methods, micromechanical models are becoming a viable alternative for analyzing the phenomena governing the behavior of soils. For any given problem, the quality of the numerical simulation depends on not only the selection of the physical model and the methods of identifying its parameters, but also the numerical process of integrating and solving the model's equations, which should satisfy the concerns of efficiency and rapidity.

Jean Biarez was one of the first in France to recognize the potential of the finite element method. In the 1970s, he encouraged a certain number of his students, namely Denis Aubry, Philippe Descrois, and Jean-Claude Hujeux to study this new method. Under the dynamic direction of Biarez, the laboratory of soil, structure and material mechanics (LMSSMat) at *Ecole Centrale de Paris* (ECP) acquired a rich and unparalleled experience in studying soil behavior. The ECP elastoplastic model, best known as the "Hujeux model", came out of the research studies of those days. Even today, it is one of the most efficient models available, especially for modeling cyclic loading.

A family of models incorporating the same basic mechanisms but serving different purposes now exists. At first, the multimechanism and cyclic version of the model presented by J.C. Hujeux [HUJ 85] introduced the notion of domain of behavior. After years of research in both mechanics and numerical modeling, the *GEFDyn* software was completed under the supervision of Denis Aubry [AUB 86, AUB 96]. Since then, other versions of this model have been developed, such as:

– multimechanism elasto-visco-plastic model incorporating the work of Kodaissi [AUB 91, KOD 83];

– model for asphalt concrete [AUB 87, MOD 89b];

– elastoplastic cyclic interface model [AUB 90];

– elastoplastic model for unsaturated soils [ABO 92, ABO 95, MOD 94a, MOD 94b, MOD 96];

– thermo-elasto-plastic or thermo-elasto-viscoplastic model [LAL 95, MOD 97];

– thermo-elasto-plastic model for unsaturated soils [MOD 95];

– elastoplastic model with damage mechanism [MOD 03];

– plastic model with oriented criteria [MOD 03, PIC 91, VOS 01].

The ECP elastoplastic model is based on a relatively simple formulation that nevertheless manages to describe the elementary mechanisms of soil behavior. The particular advantage of this model is that it has already been used over decades for calculating geotechnical structures and, thus, the experience it has mounted by now is quite considerable. This experience is manifold, from determining the model parameters for sands [DAG 08, ELM 03, HUJ 85, KOR 95, LOP 03a, LOP 03b]; and clays: [HAJ 84, LOP 03a, LAL 95, MIC 94] to its numerical integration or stability conditions [HAM 06, HAM 08a, VOS 01] and [HAM 08b]. Moreover, the model has been extended to unsaturated soils ([ABO 95, MOD 96]) and nonisothermal loading [LAL 95, MOD 97, MOD 95]. The surface version [AUB 90] is possible to use for modeling surfaces of kinematical discontinuity within a soil mass [LOP 03a] or rock [MOD 89a] or between two different materials (soil-pile [DAG 08]).

Hence, the *GEFDyn* software has been used and validated for various applications, and it is undoubtedly one of the most complete pieces of software in its field of application. Since policy and decision-makers need to have calibrated and validated numerical tools, it is essential to provide them with qualitative and quantitative knowledge about the behavior of constructions, so that the optimal design can be created out of testing different scenarios. With this in mind, a few examples of practical problems related to design and geo-structures will be presented towards the end of this chapter.

This chapter has been organized into four parts.

At first, in section 9.3, we will briefly describe the ECP elastoplastic model. We will show how this simplified model can take into account the main features of soil behavior observed in triaxial paths. In section 9.4, after generalizing this model to a multimechanism formalism to describe anisotropy and intermediary stress, we will discuss some aspects relative to the numerical integration of elastoplastic and viscoplastic laws. In section 9.5, we will describe the strategy for identifying the model parameters based on the concept of *reference behavior*, a notion highly valued by Jean Biarez. The importance of properly identifying the model parameters cannot be overstated. Special attention, therefore, will be paid to concepts such as critical state and reference behavior. In section 9.6, we conclude this chapter by illustrating the influence of the principal characteristics of soil behavior on the reaction of geo-structures. In most cases, we are concerned with granular soils which do not present any artificial or natural "glue". With the illustration of several construction cases, we will show how the constitutive law is important in analyzing the behavior of geo-structures.

9.3. Modeling soil behavior

9.3.1. *Main characteristics of the soil's mechanical behavior*

The study of the mechanical behavior of soil within the context of continuum mechanics has been the subject of many publications in international literature. The commonly termed Green Book [BIA 94] (the cover of the English edition is gray), includes an excellent discussion on the basic behavior of disturbed soil. A good synthesis can also be found in *Constitutive Modeling of Soils and Rocks*, edited by P.Y. Hicher and J.F. Shao [HIC 02].

Let us briefly recall the essential mechanical characteristics of soil behavior:

1) In the presence of water, the total stress alone cannot adequately describe the behavior of soils. This can only be described by the effective stresses that are obtained from the total stress and pore pressures. In undrained conditions and saturated soil, the behavior conforms quite well to the principle of Terzaghi's effective stress, assuming that the grains are incompressible. By working in effective stress, we can also eliminate the apparent cohesion for many soils. Real cohesion, as far as we are concerned, is the shear strength (in Mohr's plane) of a saturated soil under zero mean effective stress and not the linear extrapolation of strength at the origin, obtained from non-zero confining stresses. In unsaturated soils, capillarity can create a fictive (or apparent) cohesion, which is modeled by a pore pressure below air pressure (suction). The behavior of unsaturated soils will be examined in section 9.3.5.

2) Behavior of soils is highly non-linear. This non-linearity is due to four main mechanisms:

- the reversible deformation of granular particles at contacts,
- irreversible shear at intergranular contacts,
- the irreversible deformation of particles themselves,
- the irreversible deformation due to the breakage of particles.

In current applications of soil mechanics, due to the intrinsic rigidity of the soil particles, the third type of deformation is negligible. This is not always the case for some granular materials.

3) Effective confinement influences significantly both the stiffness and the soil strength.

4) Porosity (or void ratio or density), at a given level of confinement, also modifies the behavior of a soil. This change can be observed at the stiffness level, both for the initial value and during the loading (Figure 9.1).

5) Dilatancy refers to the ability to produce volumetric strains under the effect of shear, even at constant mean effective stress. It is dependent on both the porosity and the ratio between the average shear and mean effective stress.

6) During shearing, the material reaches a state where, in drained conditions, the deformation takes place under constant stress and without volume change; in undrained conditions, the pore pressure no longer changes and deformation occurs under constant effective stress. This state, called the *critical state*, occurs at a void ratio which depends only on the stresses. Triaxial tests show that the relationship between the critical void ratio and confinement is unique for a given material.

7) Although sands and clays have very different morphologies and structures, their macroscopic mechanical behavior is very similar. It must be said, however, that soil mechanics have not always expressed unanimous agreement concerning this similarity.

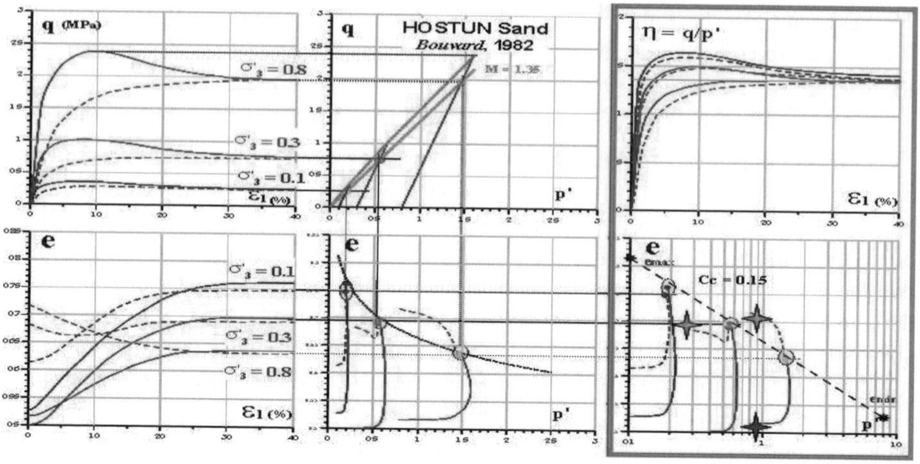

Figure 9.1. *Reference behavior based on results obtained by of Bouvard [BOU 82] from [SAI 97]*

Roscoe, Schofield and Wroth, the initiators of the critical state concept (Cambridge School [ROS 58]), first demonstrated it from the results obtained on clays (Cam Clay) in the 1960s. Previously, researchers who studied the stability of slopes composed of fine sands noticed the importance of the *in situ* void ratio on the liquefaction potential. They assumed that liquefaction could only occur in materials with a void ratio higher than a certain value (Casagrande [CAS 38], Geuze [GEU 48] and Koppejan *et al.* [KOP 48]). Casagrande and collaborators formulated the concept of critical void ratio in design methods. This method, known as the *steady state* or *collapse surface*, is described in detail by Castro *et al.* [CAS 82]. It consists of the limit analysis of potentially liquefiable soil and is based on the concept of "steady state of deformation", defined by Poulos [POU 81], as a state in which a mass of particles deforms continuously with constant volume, effective normal stress, distortion, and speed. As can be seen, these two concepts are very similar and the only point of difference seems to be the requirement that constant speed be included in the concept of *steady state* [SLA 85]. In Cam Clay theory, it is assumed that the location of all final states of a material having reached perfect plasticity may be defined by a curve. For sands, it is very difficult or impossible to obtain this line of isotropic consolidation. Nevertheless, for a given sand, we can evaluate its capacity for contraction or dilation from its initial state $(e_0 - p'_0 - q_0)$ and its position compared to the critical surface.

8) The range of reversible deviatoric behavior of soils is very small. During cyclic loading, we find that for larger amplitudes of shear strain, the hysteresis loops of stress-strain are changed at each cycle, either due to the densification of the material in drained conditions or because of changes in pore pressure in undrained conditions. Classical (G-γ) and (D-γ) curves are essentially explained by the creation of irreversible strain, which leads to phenomena such as liquefaction under undrained conditions.

a) the level of shear plastic deformation is an indicator of the type of behavior which has a major influence on cyclic properties and, therefore, seismic analyses. It varies greatly from place to place,

b) experimental results show that when the cyclic shear strain is less than 10^{-5} approximately ($10^{-3}\%$), the response of the soil is almost reversible and the non-linear elasticity related to the effects of confinement describes the behavior fairly well,

c) between 10^{-5} and 10^{-3}, irreversible strains appear, but the cycles are quite stable. Deformation occurs with no significant volume change under drained conditions (pore pressure changes slightly when undrained). Here, an approach based on the equivalent linear method with hysteretic damping provides good results (computational softwares such as [SEE 72], Flush [LYS 75], Quad-4 [IDR 73] or CyberQuake [CYB 98]),

d) for cyclic deformations larger than 10^{-2}, only a cyclic elastoplastic constitutive law can provide well-grounded results, even from a qualitative point of view.

9) The intermediate principal stress plays a role in the behavior of soils and the yield surface is not regular [LAD 77].

10) In the case of unsaturated soils, suction plays an important role in the mechanical behavior. The analysis of the behavior of unsaturated soils under various stress paths (suction controlled or total stress controlled) leads to the following findings:

a) in drainage paths under constant total stress, for both the pastes fabricated by mixing clay with water and the normally consolidated soils, the material becomes rigid when the suction reaches a certain value (Figure 9.2),

b) this rigidity is very similar to that observed during a wetting or an unloading,

c) this stiffening of unsaturated soil is also present during the suction controlled consolidation paths,

d) during suction imposed triaxial tests, the final points show an *apparent cohesion*. This apparent cohesion varies non linearly with suction and reaches an asymptote at high suctions.

11) The role played by time depends on the soil type and the target application. We can consider that in the field of earthquake engineering, the frequency factor has a rather low impact, although it can modify strain rates. Obviously, the time factor in the generation of pore pressures and its dissipation (consolidation) is taken into account by the use of a coupled calculation. However, some clays or rocks exhibit behavior that is dependent on the speed of loading. Viscous behavior has been observed even for certain sands. The effect of viscosity of the soil may lead to changes in the stability of underground construction works over time or with modification of cyclical behavior. If this is the case, the model must take into account the strain rate parameter. Tatsuoka *et al.* [TAT 08] describe four types of viscosity observed in geomaterials and recommend that a mathematical model simulating these different behaviors for triaxial drained tests be used.

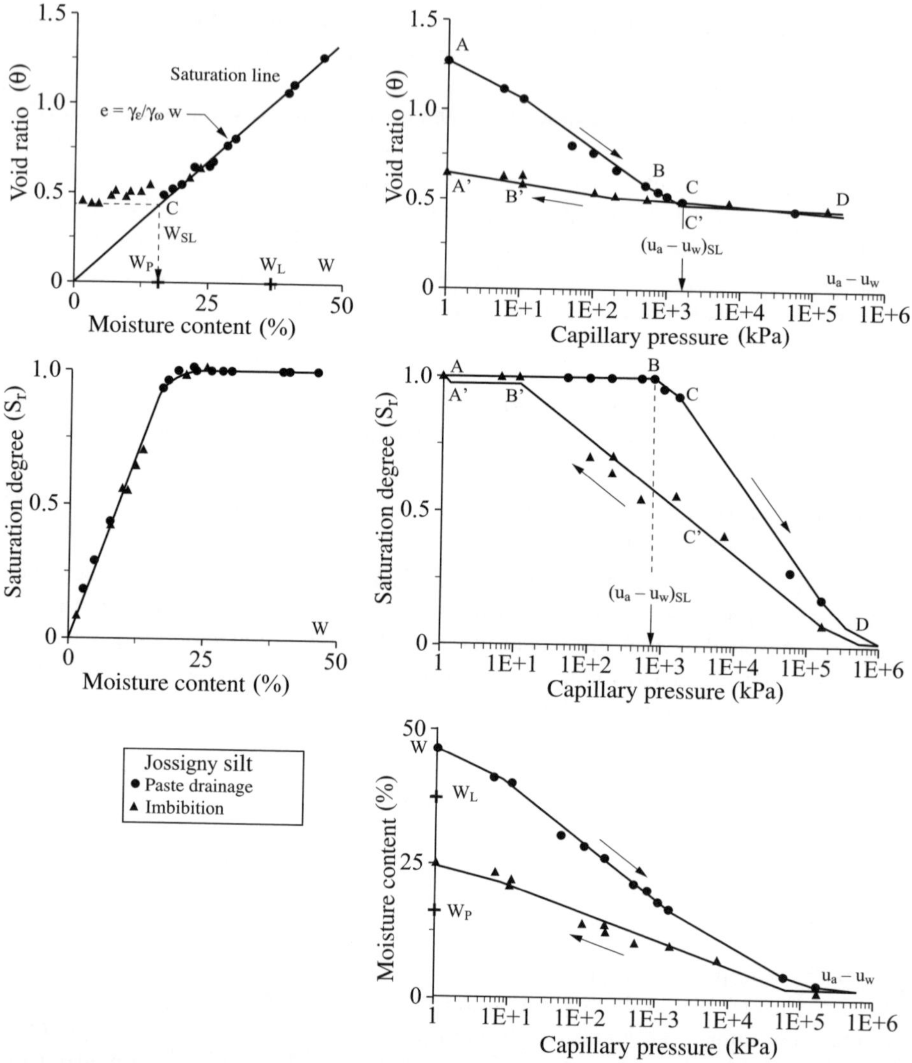

Figure 9.2. *A cycle of drainage-imbibition on normally consolidated Jossigny silt [IND 91]*

12) Other aspects, such as changes in behavior due to time, known as aging, are hereby acknowledged, but fall beyond the scope of this chapter.

According to experimental results [BIA 08], certain features, independent of the nature of constituents, can be distinguished as common to the behavior of soils. A constitutive model valid for a wide range of materials can thus be developed. It is possible to speak of *reference behavior* (Figure 9.1), which J. Biarez divided into two states:

– the pre-peak behavior;

– the post-peak behavior.

The post-peak behavior, which determines the strength of a material, depends on test conditions. When experimental precautions are taken to obtain a homogeneous state of stresses and/or strains to avoid strain localization, critical state or perfect plasticity can presumably be reached, as represented by two lines in the $(q-p')$ and $(e-\log p')$ planes with M and λ slopes, respectively. When the test is no longer homogeneous, kinematic discontinuities will appear. The pre-peak behavior is governed by an additional parameter which shows the distance compared to the critical state. For granular materials, this additional parameter represents, simultaneously, the role of density and the confinement on behavior. For clays, it describes the degree of over-consolidation and confinement. The space is thus decomposed into a zone of dilative behavior and an area of contractive behavior. Friction mobilization is thus connected to the contractive or dilative behavior of soil.

9.3.2. *Constitutive models used for computation*

Throughout its history, soil behavior has been modeled under different approaches. The earliest models were based on theories of elasticity and plasticity. From a practical point of view, seeing that it is easy to use and to apply, despite its shortcomings, the non-linear model developed by Duncan and Chang [DUN 70] was one of the most widely used models for certain applications, such as earth dams. However, unable to take into account such aspects as volumetric or cyclic behavior, its use very quickly became limited to the estimation of settlement.

The theory of plasticity provided a solid and practical framework for developing constitutive models. From Drucker's work in 1949 [DRU 49], which established the concepts of modern plasticity, many models have appeared. The first important contribution in the field of soil mechanics was inclining the yield surface from the origin downward and across the mean stress axis and, then, the introduction of the concepts of critical state and density-hardening. Gradually, it was found to be necessary to introduce deviatoric strain hardening and non-associated flow laws.

The need to generate plastic deformation at the beginning of loading resulted in generalized plasticity models, such as the concept of *bounding surface* [DAF 82, PAS 85]. Moreover, the introduction of multiple yield surfaces and kinematic hardening [MRO 67] increased the ability of models to describe cyclical phenomena. Thus, multisurface models have emerged [AUB 82, HUJ 79, MRO 81] and the theoretical elastic domain has decreased. The particularity of the elastoplastic models is that the elastoplastic tensor connecting the elastoplastic strain rate (increment) to the stress rate (increment) is dependent on the direction of loading. In the viscoplastic models, this condition, based on Perzyna's theory [PER 66], has been eliminated. This constraint is also absent in incremental and hypoplastic models [DAR 82, DAF 86]. In these incremental models, the different terms of the elastoplastic tangent tensor are defined from the test results. Their generalization to load paths, different from those for which the terms of the tensor were evaluated, is indirect. Although the behavior when loading and unloading can be modeled, they are not yet adapted to cyclic loading [KOL 00]. A vast literature on different types of models for soil behavior and their formulations can be found in the following documents: [CAM 00, HIC 02, MAN 87, MES 93, ZIE 99].

The models presented in this chapter belong to the first category and are situated within the framework of incremental plasticity. This choice is further justified by the fact that these models were destined to be integrated into computing finite element software entirely suited to this type of modeling. Moreover, since the displacements are part of the unknown variables of the system to be solved, we must choose a formulation based on the imposed strains. We proceed now to the mathematical formulation of a generic, simplified law that contains all the important features of the behavior described above.

9.3.3. *Simplified model*

As previously mentioned, the mean stress and shear (or deviatoric stress) are factors that govern the behavior of soils. Very often axisymmetrical triaxial tests, easy to perform, are used to characterize the behavior of geomaterials. Constitutive laws of isotropic materials are formulated in terms of invariants such as the mean stress p', the deviator stress q, the compressibility ϵ_v and deviatoric strain $\underline{\underline{\overline{\epsilon}}}$ or deviatoric strains ϵ given by

$$p' = \mathrm{tr}(\underline{\underline{\sigma}}')/3; \quad \underline{\underline{s}}' = \underline{\underline{\sigma}}' - p'\underline{\underline{I}}; \quad q = (\tfrac{3}{2}\mathrm{tr}(\underline{\underline{s}}' \otimes \underline{\underline{s}}'))^{1/2}$$

$$\varepsilon_v = \mathrm{tr}(\underline{\underline{\varepsilon}}); \quad \underline{\underline{\overline{\varepsilon}}} = \underline{\underline{\varepsilon}} - \tfrac{\varepsilon_v}{3}\underline{\underline{I}}; \quad \epsilon = (\tfrac{2}{3}\mathrm{tr}(\underline{\underline{\overline{\varepsilon}}} \otimes \underline{\underline{\overline{\varepsilon}}}))^{1/2}$$

The behavior of soils is also dependent on the third invariant, but as we shall see in the formalism of the multimechanism model we have adopted, even though the stress orientation has been taken into account, the third invariant does not come into play explicitly. Given the fact that during the triaxial test, the two quantities p_0 and q evolve, it is difficult to specify the role played by each factor. From a pedagogical point of view, it would be more convenient to reason in terms of normal stress and shear stress. Unfortunately, the laboratory tests used to study these quantities are often difficult to perform, or they are tests in which the stress and strain fields are not homogeneous, hence unsuitable for establishing constitutive laws.

Before we proceed to a presentation of the multimechanism model, and for the sake of simplicity and pedagogical clarity, we will discuss the isotropic elastoplastic model formulation which is only a generalization of the Cam Clay model. This comes down to writing the model for an axisymmetrical triaxial test.

The main features of the simplified model are essentially those stated in the previous section. In what follows, we will describe how these aspects are formulated under the theory of incremental plasticity:

1) We write the constitutive law in terms of Terzaghi's effective stresses [TER 67] in the case of saturated soils. We will see in section 9.3.5, the generalization of the effective stress concept in unsaturated soils and its role in describing the behavior of partially saturated soils.

2) Using the incremental theory of plasticity:

a) we assume that the strain rate (or increment) can be decomposed into an elastic part $\underline{\underline{\dot{\varepsilon}}}^e$ and a plastic part $\underline{\underline{\dot{\varepsilon}}}^p$:

$$\underline{\underline{\dot{\varepsilon}}} = \underline{\underline{\dot{\varepsilon}}}^e + \underline{\underline{\dot{\varepsilon}}}^p \qquad [9.1]$$

b) the stress tensor rate depends on the rate of elastic strains:

$$\underline{\underline{\dot{\sigma}}}' = \lambda \dot{\varepsilon}_v^e \underline{\underline{I}} + 2\mu \underline{\underline{\dot{\varepsilon}}}^e \qquad [9.2]$$

or:

$$\dot{p}' = K \cdot \dot{\varepsilon}_v^e \qquad [9.3]$$

$$\underline{\underline{\dot{s}}} = \mu \cdot \underline{\underline{\dot{\varepsilon}}}^e \qquad [9.4]$$

where λ, $\mu = G$, K, E and ν are elasticity parameters, so that:

$$\lambda = \frac{\nu E}{(1+\nu)(1-2\nu)} \qquad [9.5]$$

$$\mu = G = \frac{E}{2(1+\nu)} \qquad [9.6]$$

$$K = \frac{E}{3(1-2\nu)} \qquad [9.7]$$

3) The role played by the confinement is multiple:

a) in the elastic range, the results obtained by applying the Hertz theory to assemblies of spherical beads under elastic isotropic loading can be generalized and applied to granular materials [BIA 94]. This theory, which takes into account the increase of the contact surface during loading, results in a non-linear relationship between stresses and strain given by a stiffness modulus of the form:

$$E = \alpha(p')^n \qquad [9.8]$$

where α is a coefficient, which depends on the elastic properties of the spherical particles and the form of assembly. The exponent n that determines the degree of non-linearity is equal to $1/3$. Hardin [HAR 78] generalized this theory to the case of non-isotropic loadings, studying the shear wave propagation velocity. Having observed a linear variation between these velocities and the void ratio (e), he proposed the following form for the elasticity modulus:

$$E = A\frac{(B-e)^2}{1+e}(\sigma'_0)^n \qquad [9.9]$$

where A and B are two constants to be calibrated experimentally for each material. σ'_0 is the normal stress in the direction in which the measurement was made. The modulus is thus independent of the normal stresses in orthogonal directions. Tatsuoka [TAT 01] completed this result and generalized it to anisotropic materials:

$$\frac{E_v}{E_h} = \frac{(E_v)_0}{(E_h)_0}\left(\frac{\sigma'_v}{\sigma'_h}\right)^n \qquad [9.10]$$

where the indices v and h denote the horizontal and vertical directions. $(E_v)_0$ and $(E_h)_0$ are the vertical and horizontal Young's moduli obtained for $\sigma'_h = \sigma'_v$. He also found that the Poisson's ratio increases with value of $\frac{\sigma'_v}{\sigma'_h}$. Experience has shown that the coefficient n of relationship [9.8] is rather close to 0.5 for sands and can reach higher values for clays [BIA 94]. For a non-linear elastic law to be thermodynamically admissible, it must not introduce any dissipation. Therefore, it should derive from a potential (specific free energy). The relationship between stress increment and strain increment can be written as

$$d\underline{\underline{\varepsilon}}^e = \left(\frac{1}{3K(p',q)} - \frac{1}{2G(p',q)}\right)\frac{\mathrm{tr}\underline{\underline{\sigma}}'}{3}\cdot\underline{\underline{I}} + \frac{1}{2G(p',q)}d\underline{\underline{\sigma}}' \qquad [9.11]$$

where the coefficients could depend on p' and q. In order to have a reversible law:

$$\frac{\partial\varepsilon^e_{rs}}{\partial\sigma_{kl}} = \frac{\partial\varepsilon^e_{kl}}{\partial\sigma_{rs}} \quad \forall(k,l) \quad \forall(r,s) \qquad [9.12]$$

this condition can be formulated as

$$\left(\frac{p'}{K^2}\frac{\partial K}{\partial q} - \frac{q}{3G^2}\frac{\partial G}{\partial p'}\right)(\sigma'_{rs}\delta_{kl} - \sigma'_{kl}\delta_{rs}) = 0 \quad \forall(k,l),\forall(r,s) \qquad [9.13]$$

For an isotropic stress field ($r = s$ and $k = l$), the above condition is satisfied. But for any stress field, it is necessary that the first term on the left vanishes. The expression, which is not thermodynamically admissible, introduces dissipation even in the elastic domain. This error may be quantified [PIC 91] and we notice that its value is very low in current applications. However, in cases where the model is subjected to a large number of loading/unloading cycles, the accumulation of this error can lead to very inaccurate results. From a practical point of view, only the dependency of Young's modulus with the mean stress is regularly considered. This non-linearity results in a stiffness that increases with depth,

b) the effect of confinement on the strength of soils is double. Its effect is manifested both at the level of perfect plasticity and intermediate behavior. This role is mainly due to the friction characteristics of the granular media and is reflected by the fact that the behavior of soils at perfect plasticity is governed by the Coulomb criterion. The yield surface has the following form:

$$f(q,p',\alpha) = q - \eta(\alpha) \cdot p' \leq 0 \qquad [9.14]$$

where η is the mobilized shear (q/p'). It reaches its final value M at perfect plasticity where

$$\mathrm{M} = \frac{6 \sin \phi'_{pp}}{3 - \sin \phi'_{pp}} \qquad [9.15]$$

In this equation, α denotes the set of hardening variables due to shear and volume change that we will specify later.

4) To introduce the influence of strain hardening in shear, it is assumed that the plastic shear strain ϵ^p–since the last load reversal for cyclic conditions–controls the mobilized strength:

$$\eta(r) = r \cdot \mathrm{M} \qquad [9.16]$$

$$r(\epsilon^p) = r_{ela} + \left\langle \left(\frac{\epsilon^p}{a + \epsilon^p} \right)^{n_p} - r_{ela} \right\rangle^+ \qquad [9.17]$$

here $< a >^+ = (a + \mid a \mid)/2$.

The first parameter of strain hardening (r) that we call the degree of mobilization of plastic shear is introduced here. It moves between its initial value (r_{ela}), which sets the size of the domain where shear does not introduce strain hardening (which may be called the elastic deviatoric or shear field), to unity at perfect plasticity. Very often, a hyperbolic evolution $(n_p = 1)$ is chosen similarly to the hyperbolic curves observed for stress-strain (the hyperbolic law of Duncan is based on this observation). This hardening can be adjusted according to experimental results. Parameter a, which is one of the key parameters of the model, controls the evolution of this strain hardening.

5) The role of density is introduced by use of a Cam Clay type relation:

$$\eta(\epsilon^p, \varepsilon_v^p) = r(\epsilon^p) \cdot F(p', p_c(\varepsilon_v^p)) \cdot \mathrm{M} \qquad [9.18]$$

$$F(p', p_c(\varepsilon_v^p)) = 1 - b \cdot \ln \frac{p'}{p_c(\varepsilon_v^p)} \qquad [9.19]$$

with

$$p_c(\varepsilon_v^p) = p_{c_0} \exp(\beta \varepsilon_v^p) \qquad [9.20]$$

where p_{c_0} is the initial critical pressure corresponding to the initial void ratio, β is the plastic compressibility introducing the influence of densification or unpacking of the material on its mechanical strength. We can thus have a material that is more or

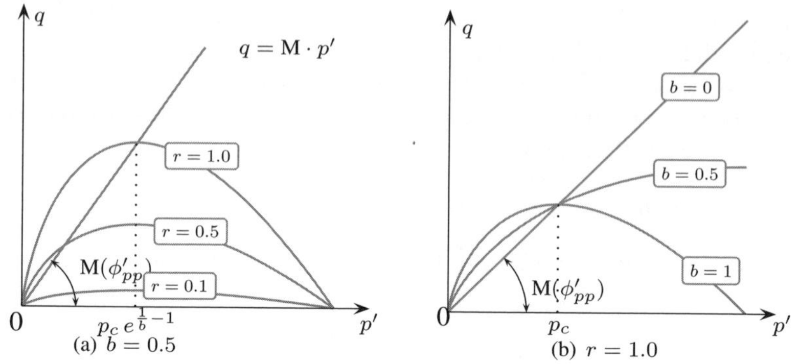

Figure 9.3. *Yield surfaces*

less resistant, as predicted by Coulomb's law, according to its position relative to the critical state line. Parameter b handles the influence of density or overconsolidation on strength and adjusts the shape of the yield surface. For clays, its value is close to 1, but for sands, it can be very low. In Figure 9.3, some yield surfaces are represented in the the $(p' - q)$ plane for different values of the parameter b and different levels of hardening (r). Note that this critical stress introduces an elastic zone in terms of normal (or spherical) stresses and strains. While the mean effective stress is less than $\exp^{1/b} \cdot p_{c_0}$, no volumetric strain may occur under spherical loading.

6) The general form of the yield surface is

$$f^s(q, p', \epsilon^p, \varepsilon_v^p) = q^s - p' \cdot r^s(\epsilon^p) \cdot F(p', p_c(\varepsilon_v^p)) \cdot \mathrm{M} \leq 0 \qquad [9.21]$$

where for a monotonic loading, $s = m$, $r^m = r$ is defined by relation [9.17] and $q^m = q$. It is found that the hardening variables introduced are r^m (or ϵ^p) and p_c (or ε_v^p). Both variables introduce isotropic hardening. Before giving their laws of evolution, we will develop the generalization of this function to the case of cyclic loading.

7) The cyclic behavior is easily developed by re-centering the yield function from the values of $\underline{\underline{s}}$ and $\underline{\underline{\varepsilon}}^p$ from the last change in the loading direction, which we denote $\underline{\underline{s}}_h$ and $\underline{\underline{\varepsilon}}_h^p$:

$$r^c(\epsilon^p) = r_{ela} + < \left(\frac{|\epsilon^p - \epsilon_h^p|}{2a + |\epsilon^p - \epsilon_h^p|} \right)^{n_p} - r_{ela} >^+ \qquad [9.22]$$

$\delta\epsilon^p = |\epsilon^p - \epsilon_h^p|$ represents the irreversible deviatoric strain since the last change of direction. Upon unloading, the evolution of work hardening is half of a monotonic loading. The yield surface of a cyclic load has the same form as in a monotonic one, but a change of reference is applied to calculate $q^s = q^c$:

$$q^c = \left(\frac{3}{2} \mathrm{tr}(\underline{\underline{s}}^c \cdot \underline{\underline{s}}^c) \right)^{1/2} \qquad [9.23]$$

with

$$\underline{\underline{s}}^c = \underline{\underline{s}} - (\underline{\underline{r}}_h - \underline{\underline{t}}_h r^c) p' \cdot F \cdot M \qquad [9.24]$$

$\underline{\underline{s}}_h$ and $\underline{\underline{r}}_h$, which give the position and size of the field at the time of change in the loading direction, are discontinuous historic parameters introducing a kinematic hardening in the model. They are given by:

$$\underline{\underline{r}}_h = \frac{\underline{\underline{s}}_h}{p'_h \cdot F(p'_h, p_c(\varepsilon^p_{vh})) \cdot \mathrm{M}} \qquad [9.25]$$

$$\underline{\underline{t}}_h = \frac{\underline{\underline{s}}_h}{q_h} \qquad [9.26]$$

We calculate them at each change of direction. As it can be seen, writing q^c in the form given in relationship [9.23] takes into account the Masing rule [MAS 26] and Bauschinger effect [LEM 09].

8) The monotonic and cyclic yield functions having been described, we now proceed to see how they can evolve, giving the evolution laws of the hardening variables:

a) it is assumed that the deviatoric plastic strain rate follows an associated law. Thus, we have:

$$\dot{\underline{\underline{\varepsilon}}}^p = \dot{\lambda}^p \underline{\underline{\Psi}} = \dot{\lambda}^p \frac{\partial f}{\partial \underline{\underline{\sigma}}} = \dot{\lambda}^p \frac{3\underline{\underline{s}}}{2q} \qquad [9.27]$$

where $\dot{\lambda}^p$ is the plastic multiplier which can be obtained from the consistency condition ($\dot{f}^s = 0$),

b) the evolution of ε^p_v, which is given by the dilatancy rule based on the relation proposed by Roscoe [ROS 58]:

$$\dot{\varepsilon^p_v} = \dot{\lambda}^p \Psi_v \qquad [9.28]$$

with:

$$\Psi_v = -\xi\Big(r(\epsilon^p)\Big)\left(\mathrm{M}'(\psi) - \frac{q}{p'}\right) \qquad [9.29]$$

where ψ is the dilatancy angle and ξ is a positive coefficient. This angle is in fact the friction angle, which corresponds to the state of null volumetric strain rate or phase transformation. This relation reveals that the behavior can be divided into two domains in the (p' – q) plane. Under the dilatancy line, where $\mathrm{M}' - \frac{q}{p'} = 0$, the material is contractant ($\dot{\varepsilon^p_v} > 0$). Above, it is dilatant ($\dot{\varepsilon_v}^p < 0$) and, on this line, the dilatancy is zero. Parameter ξ takes into account the fact that, under drained conditions, the volume change is zero until a threshold shear (ϵ_{hys}) and that, under undrained conditions, the pore pressure evolution (especially during the cycles) depends on the level of strain. It is defined by the behavior zones given below:

$$
\begin{array}{llll}
\xi(\delta\epsilon^p) = 0 & \text{if} & \delta\epsilon^p \leq \epsilon_{hys} & \text{pseudoelastic domain} \\
0 < \xi(\delta\epsilon^p) < 1 & \text{if} & \epsilon_{hys} \leq \delta\epsilon^p \leq \epsilon_{mob} & \text{hysteretic domain} \\
\xi(\delta\epsilon^p) = 1 & \text{if} & \delta\epsilon^p \geq \epsilon_{mob} & \text{plastic domain}
\end{array}
$$

It is due to these areas that we manage to find the conventional $(G - \gamma)$ curves used in soil dynamics. We can simulate loads from small deformations encountered in geophysics to those of strong earthquakes. These play an important role in monotonic behavior. The parameter ξ must be identified based on experimental observations. The relationship originally proposed by Hujeux [HUJ 85] is

$$
\xi(r) = \left(\frac{r - r_{hys}}{r_{hys} - r_{mob}}\right)^{m_p} \quad \text{for} \quad r_{hys} \leq r \leq r_{mob} \tag{9.30}
$$

In the case where $b = 1$ and $r = 1$, the flow rule is an associated one.

c) while studying the evolution of shear stress under constant p', we obtain the following relation:

$$
\dot{q} = \frac{p' \cdot F \cdot \mathrm{M}(1-r)^2}{a}\dot{\epsilon}^p + p' \cdot r \cdot \mathrm{M} \cdot \dot{F} \tag{9.31}
$$

The following observations may be made:

- if we assume that F = constant at the beginning of loading (which is justified as long as $\epsilon^p < \epsilon_{hys}$), the a parameter adjusts the initial slope of the response in the $(\epsilon^p$ - $q)$ plane,

- relations [9.19] and [9.20] show that, in this case,

$$
\dot{F} = b \cdot \beta \cdot \dot{\varepsilon}_v^p \tag{9.32}
$$

we can see that it is possible to obtain a positive work hardening ($\dot{q} > 0$) or a negative one ($\dot{q} < 0$) following the sign of $\dot{\varepsilon}_v^p$. Given the signs of different terms of the relation [9.31], the stress softening cannot occur unless when the material dilates ($\dot{\varepsilon}_v^p < 0$ and $\dot{F} < 0$),

- in the particular case where M = M', the null dilatancy line (characteristic line) is the same as the critical state line. We thus encounter the properties of the Cam Clay model. In this case, if the strains are high, we should verify that:

$$
q = \mathrm{M} \cdot p' \tag{9.33}
$$

which implies that $F = 1$ and, therefore, $p' = p_c$. The material has reached its critical stress or void ratio,

- knowing that it is very difficult to experimentally reach critical state because strain localization can occur when the state of stress and strain is no longer homogeneous in the sample, if we assume that $\phi'_{pp} = \psi$, we can find the friction angle at perfect plasticity from the state where the variation of volume changes sign (Figure 9.4).

NOTE: We should not confuse the angle of dilatancy ψ which we have used and which can often be equal to ϕ'_{pp} and the dilatancy which is the difference in volume between

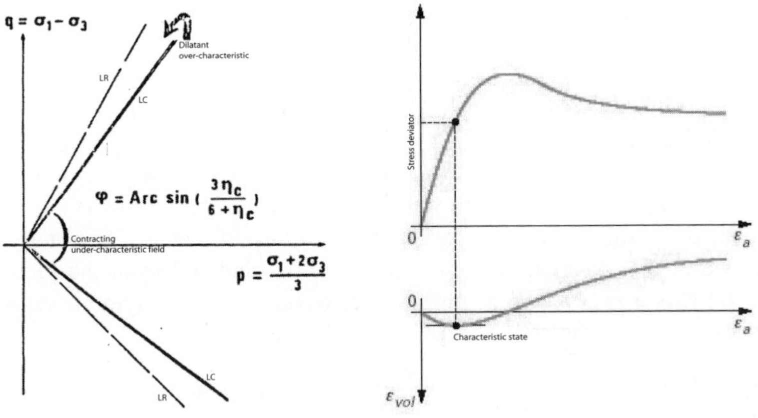

(a) (b)

Figure 9.4. *(a) Definition of the dilatancy angle and the characteristic line [LUO 80]; (b) signification of the characteristic state [MAG 91]*

a given state and the critical state line or the state of zero volume change that some researchers use in soil mechanics. In this case, we can introduce another angle, such as:

$$\begin{aligned} \mathrm{M}* &= \mathrm{M} \cdot p' - q \\ &= \mathrm{M} \cdot p' - \mathrm{M} \cdot p' \cdot F \cdot r \\ &= (1 - F \cdot r)\mathrm{M} \cdot p' \end{aligned} \quad [9.34]$$

It has been found that this dilatancy parameter integrates both the friction angle at perfect plasticity and the state of the material (its position with respect to critical state line). In this way, the role played by both the confinement and density is assembled in function F. In this chapter, we do not use the latter definition of dilatancy unless explicitly specified;

d) the evolution of r is obtained by deriving relations [9.17] or [9.22] with respect to time:

$$\dot{r}^s = \dot{\lambda}^p \frac{n_p (r^s)^{\frac{n_p - 1}{n_p}}}{a^s} \left(1 - (r^s)^{1/n_p}\right)^2 \quad [9.35]$$

The parameter a^s controls changes in r^s (or ϵ^p). This is an important parameter of the model and adjusts the plastic stiffness. Its value is double during cyclic loading ($a^m = a$ and $a^s = 2a$). It is easier to choose a constant value for a, but the parameter may also evolve according to shear deformations. Hujeux [HUJ 85] proposes the following relation which allows the two scenarios to arise, depending on the choice of a_1 and a_2:

$$a = a_1 + \xi(r)(a_1 - a_2) \qquad [9.36]$$

1) We note that this deviatoric mechanism alone cannot take into account the isotropic behavior of soils. For example, an isotropic consolidation path cannot be modeled by deviatoric mechanisms alone, unless it is in critical condition. The experimental results show that the points representing samples at a state of isotropic consolidation and those representing the samples at the perfect plasticity state in a triaxial test are positioned on two parallel lines distanced d in the $(e \; - \; \log p')$ plane and that we do not have $p' \; = \; \exp^{1/b} p_c$ which is obtained from the yield surface for $q \; = \; 0$. This aspect is taken into account by introducing an additional isotropic mechanism. The yield function is written as follows:

$$f^s_{iso}(p', r_{iso}(\varepsilon^p_v), p_c(\varepsilon^p_v)) = |p'^s| - d \cdot r^s_{iso} \cdot p_c(\varepsilon^p_v) \leq 0 \qquad [9.37]$$

This implies that in the $(p' \; - \; q)$ plane, the axis p' is crossed by a vertical line whose position changes with ε^p_v. In this mechanism: $p'^m = p'$, and $p'^c = p'^h$ with

$$p'_h \;\; = \;\; p' - r_{isoh} \cdot d \cdot p_c \qquad [9.38]$$

$$r_{isoh} \;\; = \;\; \frac{p'}{d \cdot p_c} \qquad [9.39]$$

The dilatancy law is associated and, consequently, the evolution of the degree of mobilization of the mechanism (r_{iso}) is given by

$$\dot{r}^s_{iso} = \dot{\varepsilon}^p_{v\;iso} \frac{(1 - r^s_{iso})^2}{c^s(p_c/p_{ref})} \qquad [9.40]$$

where $\dot{\varepsilon}^p_{v_{iso}}$ is the plastic volume variation due to this mechanism. Parameter c^s controls the mobilization of this mechanism, depending on whether it is a monotonic (c^m) or cyclic (c^c) one.

In the previous model, we have just summarized a way to take into account the main characteristics of soil behavior observed in the course of oedometric and triaxial tests. These formulations do not take into account the role of intermediate stress and initial or induced anisotropy. To describe the three-dimensional behavior of soils, the model should be generalized. Multimechanism plasticity is a powerful tool that can integrate these aspects.

9.3.4. *Generalizing the simplified model*

Several situations can be imagined. To better illustrate the mechanisms that govern the behavior in each situation, we introduce a tensorial projection operator, denoted $\underline{\underline{O}}_\Delta$ acting on a vector $\underline{v}$ as follows:

$$\underline{v}_\Delta = \underline{\underline{O}}_\Delta \cdot \underline{v} \qquad [9.41]$$

One part of tensor $\underline{\underline{\sigma}}$ at point $\underline{x}$ can be obtained by a double projection:

$$\underline{v}^t \cdot \underline{\underline{\sigma}}_{\Delta\Delta} \cdot \underline{w} = \underline{v}^t_\Delta \cdot \underline{\underline{\sigma}} \cdot \underline{w}_\Delta \qquad [9.42]$$

which gives

$$\underline{\underline{\sigma}}_{\Delta\Delta} = \underline{\underline{O}}^t_\Delta \cdot \underline{\underline{\sigma}} \cdot \underline{\underline{O}}_\Delta \qquad [9.43]$$

For a given plane of normal $\underline{n}$ passing through point $\underline{x}$, considering the tangent projection below:

$$\underline{\underline{O}}_\Delta = \underline{\underline{O}}_t = \underline{\underline{I}} - \underline{n} \otimes \underline{n} \qquad [9.44]$$

where $\underline{t}$ is the unit vector of the same orientation as the tangent stress in the plane perpendicular to $\underline{n}$, the stress vector and tensor can be written using the following decomposition:

– for the constraint vector:

$$\underline{\underline{\sigma}} \cdot \underline{n} = (\underline{\underline{\sigma}} \cdot \underline{n})_t + \sigma_{nn}\underline{n} \qquad [9.45]$$

with σ_{nn} being the normal stress and

$$(\underline{\underline{\sigma}} \cdot \underline{n})_t = \underline{\underline{O}}_t \cdot (\underline{\underline{\sigma}} \cdot \underline{n}) \qquad [9.46]$$

– for the stress tensor:

$$\underline{\underline{\sigma}} = \underline{\underline{\sigma}}_{tt} + (\underline{\underline{\sigma}} \cdot \underline{n})_t \otimes \underline{n} + \underline{n} \otimes (\underline{n} \cdot \underline{\underline{\sigma}})_t + \sigma_{nn}\underline{n} \otimes \underline{n} \qquad [9.47]$$

with

$$\underline{\underline{\sigma}}_{tt} = \underline{\underline{\sigma}} - \underline{\underline{\sigma}} \cdot \underline{n} \otimes \underline{n} - \underline{n} \otimes \underline{n} \cdot \underline{\underline{\sigma}} + \sigma_{nn}\underline{n} \otimes \underline{n} \qquad [9.48]$$

We can consider situations where plasticity mechanisms are written in terms of $\underline{\underline{\sigma}}_{tt}$ or in a general manner $\underline{\underline{\sigma}}_{\Delta\Delta}$. In practice, we can imagine three typical situations:

1) Stratified rocky materials, where prior to the loads applied weakness planes were already created. This could be of interest to the engineer. Further plastic deformations occur preferably along these planes distributed in space. Behavior is governed by normal and tangential stresses applied on planes of normal $\underline{n}$: σ_{nn} and σ_{nt} (or τ). The strains ε_{nn} and ε_{nt} (or γ) will also appear. The necessary number of weak planes are chosen. However, a question should be answered: are these chosen mechanisms, based on their weakness planes alone, sufficient to model the behavior of the solid or should additional isotropic mechanisms (Mohr-Coulomb or Hoek and Brown, which are often used in rock mechanics) be added, according to the microstructure of the rock mass. Therefore the mechanism is implemented, so that the user has the possibility of choosing the mechanisms in a selective manner.

2) Local tectonics and geology have imposed an initial anisotropic distribution that additional stresses induced by the construction of the structure will not change. We can imagine a summing up of the plastic mechanisms along three directions in space to be chosen based on the direction of anisotropy. This is the idea that was implemented in the multi-mechanism version of Hujeux's model [HUJ 85], in which three perpendicular directions were adopted. In this model, the mechanisms of deformation are written in terms of $\underline{\underline{\sigma}}_{\Delta\Delta}$. This decomposition into three deviatoric mechanisms has also been used by other authors to simulate three-dimensional boundary value problems [KAB 91, PRE 90]. Let $\underline{i}$, $\underline{j}$ and $\underline{k}$ be three perpendicular directions composing an orthonormal basis with unit vectors $\underline{e}_k$ with $k \in [1, 2, 3]$. The projection matrix $\underline{\underline{O}}_k$ for a plane of normal $\underline{e}_k$ can be written:

$$\underline{\underline{O}}_k = \underline{e}_{(1+mod(k,3))} \otimes \underline{e}_{(1+mod(k+1,2))} + \underline{e}_{(1+mod(k+1,3))} \otimes \underline{e}_{(1+mod(k,2))} \quad [9.49]$$

$$k \in [1, 2, 3]$$

We, therefore, have:

$$\underline{\underline{O}}_1 = \underline{e}_2 \otimes \underline{e}_3 + \underline{e}_3 \otimes \underline{e}_2 \quad [9.50]$$

$$\underline{\underline{O}}_2 = \underline{e}_1 \otimes \underline{e}_3 + \underline{e}_3 \otimes \underline{e}_1 \quad [9.51]$$

$$\underline{\underline{O}}_3 = \underline{e}_2 \otimes \underline{e}_1 + \underline{e}_1 \otimes \underline{e}_2 \quad [9.52]$$

We consider that the basic deviatoric plastic strain mechanisms, for which the hypothesis of plane strain is valid, are governed by the deviatoric stress tensor associated with each projected tensor. If we look at the tensor constraints corresponding to the proposed plane of index k, $\underline{\underline{\sigma}}_{kk}$, that for the sake of lightness, we note $\underline{\underline{\sigma}}_k$, we will have (without summation on indexes):

$$\underline{\underline{\sigma}}_k = \sigma_{ii}\underline{e}_i \otimes \underline{e}_i + \sigma_{jj}\underline{e}_j \otimes \underline{e}_j + \sigma_{ij}(\underline{e}_i \otimes \underline{e}_j + \underline{e}_j \otimes \underline{e}_i) \quad [9.53]$$

Therefore, for each mechanism we can define a mean stress p'_k, a mean strain ε^v_k, the devoiatoric stresses $\underline{\underline{s}}_k$ and the deviatoric strains $\underline{\underline{\overline{\varepsilon}}}_k$ as well as their second invariants q_k and $\overline{\epsilon}_k$.

3) The medium's memory of its geological history is short and it is its current stress state which creates the anisotropy. The three planes of projection correspond to the principal directions.

4) Frequently, surfaces appear along which displacement discontinuities can exist. These surfaces of discontinuity occur either at the interface between two materials or in the same material (such as faults or areas where deformation is localized on a thin strip). It is, therefore, very important that the volumetric constitutive law be compatible with that of the interface. Besides, experimental results show similarities between the volumetric behavior and that of interfaces. Dilatancy dependence on the average stress is a good example of this resemblance. The constitutive law of the interface is identical to the model with oriented criterion, except that instead of the strain vector components ε_n and γ, we use the displacement jump vector and its normal ($[u_n]$) and tangential ($[\underline{u}_t]$) components.

Table 9.1 summarizes the yield functions and the hardening laws for the different variants of the family of models.

9.3.5. *Mechanical behavior of non-saturated soil*

To model the mechanical behavior of a saturated soil, we define the Terzaghi's effective stress tensor [TER 67] that applies only to the skeleton:

$$\underline{\underline{\sigma}} = \underline{\underline{\sigma}}' + p_w \underline{\underline{I}} \qquad [9.54]$$

where:

– $\underline{\underline{\sigma}}$ is the Cauchy total stress tensor;

– $\underline{\underline{\sigma}}'$ is the Cauchy effective stress tensor;

– p_w is the pore pressure in the fluid; and

– $\underline{\underline{I}}$ is the second-order unit tensor.

Biot [BIO 41, BIO 57] proposed the following relation generalizing Terzaghi's postulate to materials with compressible solid phase:

$$\underline{\underline{\sigma}} = \underline{\underline{\sigma}}' + \alpha\, p_w \underline{\underline{I}} \qquad [9.55]$$

where α depends on the contrast between the stiffness of the material constituting the matrix (K_s) and that of the skeleton (K). This coefficient may vary between zero and one ($\alpha = 1 - \frac{K}{K_s}$). It is obvious that the behavior of unsaturated soil is dependent on both stresses and suction. However, the debate on whether to use one or the other of the stress tensors is still open.

The approaches used fall into two families. In the first and oldest one, our attempt is to proceed as in saturated soils. The problem, then, is how to define the effective stress tensor. Under this approach, we only partially question Terzaghi's postulate. With $\underline{\underline{\sigma}}^*$ Cauchy net stress tensor and p_g the gas phase pressure, it is assumed that the decomposition of stress tensor in the form:

$$\underline{\underline{\sigma}}^* = \underline{\underline{\sigma}} - p_g \underline{\underline{I}} = \underline{\underline{\sigma}}' + \underline{\underline{\sigma}}_f \qquad [9.56]$$

is still valid. $\underline{\underline{\sigma}}_f$ represents the portion of stresses taken up by the fluid. In the case of an unsaturated soil, we can distinguish several possible formulations of $\underline{\underline{\sigma}}_f$: the traditional formulation is to define $\underline{\underline{\sigma}}_f$ in a form close to the saturated case of the type:

$$\underline{\underline{\sigma}}_f = \overline{p}\underline{\underline{I}} \qquad [9.57]$$

where $\overline{p}$ represents the total pressure of the fluid in the media. According to the mixture theory:

$$\overline{p} = S_w p_w + S_a p_a \qquad [9.58]$$

Yield function:

$$f(\sigma_p, \sigma_q, \alpha) = \sigma_q - \mu\, \sigma_p\, F\, \rho(\alpha_d) \leq 0$$

$$F = (1 - b \ln \tfrac{\sigma_p}{\sigma_c})$$

$$\sigma_c = \sigma_{co}\, \exp(\beta\, \alpha_v)$$

$$\rho(\alpha_d) = r_{ela} + \tfrac{\alpha_d}{a + \alpha_d}$$

Flow rule:

$\dot{\alpha_v} = \dot{\lambda}^p\, \Psi_v$	with	$\Psi_v = \xi(\rho)(\mu' - \frac{\sigma_q}{\sigma_p})$
$\dot{\alpha_d} = \dot{\lambda}^p\, \Psi_d$	with	$\Psi_d = \frac{\partial f}{\partial \sigma_d}$
$\dot{\rho}(\alpha_d) = \dot{\lambda}^p\, l_\rho$	with	$l_\rho = \frac{(1-\rho)^2}{a}$

Variables				
Generic	Isotrope	Multimechanism		Interface
stresses: σ_p and σ_q	p' and q	p'_k and q_k	σ'_{nn} and $\underline{\tau}$	σ'_{nn} and $\underline{\tau}$
strains: ϵ_p and ϵ_q	ε_v and ϵ	ε_v^k and ϵ_k	ε'_{nn} and $\underline{\gamma}$	$[u_n]$ and $[\underline{u}_t]$
hardening:				
α_v	ε_v^p	${\varepsilon_v}^p$	ε'^p_{nn}	$[u_n^p]$
α_q	ϵ^p	ϵ_k^p	$\underline{\gamma}^p$	$[\underline{u}_t^p]$
$\rho(\alpha_q)$	r	r_k	r_k	r
Parameters				
μ	M	$\sin\phi'_{pp}$	$\tan\phi'_{pp}$	$\tan\phi'_{pp}$
μ'	M$'$	$\sin\psi$	$\tan\psi$	$\tan\phi$
σ_{c0}	p_c	p_c	σ_{c0}	σ_{c0}

$a = a_1 + (a_2 - a_1)\, \xi(\rho)$

$\xi(\delta\epsilon^p) = 0$	if	$\delta\epsilon^p \leq \epsilon_{hys}$	pseudo-elastic domain
$0 < \xi(\delta\epsilon^p) < 1$	if	$\epsilon_{hys} \leq \delta\epsilon^p \leq \epsilon_{mob}$	hysteretic domain
$\xi(\delta\epsilon^p) = 1$	if	$\delta\epsilon^p \geq \epsilon_{mob}$	plastic domain

Elasticity:

$$E_{max} = E_{ref} \left(\frac{\sigma_p}{\sigma'_{ref}}\right)^{n_e}$$

$$\nu = \text{cst}$$

Table 9.1. *Comparison of constitutional relations for different models of the ECP family* where S_i and p_i represent the degree of saturation and the pressure of phase i (w: water, a: air), respectively. Generally, this first approach is amended by the introduction of an additional factor, the function of S_w called the Skempton coefficient [SKE 61], and obtained experimentally, so that:

$$\underline{\underline{\sigma}}_f = B(S_w)p_w\underline{\underline{I}} \qquad [9.59]$$

Again, it is common to improve this approach by introducing experimental coefficients. The coefficient most frequently used is that of Bishop [BIS 54]:

$$\underline{\underline{\sigma}}_f = p_a\underline{\underline{I}} - \chi(S_w)P_c\underline{\underline{I}} \qquad [9.60]$$

where $P_c = p_a - p_w$ is the suction. The simplest formulation of Bishop's coefficient is

$$\chi(S_w) = S_w \qquad [9.61]$$

Experience has shown that parameter χ depends on the stress path and soil properties.

Currently, models of increasing complexity can reproduce the behavior of unsaturated soils. These models take into account different aspects of observed behavior and can be used over a long period of the structure's lifetime, if not all of it. We can divide the research on the mechanical behavior of unsaturated soils into three parts: that of the definition of effective stress, of volume change and the collapse phenomenon, and finally, of shear strength.

Historically, the first time non-saturation in the behavior of a structure was considered was through the work of Duncan and Chang [CHA 77], who calculated the behavior of a dam without explicitly taking into account the air pressure or the degree of saturation as an additional unknown variable. Under these circumstances, the modeling is done assuming a saturated or pseudo-saturated material, taking into account implicitly the reduction of the volume of air. Even though, unlike elastoplastic models, the non-linear model of Duncan and Chang does not take into account the generation of pore pressure by shear, this gap is overcome for clay by the fact that its volumetric compressibility depends mainly on the isotropic stress [POU 95]. The error regarding the pore pressure and the safety factor seems to be less than 10% in the case of clay; for sandy clay, it can be more. Ghaboussi and Kim [GHA 82] have validated this approach of variable compressibility of water by using an elastoplastic model which gives a good agreement for pressures and displacements during the construction of the Quedradona dam. These saturated or pseudo-saturated models are justified under two conditions:

– the clay is wet enough, which means that it is compacted on the wet side, so that the fluid is considered to be continuous and the principle of effective stress is applicable;

– the compressibility of the fluid is adapted to reproduce the compressibility of the mixture of water and air.

It should be noted that this approach can be used to model the construction phase. If we are interested in the other phases of a dam's lifetime, the approach is not useful. If the gas phase is continuous and the liquid phase discontinuous, or rather unbalanced in these conditions, capillarity plays an important role in the hardening of soil, essential to keep in mind during modeling. Other classes of models are the non-linear elasticity models with state surfaces. In these models, which were applied for the first time to a dam by Lloret and Alonso [LLO 85] in a partially coupled manner, the state surface defines the evolution of the degree of saturation and void ratio on a humidification path in terms of total stress and capillary pressure ($p_a - p_w$). The third category of models consists of elastoplastic models written either in terms of total or effective stress. If no hardening due to unsaturation is introduced, these models cannot be used if the gas phase is discontinuous. When the latter is continuous and the capillary pressure influences the mechanical behavior, it is necessary for the model to take it into account. The model of Alonso and Gens written in terms of total stress is a pioneering elastoplastic model obtained from the generalization of the Cam Clay model. The major drawback of these traditional approaches is that they do not describe certain phenomena observed in unsaturated soils. Based on work conducted by J. Biarez and J.M. Fleureau [BIA 93] and the microstructural model composed of particles [TAI 94], we can introduce the concept of generalized effective stress into unsaturated soils (see Chapter 6 of this volume). A capillary stress can be defined, so that the effective stress becomes:

$$\underline{\underline{\sigma}}' = \underline{\underline{\sigma}} + p_a \underline{\underline{I}} - \underline{\underline{\sigma}}_c \qquad [9.62]$$

where $\underline{\underline{\sigma}}_c$ is a function of capillary pressure as well as the density and grain-size of the material. The capillary stress, depending on the geometry of pores, is not necessarily spherical. However, for the sake of simplicity and lack of an experimentally validated model, only the spherical part of this tensor is considered and we assume that $\underline{\underline{\sigma}}_c = \pi_c \underline{\underline{I}}$. However, the same terminology has been kept to avoid confusion with the capillary pressure (P_c). A given density and particle-size, this capillary stress must comply with the conditions obtained from microstructural models and experimental observations. The deduction made from the intrinsic curves obtained for unsaturated samples [BIA 93] shows that, for a given density and particle-size, the variation of capillary stress as a function of capillary pressure reaches a plateau for high values of capillary pressure. This deduction was made by Taibi [TAI 94] in considering a microstructural model. For an assembly of deformable particles in contact, he gives the following relationship:

$$\pi_c = \pi_c^{max}(1 + \frac{3(3T - \sqrt{8 \cdot T \cdot R \cdot P_c + 9T^2})}{4R \cdot P_c}) \qquad [9.63]$$

where π_c^{max} characterizes the level of maximum capillary pressure, R is the radius of the particles and T is the superfacial tension of the air-water interface. The influence of density and of particle-size on the capillary pressure is taken into account in the parameter π_c^{max} which is written as follows:

$$\pi_c^{max} = \frac{2\pi T}{K(e)R} \qquad [9.64]$$

where $K(e)$ is a function of void ratio e given by: $K(e) = 0.32e^2 + 4.06e + 0.11$. Taibi shows that this relationship can be generalized to soils by replacing parameter R by d_{10} (maximum diameter of the particles with 10% of the total weight) which appears to be the most representative parameter. Recent results on Perafita sand [HAR 02] show that, in the presence of a large quantity of fines, parameter R may be less than d_{10} (ratio of 1/20). We also impose a continuity between the pore pressure and capillary stress at the limit of saturated and unsaturated domains.

This condition deduced from the results with a high degree of saturation allows us to use a single framework to model saturated and unsaturated geomaterials. Indeed, the same modeling tools can be used for analyzing actual cases where the state of the *in situ* material varies with hydrological and hydroclimatical conditions. For example, we give an expression of function $\pi_c(P_c)$, which ensures that the latter can be retained for a first approximation:

$$\pi_c = \pi_c^{max} \tanh(\frac{P_c - p_{c_d}}{\pi_c^{max}}) \qquad [9.65]$$

This simple mathematical relationship ensures the continuity of pore pressure and capillary stress in saturated and unsaturated cases, but may prove to be too stiff or not valid over the entire range of pressures. It must be adapted to suit the material and the range of pressures considered. Coussy and Dangla [COU 02a] extending the energy approach of non-linear poroelasticity in Biot's saturated media [BIO 41] to the case of unsaturated soils give the expression of the capillary stress which reduces to a pressure (π):

$$\pi = S_w p_w + S_a p_a - U(S_w, T) \qquad [9.66]$$

where U is the free energy stored in interfaces and T is the superficial tension.

In some studies, the existence of an effective stress is rejected and the behavior is defined by either using a state surface which connects the total stress, void ratio and capillary pressure (Bishop and Blight [BIS 63], Matyas and Radhakrishna [MAT 68], Fredlund and Morgenstern [FRE 77], Lloret and Alonso [LLO 85]) or elastoplastic models written in terms of total stress (Alonso *et al.* [ALO 90] and Toll [TOL 90]).

The debate over the use of an effective stress tensor to model the behavior of unsaturated soils remains open. However, many recent publications support the idea

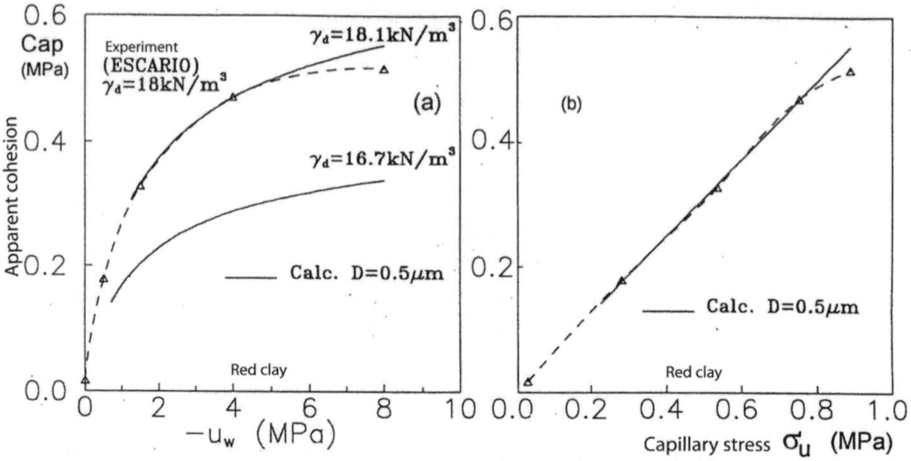

Figure 9.5. *Comparison of measured and computed apparent cohesion of a clay as a function of: (a) negative pore pressure ($-u_w$); and (b) capillary pressure [TAI 94]*

of using this notion [COU 02b, KHA 98]. Our experience shows that the choice of a generalized effective stress tensor, associated with a constitutive law taking into account the capillary effect, leads to acceptable results.

The choice of formulation in generalized effective stresses provides a natural transition between the dry, unsaturated and saturated states. Moreover, the constitutive model is obtained from extending the Hujeux model, already widely used and validated for saturated or dry soils. This also allows us to benefit from the experience already gained from model parameters and their identification. Therefore, only a few parameters dealing with the influence of non-saturation on behavior remain to be characterized.

This model is essentially governed by the influence of capillary pressure on the void ratio. Thus, changes in the capillary pressure will change the consolidation (critical) pressure. This is explicitly taken into account in the isotropic mechanism.

In terms of plasticity, this is equivalent to the expansion of the elastic domain considered by introducing a kind of isotropic capillary *cohesion* resulting from the increase of overconsolidation pressure owing to capillary forces. The isotropic yield function becomes:

$$f_{iso} = p' - d \cdot r_{iso} \cdot p_c^{ns}(\varepsilon_v^p, \pi_c) \qquad [9.67]$$

with

– p': the mean effective stress estimated from relations [9.54] or [9.62], depending on whether the gas phase is discontinuous or continuous, respectively:

$$\begin{array}{lll} p' = p - (p_w - p_a) & \text{if} & p_w - p_a > p^*_{sat} \\ p' = p - \pi_c & \text{if} & p_w - p_a \leq p^*_{sat} \end{array}$$

where p^*_{sat} is the pressure at which the meniscus forms or disappears in the soil. It represents the limit between the saturated and unsaturated behavior. It may have different values, under drainage and imbibition paths: ($p^*_{sat} = p_{desat}$ if $p_w - p_a < 0$) or imbibition ($p^*_{sat} = p_{resat}$ if $p_w - p_a > 0$);

– $p_c^{ns} = p_{c_0} \exp(\beta \varepsilon_v^p) + \Re(\pi_c)$: the critical pressure for unsaturated behavior which is a function of the plastic volumetric strain ε_v^p, plastic compressibility β, critical pressure at initial state p_{c_0} and of $\Re(\pi_c)$, which is a function of capillary stress controlling the extension of elasticity of the unsaturated soil;

– d is the parameter measuring the distance between the critical state line and the isotropic normally consolidation (virgin) state in $(e - \ln p')$ plane for a saturated soil.

NOTE: The influence of temperature on the behavior is governed by this parameter which can evolve with temperature [LAL 95, MOD 97]. As the isotropic consolidation line moves to the right side of the perfect plasticity (critical state) when the temperature (T) increases, the following relation can be considered for d: $d(T) = d_0 \exp^{-\beta\varsigma(T)}$ where ς represents the variation of the void ratio with temperature.

The deviatoric mechanisms remain unchanged except that a modified critical pressure is introduced;

$$f_k = q_k^s - p'_k \cdot r_k^s \cdot (1 - b \ln \frac{p'}{p_c^{ns}}) \cdot \sin \phi'_{pp} \qquad [9.68]$$

Therefore, the variation of the friction angle due to the nonsaturation is taken into account by p_c^{ns}, which couples the two types of mechanisms. As mentioned earlier, the isotropic mechanism is assumed to be associated as well as the deviatoric mechanisms for their deviatoric part. For the volumetric strain associated with these mechanisms, the same dilatancy rule has been used. The coupling between mechanisms is done by p_c^{ns} and leads to expansion or narrowing of the isotropic and deviatoric yield surfaces. During a thermal loading, if the stress point remains on the yield surface, as the mean effective stress is not modified, this coupling leads to increasing deviatoric strength and thermal hardening, which is even more pronounced for unsaturated soil. Assuming that the friction angle varies with temperature, this effect can be counterbalanced. The laws governing the evolution of hardening variables remain unchanged.

9.3.6. *Loading/unloading definition in plasticity*

In the case of general plasticity, the calculation of the elastoplastic yield function is done according to the consistency or compatibility condition expressed by $\dot{f} = 0$. This relationship helps to establish the evolution of the plastic multiplier $\dot{\lambda}^p$. The two following conditions can be distinguished for plastic loading or elastic unloading, respectively:

$$\left\{\begin{matrix} \dot{\lambda}^p > 0; & \dot{f} = 0, \\ \dot{f} < 0; & \dot{\lambda}^p = 0. \end{matrix}\right\}$$

These conditions can be assembled:

$$\dot{\lambda}^p \cdot \dot{f} = 0 \qquad [9.69]$$

In a general manner, five states of behavior can be distinguished (Figure 9.6) according to the expression of the yield function in terms of stresses (f) or strains($\mathcal{F}$):

	Conditions in stresses	Conditions in strains
loading *(hardening)*	$\partial_{\underline{\underline{\sigma}}} f : \underline{\underline{\dot{\sigma}}} > 0$	$\partial_{\underline{\underline{\varepsilon}}} \mathcal{F} : \underline{\underline{\dot{\varepsilon}}} > 0$
loading *(perfect plasticity)*	$\partial_{\underline{\underline{\sigma}}} f : \underline{\underline{\dot{\sigma}}} = 0$	$\partial_{\underline{\underline{\varepsilon}}} \mathcal{F} : \underline{\underline{\dot{\varepsilon}}} > 0$
loading *(softening)*	$\partial_{\underline{\underline{\sigma}}} f : \underline{\underline{\dot{\sigma}}} < 0$	$\partial_{\underline{\underline{\varepsilon}}} \mathcal{F} : \underline{\underline{\dot{\varepsilon}}} > 0$
neutral loading	$\partial_{\underline{\underline{\sigma}}} f : \underline{\underline{\dot{\sigma}}} = 0$	$\partial_{\underline{\underline{\varepsilon}}} \mathcal{F} : \underline{\underline{\dot{\varepsilon}}} = 0$
unloading	$\partial_{\underline{\underline{\sigma}}} f : \underline{\underline{\dot{\sigma}}} < 0$	$\partial_{\underline{\underline{\varepsilon}}} \mathcal{F} : \underline{\underline{\dot{\varepsilon}}} < 0$

Let us note that the conditions of loading and unloading may not be properly defined merely by conditions on stresses, whereas conditions on strains are not all ambiguous. However, as we work with yield functions defined in the stress space, we must find the relationship between $\partial_{\underline{\underline{\varepsilon}}} F$ and $\partial_{\underline{\underline{\varepsilon}}} \mathcal{F}$. With a linear elasticity assumption, we can write:

$$\underline{\underline{\dot{\sigma}}}' = \underline{\underline{\underline{\underline{C}}}}(\underline{\underline{\dot{\varepsilon}}} - \underline{\underline{\dot{\varepsilon}}}^p) \qquad [9.70]$$

In this case, the yield function in terms of stresses can be obtained from that expressed in terms of strains:

$$\mathcal{F}(\underline{\underline{\varepsilon}}, \underline{\underline{\varepsilon}}^p, \alpha) = f(\underline{\underline{\sigma}}', \underline{\underline{\varepsilon}}^p, \alpha) = f(\underline{\underline{\underline{\underline{C}}}}(\underline{\underline{\varepsilon}} - \underline{\underline{\varepsilon}}^p), \alpha) = 0 \qquad [9.71]$$

replacing $\partial_{\underline{\underline{\varepsilon}}} \mathcal{F}$ by $\underline{\underline{\Phi}} : \underline{\underline{\underline{\underline{C}}}}$ where $\underline{\underline{\Phi}} = \partial_{\underline{\underline{\sigma}}} f$. Therefore, in the case of yield functions written in terms of stresses, the following conditions on the yield function and plastic multiplier $\dot{\lambda}^p$ are required to define the various states:

$$\begin{array}{llllllll} if & f = 0; & \dot{f} = 0; & \underline{\underline{\Phi}} : \underline{\underline{\underline{\underline{C}}}} : \underline{\underline{\dot{\varepsilon}}} > 0 & and & \dot{\lambda}^p > 0 & \text{then} & \text{loading} \\ if & f = 0; & \dot{f} = 0; & \underline{\underline{\Phi}} : \underline{\underline{\underline{\underline{C}}}} : \underline{\underline{\dot{\varepsilon}}} = 0 & and & \dot{\lambda}^p = 0 & \text{then} & \text{neutral loading} \\ if & f = 0; & \dot{f} < 0; & \underline{\underline{\Phi}} : \underline{\underline{\underline{\underline{C}}}} : \underline{\underline{\dot{\varepsilon}}} < 0 & and & \dot{\lambda}^p = 0 & \text{then} & \text{unloading} \end{array}$$

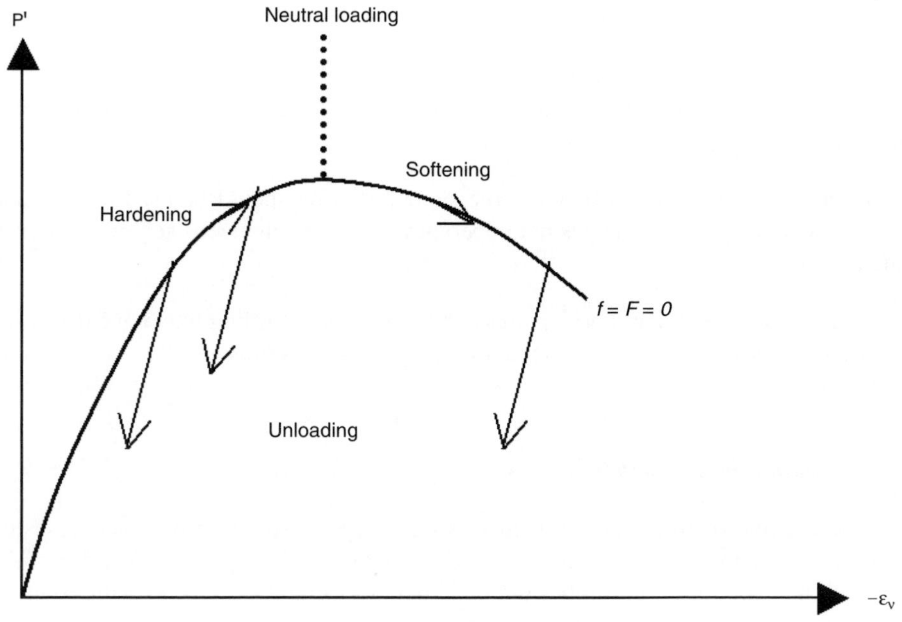

Figure 9.6. *Five states of behavior*

9.3.6.1. *Capacity of the unsaturated model to simulate the collapse phenomenon*

In unsaturated soils, the material sometimes contracts when the suction and effective stress decrease. This is apparently the argument used to reject the use of effective stress for describing the behavior of unsaturated soils. We demonstrate that these two aspects are not contradictory and the developed constitutive model is capable of taking into account this phenomenon, despite its formulation in terms of effective stress. In fact, as we have seen in the previous subsection, a decrease in stress may be accompanied by a generation of plastic strains.

The consistency condition ($\dot{f} = 0$) of the isotropic mechanism results in:

$$A\dot{\lambda}^p_{iso} = \dot{B} - \dot{C} - \dot{D} \quad [9.72]$$

where

$$\begin{aligned} A &= K - \partial_{r_{iso}} f_{iso} \cdot h_{iso} - \partial_{\varepsilon^p_v} f_{iso} \\ \dot{B} &= K \cdot \dot{\varepsilon}_v \\ \dot{C} &= \partial_{\pi_c} f_{iso} \dot{\pi}_c = -d \cdot r_{iso} \cdot \partial_{\pi_c} \Re \cdot \dot{\pi}_c \\ \dot{D} &= \partial_T f_{iso} \cdot \dot{T} = -r_{iso} \cdot dp^{ns}_c \cdot \varsigma \cdot \beta \cdot \dot{T} \end{aligned} \quad [9.73]$$

Given that A is always positive, it is clear that the loading condition ($\dot{\lambda}^p_{iso} > 0$) requires $\dot{B} - \dot{C} - \dot{D} > 0$. As can be seen under isothermal condition ($\dot{D} = 0$), during a humidification path for which $\dot{C} > 0$, material will collapse so long as condition $\dot{B} > \dot{C}$ is fulfilled. Otherwise, an unloading is detected and the material swells elastically.

From a theoretical point of view, the same approach should be used to evaluate the plastic strains due to changes in temperature or the simultaneous action of the two phenomena.

It should be noted that thermal stress or a mechanical loading may have the same effect, which is to make an unsaturated soil contract during a subsequent humidification [MOD 95].

9.3.7. *Multimechanism model*

The multimechanism approach involves the superpositioning of several plastic strain mechanisms, each assigned by its own yield function (f_k), its own flow rule $\underline{\underline{\Psi}}_k$ with its own plastic multiplier ($\dot{\lambda}^p_k$). Hardening variables can be specific to each mechanism (α_k): e.g. degrees of mobilization of mechanisms (r^s_k) or possibly ε^p_n, where the mechanisms are uncoupled, or be common to several mechanisms (η) such as plastic volumetric strain (ε^p_v) in the case of a unique critical density. The hardening variables may then evolve if a mechanism is not active. The flow rule of each mechanism can be written as follows:

$$\dot{\underline{\underline{\varepsilon}}}^p_k = \dot{\lambda}^p_k \underline{\underline{\Psi}}_k \tag{9.74}$$

where the volumetric part of the plastic strains generated by the k mechanism is given by the dilatancy law:

$$\dot{\varepsilon}^p_{v_k} = \dot{\lambda}^p_k \Psi_{v_k} \tag{9.75}$$

The following relations are also applied:

$$\dot{\lambda}^p_k \geq 0, \quad f^s_k(\underline{\underline{\sigma}}', \alpha_k, \eta) \leq 0; \quad \dot{\lambda}^p f^s_k(\underline{\underline{\sigma}}', r_k, \varepsilon^p_v) = 0$$

The rate of change for hardening variables is given by their evolution rule:

$$\dot{\alpha}_k = \dot{\lambda}^p_k L_{\alpha_k} \tag{9.76}$$

$$\dot{\eta}_k = \dot{\lambda}^p_k L_{\eta_k} (\text{in our case: } \eta_k = \varepsilon^p_v \quad \text{and} \quad L_{\eta_k} = \Psi_{vk}) \tag{9.77}$$

with

$$\eta = \sum_k \eta_k \qquad k \in M^p_{act} \tag{9.78}$$

where M^p_{act} designates the set of all active plastic mechanisms. In the set of all potentially active plastic mechanisms (M^p_{pact}), the active ones (M^p_{act}) are

$$M^p_{pact} = \{a \mid f_a = 0\} \quad [9.79]$$

$$M^p_{act} = \{b \in M^p_{pact} \mid \dot{\lambda}^p_b > 0\} \quad [9.80]$$

In the case of multimechanism plasticity, when all mechanisms are active, a total loading state is reached. In a partial loading state, only some of the mechanisms are active. The elastic behavior takes place when none of the mechanisms are activated.

The strain rate is given as:

$$\dot{\underline{\underline{\varepsilon}}} = \dot{\underline{\underline{\varepsilon}}}^e + \sum_k \dot{\underline{\underline{\varepsilon}}}^p_k \quad [9.81]$$

The active plastic multipliers are the solution to the following non-linear system of equations obtained writing the consistency condition ($\dot{f}_b = 0$) of the potentially active mechanisms ($a,\ b\ \in M^p_{pact}$):

$$\sum_a (\underline{\underline{\Phi}}_b : \underline{\underline{\underline{\underline{C}}}} : \underline{\underline{\Psi}}_a - \partial_{\alpha_b} f_b : L_{\alpha_b} \delta_{ab} - \partial_\eta f_b : L_{\eta_a}) \dot{\lambda}^p_a = \underline{\underline{\Phi}}_b : \underline{\underline{\underline{\underline{C}}}} : \dot{\underline{\underline{\varepsilon}}} \quad [9.82]$$

$$\forall a\ , b\ \in M^p_{pact}$$

9.4. Parameter identification strategy for the ECP model

The essential ingredients of computational geotechnics are geotechnical data and structure data in order to build the constitutive model and its parameters with emphasis on aspects related to the validity and justification of the model for each application. The particularity of the models we have presented is that they have been designed to describe the behavior of geomaterials under a wide range of deformations. As a result, a certain number of introduced parameters may not be into practice by certain engineers. This difficulty is compounded by the fact that the required test results are not always available. Furthermore, even when these tests are available, they may not always be usable. For example, the model is based on the concept of critical state and the parameters representing this state play a vital role. However, this critical state is very difficult to reach; if particular precautions are not taken experimentally, kinematical discontinuities may develop in the sample, especially in the case of dense sands and heavily overconsolidated clays. It is very difficult to know the position of the critical state line and its distance compared to the line of virgin consolidation in $(e - \ln p')$ space or the angle of friction at perfect plasticity. Similarly, the small size of the elastic domain for granular materials makes it very difficult to determine their elastic properties. On the other hand, despite all the experimental difficulties, if we do manage to obtain consistent and reproducible results on samples made from remolded materials prepared in the laboratory, it may still not be possible to obtain similar results from unremolded soil sampled *in situ*. Therefore, opting for the strategy of identifying model parameters, we used the results of many studies for their correlations between intrinsic characteristics of the soil, such as the elasticity modulus, or those related to the critical state and other characteristics often used in soil mechanics to classify soils, such as Atterberg limits and the consistency index for clays and characteristics related to the grain-size distribution for sands. Although these correlations help identify parameters, a good knowledge of the physics contained in the model is absolutely indispensable.

Given the non-linear character of the latter and the high number of parameters, we are not sure to obtain a single set of parameters after calibration, especially if only one type of stress path is considered. Using correlations and understanding the physical model and its parameters is the only way to ensure proper calibration. To do this, we must add that the mechanisms are directly related to the loading path experienced by the soil. We first present the strategy of identifying parameters for remolded materials for which many experimental results are available. Moreover, several authors have published reference curves for different materials by conducting the synthesis of a large number of experimental results. In practice, different types of tests are performed in the laboratory to characterize soils. They concern essentially triaxial compression or extension or shear tests under drained or undrained conditions. Resonant column tests are performed for geotechnical earthquake engineering applications. Biarez and colleagues propose reference curves for Hostun sand, depending on its density, and also clays under monotonic loading. For cyclic behavior, the curves given by Vucetic and Dobry [VUC 91] for clays and those of Seed and Idris [SEE 70] for sands are widely used. In presenting the parameters' identification strategy, we try to be as general as possible. Where the parameters must be identified by curve-fitting, the reference curves are used. This strategy will then be applied to soils of well-studied sites to justify its efficacy for the study of real problems.

9.4.1. *Classification and identification of the ECP model parameters*

In this study, we focused on aspects of parameter identification for the simulation of the behavior under cyclic loads. In fact, although this type of loading is routinely applied in earthquake engineering, geo-structures regularly undergo loading and unloading. As an example, we can consider an earth dam during its life cycle which, after construction, traverses several cycles of water impounding and drainage.

As we shall see in section 9.5.2, the objective is to identify the parameters using a minimum amount of laboratory and field data, and to make the most of existing correlations. We thus seek a set of consistent parameters for sandy and clayey soils using parameters such as Atterberg limits for clays and relative density and the grain-size distribution for sand.

The model parameters can be classified differently depending on the objective. As the model is a plastic one, a preliminary classification of obvious parameters gives:

– elastic parameters:	E, ν, n_e, P_{ref}
– plastic and critical state parameters:	ϕ'_{pp}, β, b, d
– hardening parameters:	
	$\psi, a_1, a_2, n_p, c, c_{cyc}$
	$r_{ela}, r_{hys}, r_{mbl}, r_{iso}, m_p, n_p$
– initial state parameters:	p_{c0}

to these parameters we should add those which govern the unsaturated behavior:

– non-saturation parameters: $p^*_{sat}, \kappa_{ns}, d_{10}$

We classify model parameters according to the role they play in soil behavior. The parameters can be divided into three categories: those that contribute directly to the stiffness (stress strain), those which are associated to a state independent of the loading history (τ/σ'_n, q/p' or q_k/p'_{l_0} for instance) and, finally, those which govern the evolution of hardening variables. For the identification process, we suggest a third type of classification, according to the method of their determination (see Table 9.2).

There are parameters that are directly measured from the quantitative analysis of observed results, such as parameters of elasticity and perfect plasticity. There are thus parameters which are not directly measurable and which must be obtained through calibration. Let us note that the equivalent parameters (r_{ela},r_{hys}) and ($\gamma_{ela}, \gamma_{hys}$) because of the parameters (a_1,a_2) belong to the not directly measurable and directly measurable categories, respectively. Parameter p_{c0} defining the initial state, depending on the initial density for sands and the degree of overconsolidation for clays is directly measurable. There are correlations in the literature which permit all the directly measurable parameters to be quantified. Biarez and Hicher [BIA 94] provide a summary of the major directly measurable soil parameters. They have classified parameters governing the behavior of soils into two categories:

	Stiffness	State	Hardening
Directly measurable	Elasticity (V_s, V_p) (G, K) (E, ν) n_e Plasticity β	ϕ'_{pp}, ψ d p_{c0} or σ_{c0}	$\gamma_{el}, \gamma_{hys}$
Not directly measurable	a_1, a_2	b	$r_{ela}, r_{hys}, r_{mbl}, \gamma_{mbl}$ n_p, m_p r_{iso}, c, c_{cyc}

Table 9.2. *Classification of the ECP's constitutive model parameters*

I) parameters independent of the assembly of grains:

1) geometric parameters:

a) granulometric description: d_{10}, d_{60}/d_{10} and F (% < 0.8 mm), $\cdots$

b) shape of grains: angular or rounded,

c) texture of grain surface;

2) mechanical parameters (independent of loading history):

a) perfect plasticity: ϕ'_{pp}, one point and the slope of critical state line in the $(e \; - \; \log p')$ plane: (C_c and $\Gamma(p_\Gamma)$) or two points of the line: $(e_{max}, p_{c_{max}})$ and $(e_{min}, p_{c_{min}})$,

b) normally consolidated behavior: C_c and C_d (slope of a p'=cst test in $(e \; - \; \log q)$) plane;

II) parameters depending on the grain assembly:

1) geometrical arrangement: e and the tangent planes of contact between grains;

2) history of loading:

a) normally consolidated: $\underline{\underline{\sigma}}'_c$ (p'_{ic} isotrope, σ'_{vc} oedometer),

b) overconsolidated: OCR.

In addition to these parameters, which describe the state of a soil, the state of anisotropy and the applied stresses should be mentioned.

The parameters of unsaturated behavior depend mostly on the arrangement and are classified in the latter category. We present the strategy developed for both clays and sands:

1) case of clays: Atterberg limits are parameters that are very easy to obtain and they have great significance in the behavior of clayey soils [BIA 72, BAR 97, LAM 79]. Therefore, the limit of liquidity (w_L) and plasticity index ($I_p \;=\; w_L \;-\; w_P$) are very useful for determining the parameters. Furthermore, existing correlations

between these two quantities reduce the correlation parameters to a single one [BIA 72, CAS 58, SKE 53]. Chapter 5 in this volume provides interesting comparisons on these correlations for different clays;

2) case of sands: the relative density $D_r = \frac{e_{max}-e}{e_{max}-e_{min}}$ or the density index $I_D = 1 - D_r$ and the grain-size distribution together with the shape of grains are basic parameters from which correlations are established for sands [BIA 94].

9.4.2. *Directly measurable parameters*

9.4.2.1. *Elastic parameters*

9.4.2.1.1. Clay

We can use the results of geophysical measurements; however, there are correlations for the shear modulus of clays that link this parameter to the void ratio (e), the state of overconsolidation (OCR) and mean stress (p'_0). Certain authors suggest using the plasticity index I_p and consistency index ($I_c = \frac{w_P-w}{w_L-w_P}$). As the void ratio and the plasticity index in turn are correlated, these two approaches are identical. Seed and Idriss [SEE 70] connect the shear modulus G_{max} to maximum undrained strength (C_u), which also takes into account the influence of the initial state. A number of correlations of the literature are shown in Figure 9.7 and Tables 9.3 and 9.4.

The Poisson's ratio varies between 0.2 and 0.3. The elasticity exponent n_e varies between 0.5 and 0.9 for clays.

9.4.2.1.2. Sands

As for clay, correlations based on p'_0 and void ratio e are given for sand. Hardin and co-authors [HAR 63, HAR 68, HAR 78] give the same relationship for low plasticity clay and Monterrey sand. Exponent n_e is about 0.5 for sand. According to Brul, in unsaturated soils, the shear modulus increases as a function of suction [BRU 80]. However, we can expect that this increase is limited by the maximum capillary stress. Furthermore, Coronado [COR 05] and co-authors [COR 07] have shown that a unique relationship can be found between the evolution of the elastic modulus and the generalized effective stress for different water contents. In the Cam Clay model, a linear evolution between the void ratio and the logarithm of the mean effective stress is assumed during an isotropic unloading path (parameters κ or swelling index C_s), whereas, in our model, the elastic modulus satisfies a power law. It is only in the special case where $n_e = 1$ that the two laws are equivalent. Although the value of n_e can approach unity for some clays, an exponential increase is not reasonable for sands.

9.4.2.2. *Plastic compressibility (β)*

Experience has shown that, at the first approximation, the relationship between void ratio and mean stress at the logarithmic scale, in an isotropic compression, is a straight

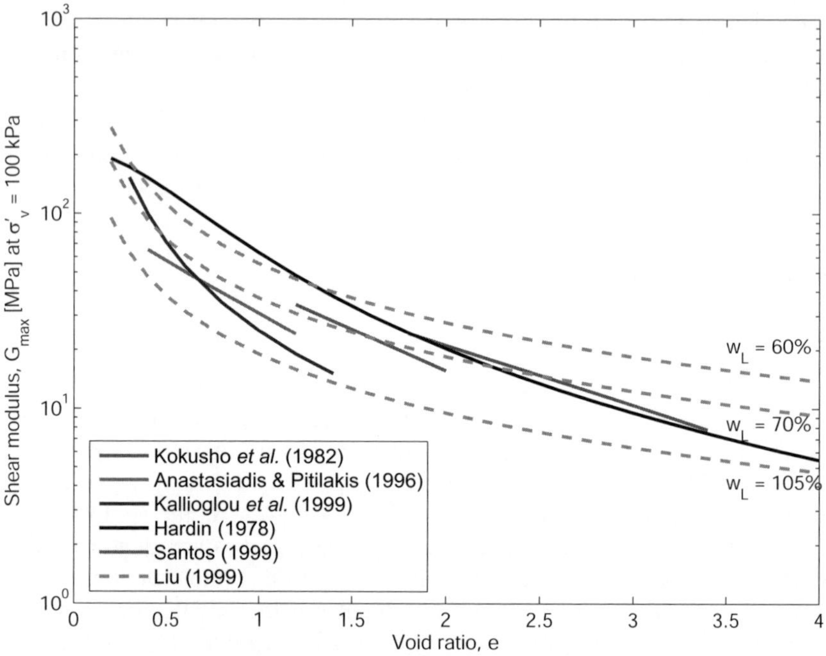

Figure 9.7. *Comparison of some empirical relations given for G_{max} at a mean pressure of 200 kPa*

line (compressibility index C_c or λ). When a normally consolidated soil passes from a state (e_0, p'_0, p_{c0}) to (e, p', p_c), the following relations are valid:

$$\varepsilon_v = \varepsilon_v^e + \varepsilon_v^p = -\frac{e-e_0}{1+e_0} = \frac{\lambda}{1+e_0}\ln\frac{p_c}{p_{c0}}$$

$$\text{if} \quad n_e = 1 \quad \text{then} \quad \frac{\lambda}{1+e_0}\ln\frac{p_c}{p_{c0}} = \frac{\kappa}{1+e_0}\ln\frac{p_c}{p_{c0}} + \frac{1}{\beta}\ln\frac{p_c}{p_{c0}}$$

$$\text{if} \quad n_e \neq 1 \quad \text{then} \quad \frac{\lambda}{1+e_0}\ln\frac{p_c}{p_{c0}} = \frac{{p'_0}^{1-n_e} - {p'}^{1-n_e}}{(1-n_e)K_{ref}^{1-n_e}} + \frac{1}{\beta}\ln\frac{p_c}{p_{c0}}$$

$$= \frac{1}{K^*}\ln\frac{p_c}{p_{c0}} + \frac{1}{\beta}\ln\frac{p_c}{p_{c0}}$$

$$= \frac{k*}{1+e_0}\ln\frac{p_c}{p_{c0}} + \frac{1}{\beta}\ln\frac{p_c}{p_{c0}}$$

where K^* and k^* are equivalent moduli giving the same elastic strain with a logarithmic law in the range of the considered pressure.

Type of soil	G_{max}	Reference	Test
Clays			
Low plasticity $e = [0.6 - 1.5]$ $\sigma'_0 = [240 - 700]$ kPa	$323\ p_a \frac{(2.97-e)^2}{1+e} (\frac{\sigma'_0}{p_a})^{0.5}$	[HAR 68]	RC
Kaolinite $I_p = 35\%$ $e = [1.1 - 1.3]$ $\sigma'_0 = [70 - 550]$ kPa	$450\ p_a \frac{(2.97-e)^2}{1+e} (\frac{\sigma'_0}{p_a})^{0.5}$	[MAR 72]	RC
Bentonite $I_p = 60\%$ $e = [1.8 - 2.4]$ $\sigma'_0 = [70 - 550]$ kPa	$45\ p_a \frac{(4.4-e)^2}{1+e} (\frac{\sigma'_0}{p_a})^{0.5}$	[MAR 72]	RC
Sands			
Toyoura $e = [0.59 - 0.81]$ $\sigma'_0 = [25 - 200]$ kPa	$90\ p_a \frac{(2.17-e)^2}{1+e} (\frac{\sigma'_0}{p_a})^{0.4}$	[IWA 78]	RC CTS
Hostun $e = [0.59 - 0.81]$ $\sigma'_0 = [25 - 200]$ kPa	$90\ p_a \frac{(2.17-e)^2}{1+e} (\frac{\sigma'_0}{p_a})^{0.4}$	[IWA 78]	RC CTS
Ottawa $e = [0.37 - 0.79]$ $\sigma'_0 = [24 - 287]$ kPa	$69\ p_a \frac{(2.17-e)^2}{1+e} (\frac{\sigma'_0}{p_a})^{0.5}$	[HAR 63]	RC CTS
Monterrey N.0 $e = [unknown]$ $\sigma'_0 = [25 - 200]$ kPa	$323\ p_a \frac{(2.973-e)^2}{1+e} (\frac{\sigma'_0}{p_a})^{0.5}$	[HAR 72] [DOB 82]	RC CTS
p_a – atmospheric pressure, G_{max} in the same unit as p_a			

Table 9.3. *Empirical relationships to estimate $G_{max} = \frac{E_{max}}{2(1+\nu)}$ in remolded soil [LOP 03a]*
Key: RC: resonant column; CTS: cyclic torsion test

9.4.2.2.1. Clay

For normally consolidated clays, the variation of void ratio e with the mean stress in an isotropic compression test or the vertical stress during an oedometeric test can be expressed in terms of the compression index C_c. Several authors have suggested correlations between this parameter and the liquid limit w_L or the plasticity index I_p given in Table 9.5 [BIA 72, TER 67]. According to Bardet [BAR 97], C_c varies from a value of 2.6 for montmorillonite down to 0.08 for "till" clays while C_s varies between 0.01 and 0.51. We note that the compression index given by equation (a) of Table 9.5 is greater than that given by relationship (b). Biarez attributes this difference

Type of soil	G_{max}	Reference	Test
Clays			
Silty (Portugal) $I_p = [40 - 42\%]$ $e = [1.32 - 2.0]$ $\sigma'_0 = [20 - 200]$ kPa	$52\, p_a \frac{(5-e)^2}{1+e} (\frac{\sigma'_0}{p_a})^{0.5}$	[SAN 99]	RC
NC alluvials (Japan) $I_p = [38 - 103\%]$ $e = [1.8 - 3.5]$ $\sigma'_0 = [20 - 1{,}000]$ kPa	$90\, p_a \frac{(7.32-e)^2}{1+e} (\frac{\sigma'_0}{p_a})^{0.6}$	[KOK 82]	CT
Clay and silty sand (Greece) $I_p = [5 - 66\%]$ $e = [0.36 - 1.36]$ $\sigma'_0 = [50 - 400]$ kPa	$262\, p_a e^{-1.505} (\frac{\sigma'_0}{p_a})^{0.623}$	[KAL 99]	RC
Clay $OCR < 3$ (Mexico) $I_p = [91 - 237\%]$ $e = [4.0 - 9.37]$ $\sigma'_0 = [100 - 350]$ kPa	$122\, p_a (\frac{\sigma'_0}{p_a})^{0.82} (\frac{1}{I_P - I_c})^{(I_P - I_c)}$	[ROM 95]	RC
Hard clays (Greece) $e = [0.4 - 1.10]$ $\sigma'_0 = [20 - 200]$ kPa	$80\, p_a \frac{(3.78-e)^2}{1+e} (\frac{\sigma'_0}{p_a})^{0.49}$	[ANA 96]	RC
Soft clays (Greece) $e = [0.6 - 1.5]$ $\sigma'_0 = [20 - 200]$ kPa	$70\, p_a \frac{1}{(0.3+0.7e)^2} (\frac{\sigma'_0}{p_a})^{0.88}$	[ANA 96]	RC
p_a = atmospheric pressure, G_{max} in the same unit as p_a			

Table 9.4. *Empirical relations for estimation of G_{max} of natural soils [LOP 03a]*
Key: RC: Resonant column;
CTS: Cyclic torsion test;
CT: Cyclic triaxial test

	Compressibility index	Swelling index	Reference
(a)	$C_c = 0.009(w_L - 10)$	$C_s = C_c/4$	[TER 67]
(b)	$C_c = 0.009(w_L - 13)$	$C_s = C_c/4$	[BIA 72]
(c)	$C_c = I_p/74$	$C_s = I_p/370$	[WRO 78]

Table 9.5. *Compressibility index, Swelling index and reference*

to the overconsolidated character of London clay samples tested by Skempton. Besides, for normally consolidated clays, Biarez and Favre [BIA 72] give indicative values of 7 kPa and 1 MPa for vertical effective stresses to reach w_L and w_P, respectively (see Figure 9.8(a)).

9.4.2.2.2. Sands

For sands, there is a correlation between the value of C_c at the critical state and void ratios and e_{min} and e_{max} which are themselves connected to the ratio d_{60}/d_{10}. According to Grésillons *et al.* quoted by Saïm [SAI 97], when D_r varies from 0 to 1, σ'_0 varies between 100 kPa and 5 MPa (see Figure 9.8(b), and Figure 9.9). The synthesis carried out by Saïm shows that the value of 5 MPa can possibly be increased to 7 MPa. Hicher and Rahman [HIC 94] propose a relationship for determining β from e_{min}, e_{max}, $U_c = \frac{d_{60}}{d_{10}}$ and $C_z = \frac{d_{30}^2}{d_{60}.d_{10}}$:

$$\beta = 2.3(1+e)\,\frac{\beta_{norm}}{I_e} \tag{9.83}$$

with $I_e = e_{max} - e_{min}$:

$$\begin{aligned}
\beta_{norm} &= \log P'c_{max} - \log P'c_{min}\\
\log P'c_{max} &= -4\log U_c - 7.5 \cdot e_{min} + 4.75 \cdot I_e + 7.1 \pm 0.3\\
\log P'c_{min} &= -12\log C_z - 4.7 \cdot e_{max} + 9.71 \cdot I_e + 2.2 \pm 0.5
\end{aligned}$$

Wherever $n_e = 1$, the plastic compressibility modulus β of the model is obtained from the parameters $\lambda = C_c/2.3$ and the elasticity stiffness K^*, which should be estimated for the range of considered p'. We note that assuming a linear relationship in the plane $(e - \log p')$ implies a value of β which varies with the void ratio:

$$\beta = \frac{1+e_0}{\lambda - k^*} \tag{9.84}$$

where k^* is the mean slope of the elastic line in the $(e - \log p')$ plane, as defined in section 9.4.2.2. Because of the higher value of plasticity compressibility with respect to the elastic one, for an initial approximation, k^* can be neglected before λ and $\beta \approx \frac{1+e_0}{\lambda}$ can be assumed.

9.4.2.3. *Friction and dilatancy angles (ϕ'_{pp} and ψ)*

In principle, these two angles are similar (e.g. Cam Clay model) unless experimental results show otherwise.

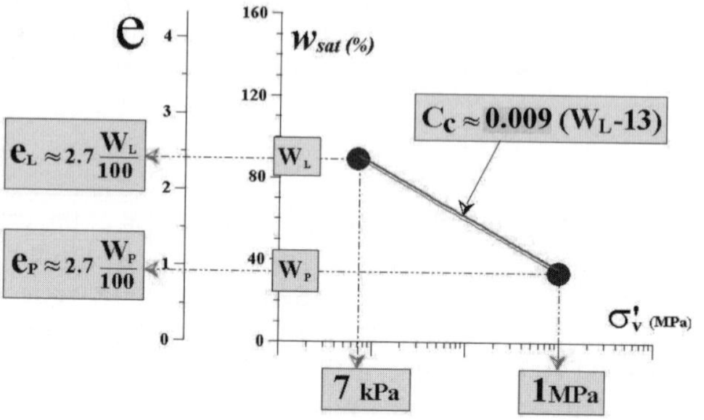

(a)

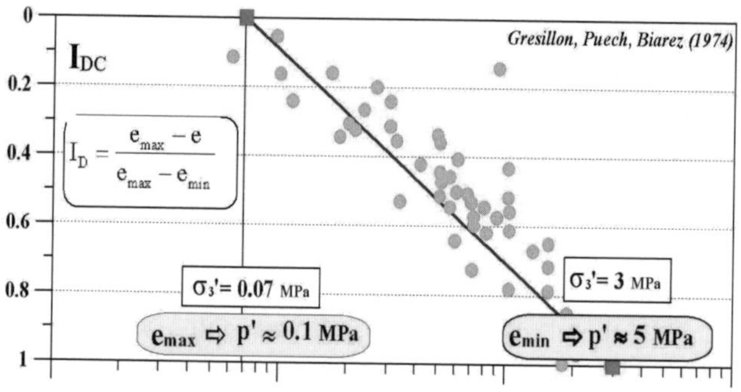

(b)

Figure 9.8. *Variation in void ratio as a function of effective stress at critical state: a) for clays [BIA 72]; and b) for sands [SAI 97]*

9.4.2.3.1. Clay

According to experimental results provided by several authors, the friction angle at critical state decreases with the plasticity index (I_p). As can be noted in Figure 9.10, for the same value of I_p, a difference of 10 degrees may exist. However, though this error may have important consequences for applications, where the deformations can be relatively great, for current applications, its impact is minor. Biarez and Favre [BIA 72] suggest a relation between ϕ'_{pp} and w_L. The friction angle varies from 32 to 20 degrees when w_L varies between 20 and 100 % (see Table 9.6).

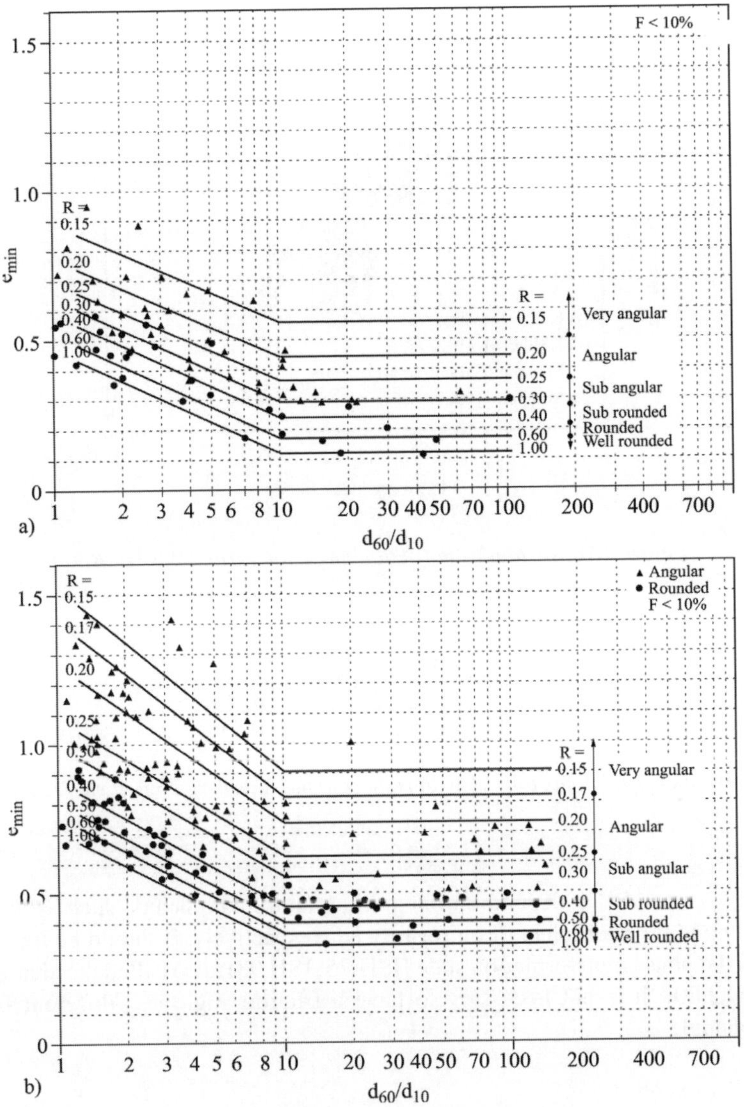

Figure 9.9. *Evolution of void ratios e_{min} (a) and e_{max} (b) with respect to the grain-size of sand [BIA 94]*

9.4.2.3.2. Sand

Acoording to [FAV 80], the friction angle ϕ'_{pp} of sands can be obtained from the following relation:

$$\phi'_{pp} = 31,5 + \varphi_D + \varphi_F + \varphi_{U_c} \quad \text{[degrees]} \qquad [9.85]$$

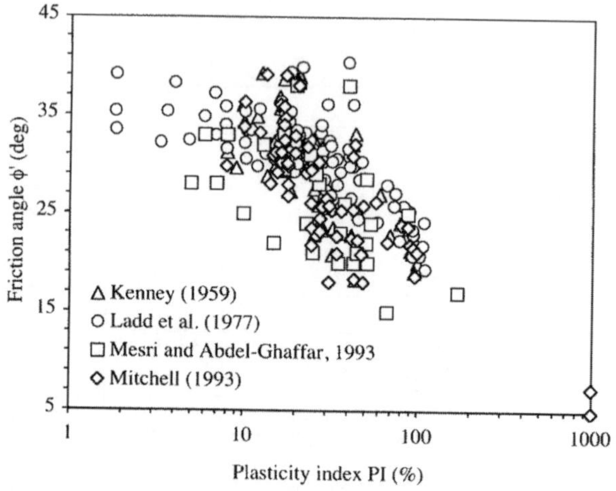

Figure 9.10. *Variations in friction angle for various I_ps [BAR 97]*

w_L [%]	20	30	40	50	60	70	80	90	100
I_p [%]	5	12	20	27	34	42	49	56	65
ϕ'_{pp}	32	30	28	26	25	24	23	22	21

Table 9.6. *Correlations for ϕ'_{pp} with respect to the Atterberg limit of clay [BIA 72]*

where φ_D, φ_F and φ_{U_c} depend on integrating nature parameters, such as grain size, angularity and the coefficient of uniformity U_c, respectively. It should be recalled that to obtain the phase transformation state [ISH 75, ISH 93], also called the characteristic state [LUO 80], it is not necessary to have a friction angle ϕ'_{pp} different from the dilatancy angle ψ.

9.4.2.4. *Parameter d*

This parameter represents the distance between isotropic consolidation and that of the perfect plasticity. Saïm [SAI 97] suggests a vertical distance of $\Delta e = 0.1$ for both sands and clays in the $e - \log p'$ plane. Knowing λ, the horizontal distance can be directly obtained. We can tie parameters d and β to parameter c_d, obtained as the slope of the loading path in the $(e - \log(1 + q^2/(M^2 p'^2)))$ plane when a constant p' test is performed on a normally consolidated material. The yield function is given as:

$$q - Mp'(1 - b \log \frac{p'}{p_c})r = 0 \qquad [9.86]$$

where

$$p_c = p_{c0} \exp^{\beta \varepsilon_v^p} \quad [9.87]$$

As the material is normally consolidated:

$$d = \frac{p'}{p_{c0}} \quad [9.88]$$

which results in:

$$\varepsilon_v^p = \frac{1}{b\beta}(\frac{q}{rMp'} - 1 + b \ln d) \quad [9.89]$$

Therefore, the volume change at constant p' depends on b, d, β and K parameters. The hardening parameters act through variable r which reaches 1 at perfect plasticity. For a given type of material (fixed b), the volume change with respect to $q/(\mathrm{M}p')$ depends on d and β parameters which are correlated to liquidity limit w_L for the clays and relative density D_r for the sands. For the elastic part, under constant p' path, K can only change with void ratio. However, the elastic part being negligible compared to plastic part, it does not influence c_d value significantly.

9.4.2.5. *Behavior domains*

γ_{ela} and γ_{hyd} can be obtained from test results. They have been the subject of numerous research studies. Vucetic [VUC 94] suggests the range of 0.01 to 0.04% for γ_{hys}. These values are based on tests performed on sands of different ($D_r \in$ [20 – 85%]), under various confining stresses $\sigma_0' \in$ [25-190 kPa] (see Figure 9.11). For clays, similar results are given. γ_{hys} is about 0.1% and hardly varies with overconsolidation.

9.4.2.6. *Unsaturated soil parameters*

Given the formulation of the model for unsaturated soil, only a few parameters depend exclusively on unsaturation. Fleureau and his co-authors [FLE 93] have shown the validity of the correlations suggested by Biarez and Favre [BIA 72] between the C_c and w_L parameters on drainage paths applied to normally consolidated clays. Indeed, the parameters necessary to model the mechanical behavior are: p_{desat}, p_{resat}, $\pi_c(pc)$ and $\Re(\pi_c)$. Zerhouni [ZER 91] suggested a relation between suction at desaturation p_{desat} and the liquidity limit (see Figure 9.12). Khalkhali and Khabbaz [KHA 98] obtained the following relation illustrated in Figure 9.13 for χ, which determines the effective stress (see relation [9.60]):

$$\chi = (\frac{p_c}{p_{c_{desat}}})^{-0.55} \quad [9.90]$$

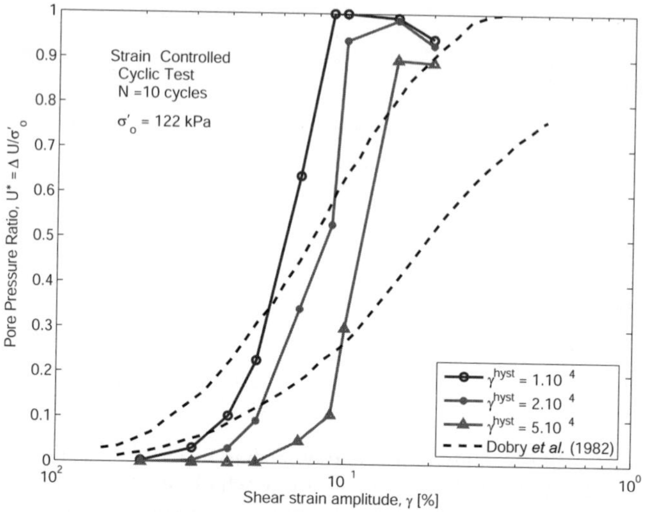

Figure 9.11. *Evolution of the hysteresis domain of sands in undrained cyclic shear tests [LOP 07] compared to the experimental results of [DOB 82]*

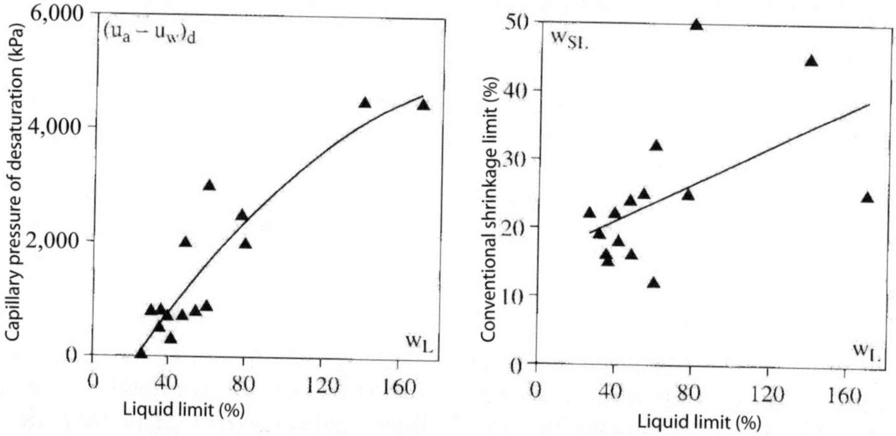

Figure 9.12. *Correlation between desaturation pressure and liquidity limit [ZER 91]*

9.4.3. *Parameters that are not directly measurable*

The parameters which cannot be directly measured need to be estimated by the best fitting of experimental curves. We may, however, obtain these correlations in the future.

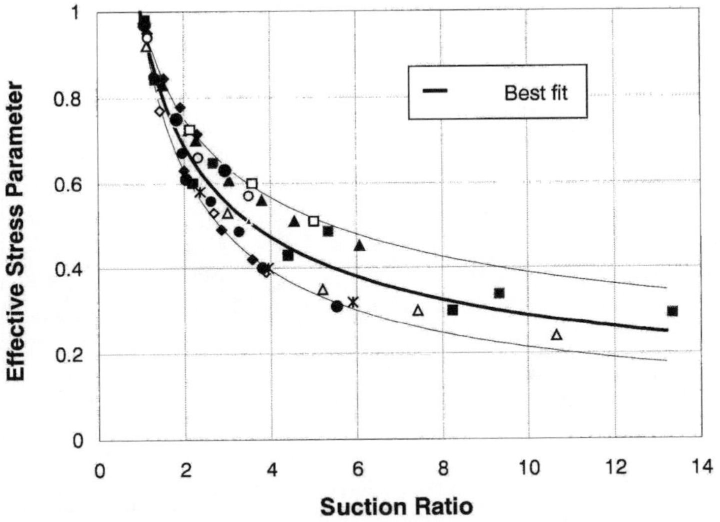

Figure 9.13. *Relation given by Khalkhali and Khabbaz [KHA 98] for Skempton's χ parameter*

9.4.3.1. *Parameter b*

This parameter controls the form of the yield surface and can vary between 0 for a Mohr-Coulomb and 1 for a Cam Clay type surface. A value equal to 1 can be considered acceptable for clays. We may choose b between 0.1 and 0.2 for sands.

9.4.3.2. *Parameters related to plastic stiffness a_1 and a_2*

These parameters which control the evolution of the yield surface play an important role.

9.4.3.2.1. Clays

Numerical simulations to calibrate the $G - \gamma$ and $D - \gamma$ given by Vucetic and Dobry [VUC 91] for clays have made it possible to establish a correlation between theses parameters and the factor $\frac{G_{max}}{\sigma'_0 OCR}$ for the range of OCR studied [LOP 03a]. As we see in Figure 9.14, the range of their variation depends on the plasticity index I_p.

9.4.3.2.2. Sands

Parameters a_1 and a_2 vary from 0.1 to 0.01 for sand. At present, we do not have a well established correlation for such materials.

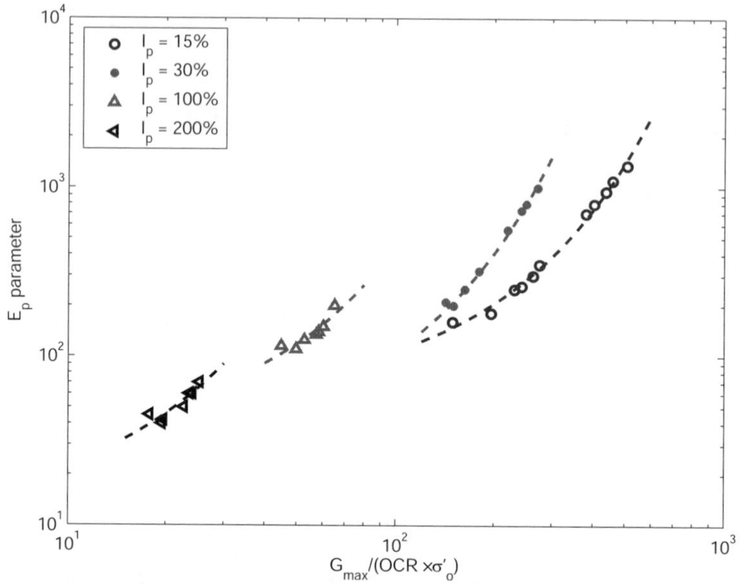

Figure 9.14. *Correlations obtained by Lopez-Caballero [LOP 03a] for* $E_p = \frac{\tan\phi'_{pp}}{a}$ *fitting G modulus and D damping -γ given by Vucetic and Dobry [VUC 91]*

9.4.3.3. *Domain of behavior*

These parameters have to be calibrated by curve fitting by using modulus variation with strain or pore pressure variation with strain in cyclic tests.

9.4.4. *Parameters defining the initial state*

Generally speaking, the initial state is represented by the void ratio, the stress state, and the maximum stress to which the material has been historically subjected (that is to say, OCR). In the framework of the constitutive model, they are defined by $\underline{\underline{\sigma}}'_0$ (σ_{c0}, γ_p, $\underline{r}_h$, $\underline{t}_h$) for the generic model and (p_{c0}, r_0, $\underline{\underline{T}}^h_k$,$\underline{\underline{t}}^h_k$) for the multimechanism model. In the case of an isotropic monotonous behavior, the initial state can be represented by both the initial stresses or its position with respect to the critical state. Therefore, the unique parameter (p'_0/p_{c0}) which takes into account the relative density and the overconsolidation (see Figure 9.15) can be used. In the case of an anisotropic stress state, we should verify that the yield surface has not been overstepped (it should be plastically admissible). This will be done by an adapted initialization of the degree of mobilization of the mechanisms. If the vertical stresses can be estimated fairly easily, this is not necessarily the case for the horizontal stresses. Moreover, for a homogeneous

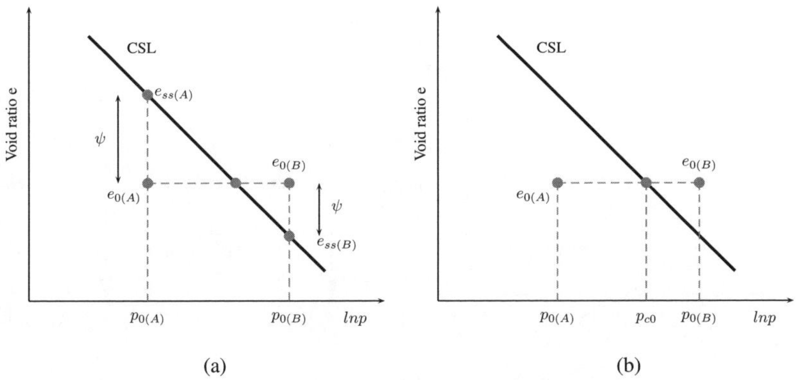

Figure 9.15. *Initial state: a) [BEE 85], definition of state parameter (ψ); b) definition of the initial state in the ECP family models. CSL - critical state line*

normally consolidated clay, as the void ratio varies with depth, the ratio σ'_{c0}/σ'_0 may be considered constant. On the other hand, in the case of sand, the change in void ratio with depth is low and, for a given D_r, σ'_{c0} remains constant.

In summary, under isotropic loading conditions, the initial state of a clay is known when we know the initial stress (or void ratio) and its OCR. For sands, knowing the void ratio (or relative density) and the initial stress is sufficient. Under anisotropic conditions, the degree of shear mobilization should be added.

We can obtain the following relation for clays:

$$\frac{\sigma'_{c0}}{\sigma'} = \frac{C_u}{\sigma'_0 \tan \phi'_{pp}} \qquad [9.91]$$

where C_u is the undrained shear strength. In practice, the ratio $\frac{C_u}{\sigma'_0}$ can be obtained from laboratory tests or by using correlations such as the one given below:

$$\frac{C_u}{\sigma'_0} = (0.23 \pm 0.04) OCR^{0.8} \quad \text{[JAM 85]} \qquad [9.92]$$

According to Been and Jefferies [BEE 85], the ratio $\frac{\sigma'_{c0}}{\sigma'}$ for sands may be obtained from coefficients such as the state index ψ (which should not be confused with the dilatancy angle) or I_s used by Ishihara [ISH 93]. The state index given by Been and Jefferies is defined as the difference of void ratios at the initial state e_0 and that at critical state e_{ss} under the same level of stress (see Figure 9.15).

In the previous sections, a methodology of identifying the ECP elastoplastic model parameters was presented. This methodology consists of two objectives. The first is to suggest an alternative to complex laboratory tests which are not always available for

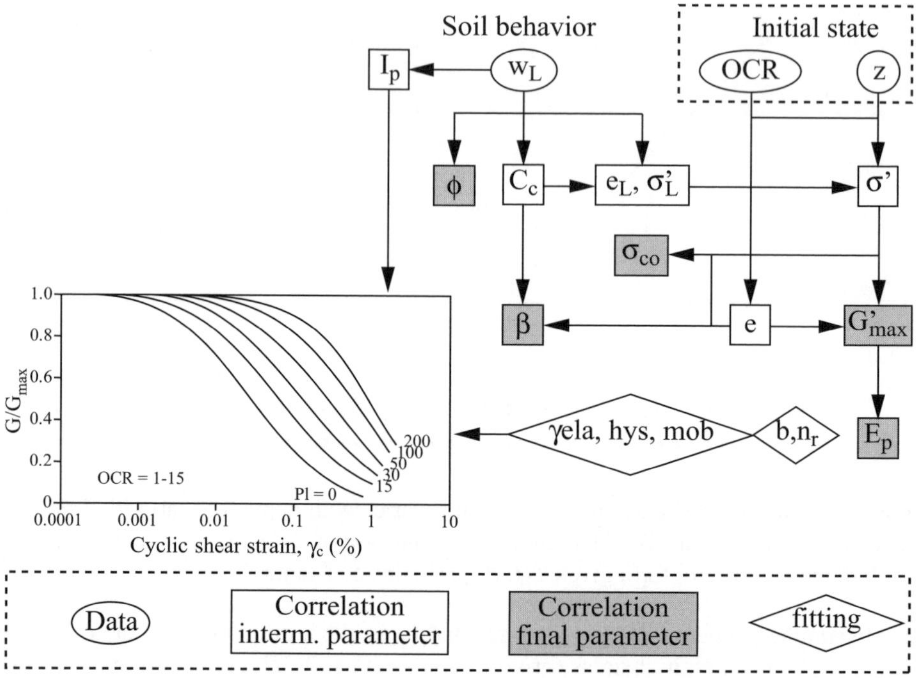

Figure 9.16. *Methodology to identify model parameters for clays [LOP 07]*

current applications. The second is to provide a rapid way to verify the coherence of data from a series of geotechnical tests.

This methodology is operational for clays and sands. Therefore, with a minimum amount of data, we can define the first set of usable parameters. A summary of this methodology applied to clays is given in Figure 9.16. The response of a clay under a given stress-path is defined by two categories of information. The first category corresponds to the determination of its initial state and its loading history. The second category corresponds to its proper behavior.

For a normally consolidated clay, knowing the initial void ratio and w_L is sufficient for determining the rest of the parameters. As can be seen in Figure 9.16, we dispose of correlations for almost all the model parameters (ϕ, G_{max}, β, E_p (or a_1 and a_2), p_{co}, behavior domains). Only a few parameters remain for which the correlations have not yet been validated and tested (b).

For an overconsolidated clay, it is necessary in addition to w_L to know, the couple initial stress-initial void ratio or OCR-void ratio. The next step for clays is to generalize

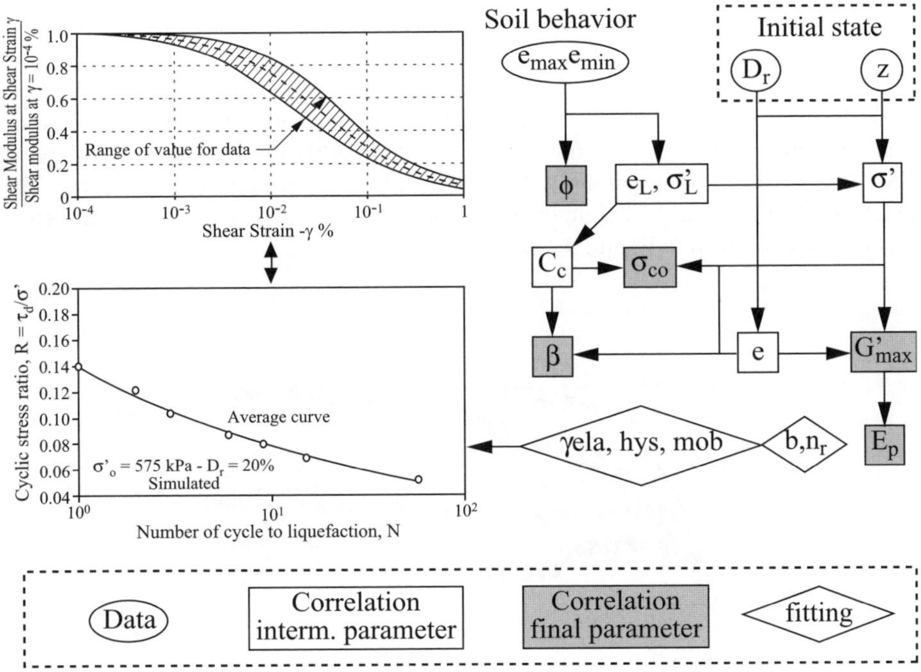

Figure 9.17. *Methodology to identify model parameters for sands [LOP 07]*

the methodology to natural soils and to continue its validation on real sites and geo-structures.

The approach for sands, summarized in Figure 9.17, has already been applied to Hostun and Toyoura sands. To be validated, it needs to be applied to sands with different physical characteristics. In particular, the determining parameters for these materials being their relative density and the likelihood to liquefaction, a more profound investigation is currently being carried out at ECP.

9.4.5. *Application of parameter identification strategy*

We hereby present this strategy as it is applied to a coarse material coming from a dike [FLE 02]. Some of the parameters are measured directly from the test results, whereas others are calibrated by using numerical simulation [NGU 05]. With this example, we want to illustrate the importance of a coherent set of parameters. Even if it is always possible to find a set of parameters that simulate the behavior for a given stress-path, it is not necessarily possible to obtain the appropriate behavior with the same set on other stress paths.

The material is composed of particles with a diameter that varies from 0.8 mm to 10 mm with e_{max} and e_{min} equal to 0.545 and 0.209, respectively. The density of the particles is 2.67. We obtain $d_{10} = 200\ \mu\text{m}$; $d_{30} = 600\ \mu\text{m}$ and $d_{60} = 4{,}000\ \mu\text{m}$ from the grain size distribution curve. A series of drained and undrained triaxial tests under different confining pressures (from 50 kPa to 200 kPa) has been performed at LMSS-Mat [FLE 02] on samples with relative densities, varying from 48 to 81%. The cylindrical samples have a height of 20 cm and a diameter of 10 cm. The experimental results are assembled in Figure 9.18.

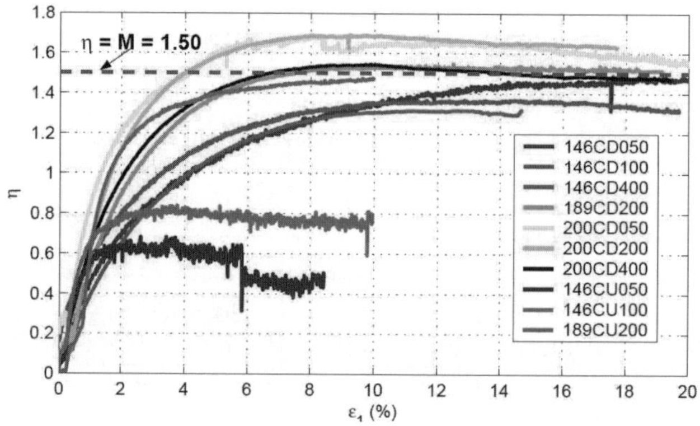

(a)

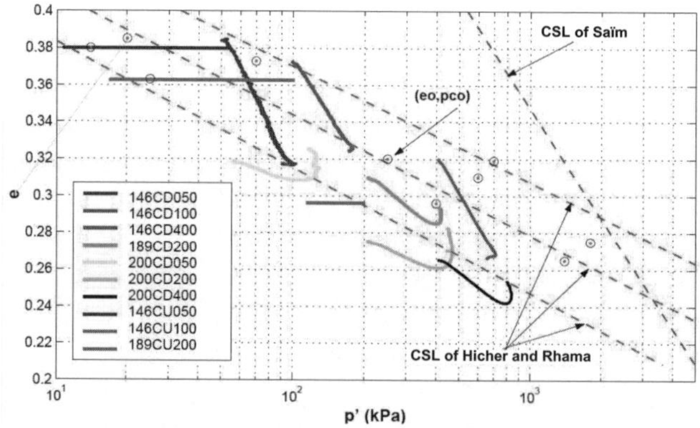

(b)

Figure 9.18. *Drained and undrained triaxial test results on coarse gravel [FLE 02]*

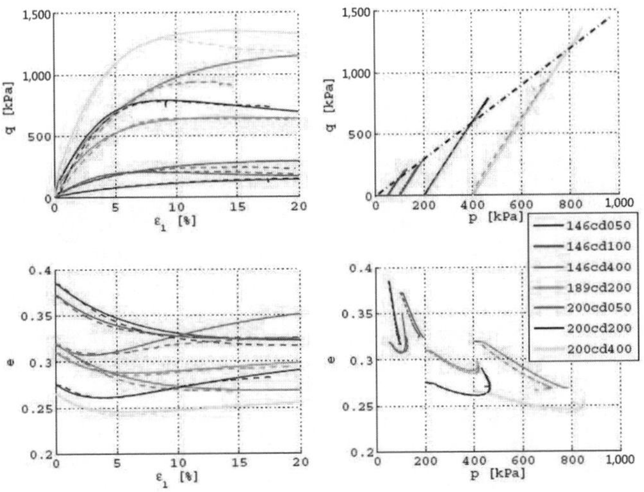

(a)

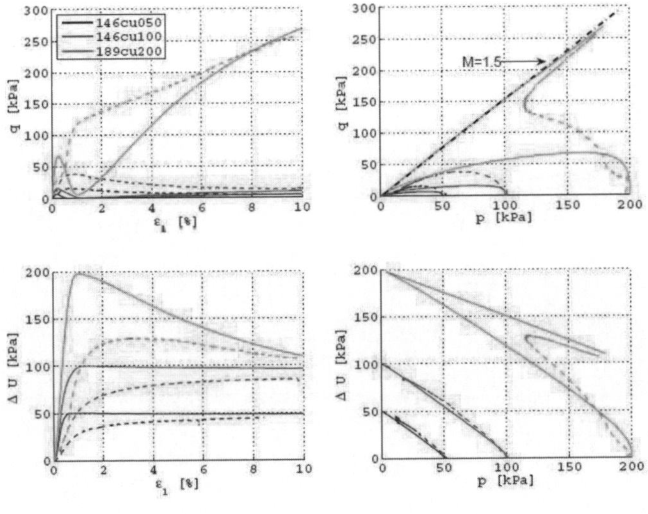

(b)

Figure 9.19. *Triaxial tests on coarse gravel-PA01; solid lines: simulation [NGU 06]; dashed lines: experience [FLE 02] data set J1 (Table [A.1] in the appendix, section 9.8) with* $A = 1{,}300$ *(relation [9.9])*

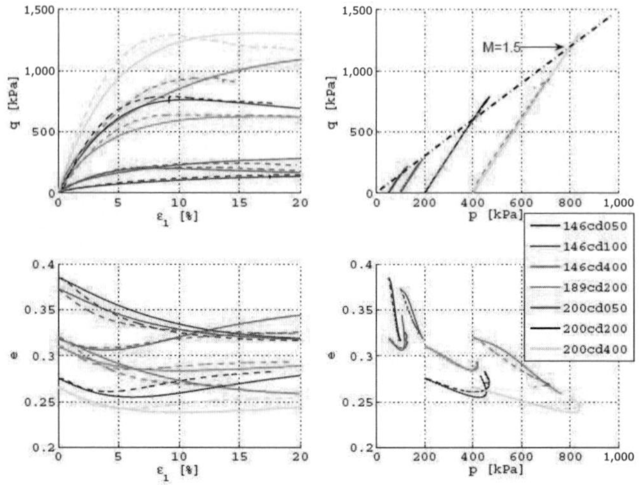

(a)

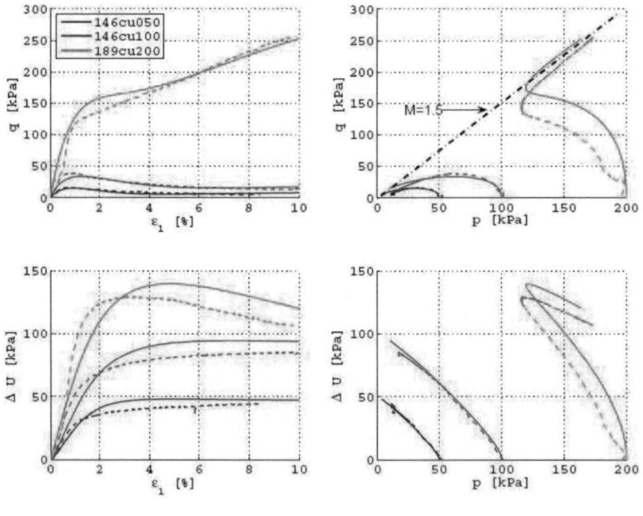

(b)

Figure 9.20. *Triaxial tests on coarse gravel-PA02; solid lines: simulation [NGU 06]; dashed lines: experiment [FLE 02] data set J1 (Table [A.1] in the appendix, section 9.8) with A = 70 (relation [9.9])*

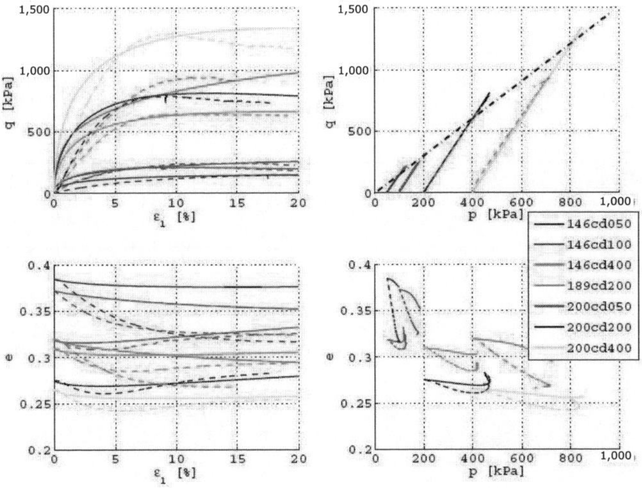

(a)

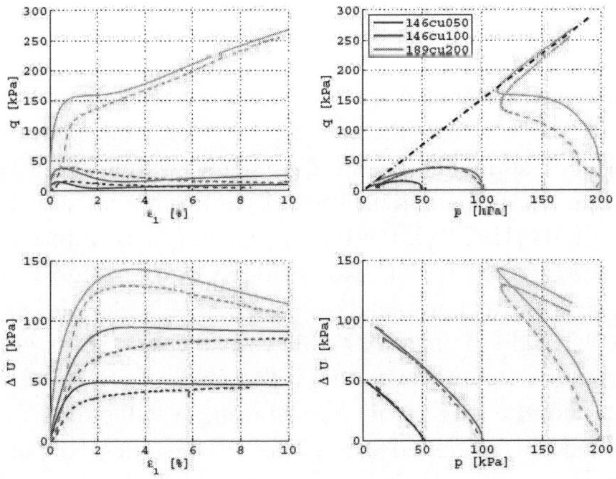

(b)

Figure 9.21. *Triaxial tests on coarse gravel-PA07; solid lines: simulation [NGU 06]; dashed lines: experiment [FLE 02] data set J2 (Table [A.1] in the appendix) with* $A = 1{,}300$ *(relation [9.9])*

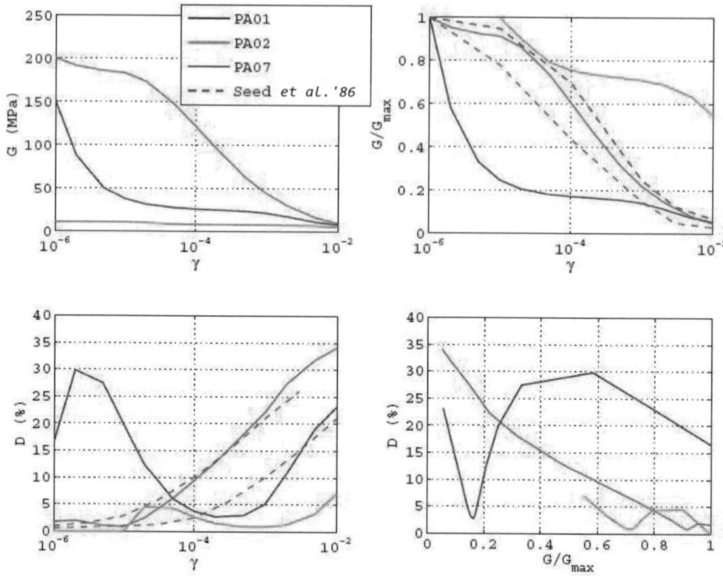

Figure 9.22. *Cyclic shear test simulations obtained with the three sets of parameters PA01: J1 set with A = 1,300; PA02: J1 set with A = 70; and PA07: J2 set with A = 1,300*

The critical state parameters (M and λ) are chosen through an analysis of the final points of the tests. We have noticed that the tests with the same confining pressure reach the same point for large strains in ($e - \log p'$) plane, which is in accordance with the critical state concept. $C_c = 0.05$ is positioned within the range given by Hicher and Rahma (relation [9.83]) [HIC 94]. On the other hand, it is different from the values given by Biarez and Hicher [BIA 94]. Kokusho [KOK 82] suggests using $A = 1{,}300$, $n_e = 0.6$ and $b = 2.17$ for parameters of relation [9.9] for the elastic modulus of rounded gravels. This value is used for a first set of parameters (J1: see Table A.1, section 9.8). We should note, however, that if this choice is satisfactory for simulating drained tests, this is not the case for undrained tests. A value of $A = 70$ improves the results of both drained and undrained tests. However, this last set of parameters results in an incoherent response for liquefaction tests. Furthermore, the $G - \gamma$ and $D - \gamma$ curves, obtained with this last set of parameters, are outside of the zone currently used. (Figure 9.22). A third set of parameters is recommended using the guidelines produced by Kikoshu ($A = 1{,}300$) but by refinement of the values of p_{c0}. Hence, a set of parameters with acceptable precision for the response of all drained, undrained, monotonous and cyclic triaxial tests is obtained. This set of parameters is given in Table A.1 in the appendix (section 9.8).

9.5. Influence of constitutive behavior on structural response

In this section, we will present a number of applications highlighting the influence of the principal characteristics of the mechanical behavior of soil on the behavior of geotechnical structures. Through these examples, the performance of the constitutive models in a wide and diverse range of applications is illustrated. We will limit ourselves to the presentation of the observed phenomena and the parameters of the behavior which influence them. We will not examine the application and justification of the computations on real constructions, which are beyond the objectives of this chapter. Readers interested in this aspect may refer to the research studies and PhD theses cited.

First, the importance of taking into account a reduced elastic range is addressed by studying the response of a retaining wall obtained by two models [VOS 01]. Then, the influence of the initial state and loading history is illustrated by studying the mechanisms of load transfer in axially loaded piles [DAG 08]. We will see how the ability to change volume influences the mobilization of friction at the pile-soil interface. The influence of the applied load level against the strength of the pile, particularly the cyclic fatigue, is discussed. We analyze the issue of identifying model parameters of materials for which the macroscopic behavior is difficult to study in the laboratory and the procedure for obtaining a predictive model, by looking at the example of the El Infiernillo dam [SIC 00, SIC 08].

9.5.1. *Retaining walls*

Retaining walls have become increasingly used in the development of urban sites, as they present a rather practical building solution. Given that the settlements on the surface and the deflection of the wall are the key parameters quantifying the damage, the finite element method, which offers the possibility of providing such information, becomes very attractive. Furthermore, it is able to take into account the history of loading and construction and/or excavation scenarios, for it can be used to simulate the behavior of the structure during its lifetime.

The importance of the constitutive model on the estimation of soil displacement in such geo-structures has been the subject of a *German Benchmark* [SCH 02]. If the behavior of a structure in service conditions is of capital importance, the notion of failure is another key parameter. Historically, the structures have been designed according to criteria based on stability and ultimate load. At present, we are seeing the emergence of the concept of ultimate limit state, which is a more general concept related to the function of the structure. It is defined as the state from which the structure may undergo large displacements without endangering human lives. Thus, the stability criteria should be verified through appropriate safety coefficients or through the deformation obtained for different behavior coefficients.

In current practice, methods for obtaining the state of a geo-structural breakdown are different from those used for determining their displacement. However, it would be very interesting to use the same methods, models, and tools to analyze these two aspects. These methods should not only detect the ultimate limit state of the structure but also the displacements that a structure could be subjected to at the time of breakdown.

Through this example, we study how the finite element method is able to detect the ultimate limit state. We illustrate the importance of the constitutive law on the evaluation of the breakdown mechanism and deformation of the soil mass. The example involves a flexible retaining wall with two inclined ties. Several criteria can be considered as precursors of a breakdown:

1) In current practice, breakdown is associated with the non-compliance of the balance between internal and external forces. In a numerical simulation by finite element method, this results in non-compliance with the convergence criteria associated with residues.

2) In the area of weak discontinuities, that is to say when only the rate of deformation is discontinuous, the Rice criterion [RIC 76], based on the analysis of bifurcation and strain localization states:

$$\underline{\underline{Q}}^{ep} \cdot [d_n \underline{v}] = 0 \qquad [9.93]$$

where [.] represents the jump. The acoustic tensor $\underline{\underline{Q}}^{ep}$ is given by

$$\underline{\underline{Q}}^{ep} = \underline{n} \cdot \underline{\underline{\underline{\underline{C}}}}^{ep} \cdot \underline{n} \qquad [9.94]$$

In these relations $\underline{v}$ is the velocity, $\underline{\underline{\underline{\underline{C}}}}^{ep}$ is the elastoplastic tensor, and $\underline{n}$ is the unit normal vector to the plane of $d_n\underline{v}$ discontinuity. This is obtained by writing the condition in such a way that this equation gives a non-trivial solution:

$$\det(\underline{\underline{Q}}^{ep}) = 0 \qquad [9.95]$$

3) If we are interested in material instability, a stable state is defined by Hill's condition [DAR 99] given by

$$d^2W = d\underline{\underline{\sigma}} : d\underline{\underline{\varepsilon}} > 0 \quad \forall d\underline{\underline{\sigma}}, d\underline{\underline{\varepsilon}} \qquad [9.96]$$

where $d\underline{\underline{\sigma}}$ and $d\underline{\underline{\varepsilon}}$ are the stress and strain increments of the material with a given history. Therefore, instability occurs when this condition is breached.

The normalized second-order work considered by Darve and Laouafa [DAR 00] is

$$d^2W_{norm} = \frac{d\underline{\underline{\sigma}} : d\underline{\underline{\varepsilon}}}{||d\underline{\underline{\sigma}}|| : ||d\underline{\underline{\varepsilon}}||} > 0 \quad \forall d\underline{\underline{\sigma}}, d\underline{\underline{\varepsilon}} \qquad [9.97]$$

For an incremental linear constitutive law, the positivity of the second-order work is a sufficient condition for the uniqueness of the response and corresponds to the fact that the elastoplastic tensor is positive definite. When the incremental stress-strain relationship is non-linear, or is piecewise linear, it becomes more difficult to prove. Nevertheless, several authors have demonstrated the positive-definite nature of the elastoplastic tensor [BIG 91, CHA 99, NOV 04], and, consequently, the positivity of the second-order work is a sufficient condition for the uniqueness of the solution or the absence of stability-loss in the boundary problem.

The multimechanism model's ability to predict different conditions of bifurcation has been studied as part of the PhD work of K. Hamadi [HAM 06, HAM 08a, HAM 08b]. It has been proved that with the Hill criterion, instability may exist before the state of perfect plasticity is reached. The state of the material and the loading path influence the instability. The analysis of the behavior of Hostun RF sand has shown that instabilities similar to those encountered in undrained paths may occur during drained triaxial paths measured on loose and dense samples. Tests at constant q have also shown the occurrence of instabilities when the volume change is maximal.

These instabilities are reached before the condition of perfect plasticity and the multimechanism model exhibits a wide range of instability domains, compared to other models. Moreover, for all the cases studied, the unstable branches form a single cone in the stress space. We can conclude that $(det(C_s^{ep}) = 0)$ is a general condition necessary for detecting the existence of bifurcation. The circular presentation of $(det(C_s^{ep}))$ and $(det(C^{ep}))$ highlights tensorial areas where the constitutive relation is linear. Two mechanisms can be distinguished when the loading level increases. One corresponds to an elastic unloading and the other to a plastic loading.

These examples confirm the model's ability to reproduce the instability and the phenomenon of strain localization. The response is consistent with experimental observations, theoretical developments, and numerical analyses in the existing literature.

As part of K.C. Vossoughi's PhD thesis [VOS 01], we examined these criteria to analyze the ability of the standard finite element method to study retaining structures. However, if from a theoretical point of view, these concepts are well developed, their application in numerical methods is not self-evident. For example, we may cite the dependence of the shear band with respect to the mesh size and orientation [PAS 95].

Here, we restrict our presentation to the role played by the constitutive model that represents soil behavior. However, we note the possibility of using a dilation second gradient model to circumvent the problem of mesh sensitivity, which has been applied to Hujeux's model in the framework of Alexandre Foucault's PhD thesis ([FOU 10]). Therefore, we compare the results obtained with a Mohr-Coulomb yield criterion and non-associated flow rule (the bifurcation can only occur if the law is not associated) with those obtained with the ECP elastoplastic model.

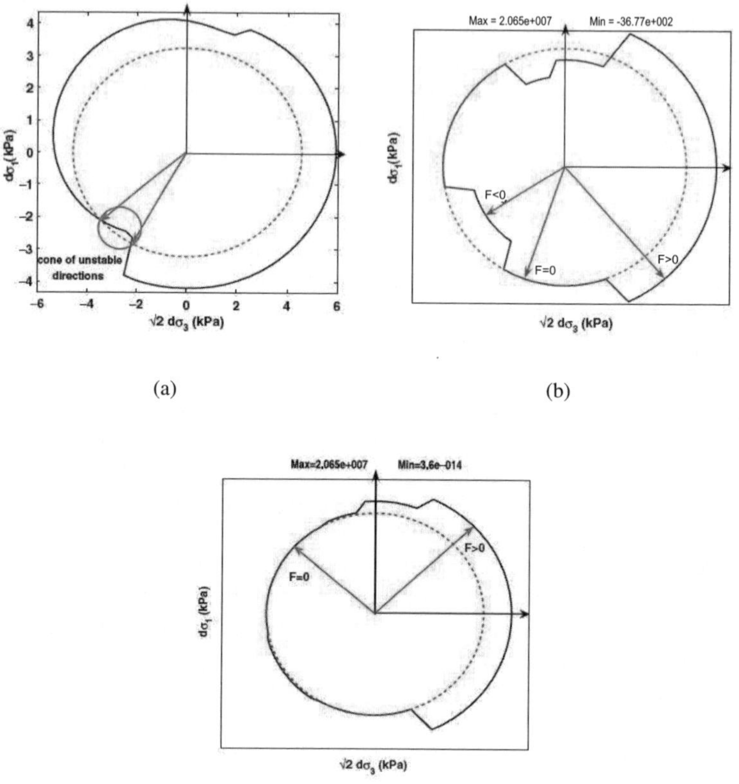

(a) (b)

(c)

Figure 9.23. *Directional variations of: (a)* d^2W*; (b)* $det(C_s^{ep})$*; and (c)* $det(C^{ep})$ *for Hostun sand at* $D_r = 40$ *with* $\sigma_3 = 100\ kPa$ *and* $\xi = \frac{\sigma_1}{\sigma_3} = 2.5$ *under axisymmetrical conditions*

In Figure 9.24, we present the deformed mesh and the displacement increments at the last acceptable stage of excavation (the balance of forces is verified). On the same figure, the potential directions satisfying the condition of Rice are indicated by lines with these directions. If we compare the two diagrams, we see that the failure mechanisms obtained by the two models are different. When the Mohr-Coulomb criterion is used, we are in the presence of an unrealistic swelling of the medium because of the elastic behavior during the excavation and the unloading of the soil. Thanks to its reduced elastic range, no swelling occurs with the Hujeux model, leading to results closer to observations [BEL 98]. The difference in the geometry of the failure surface must also be reported. With the Hujeux model, a well identified passive zone

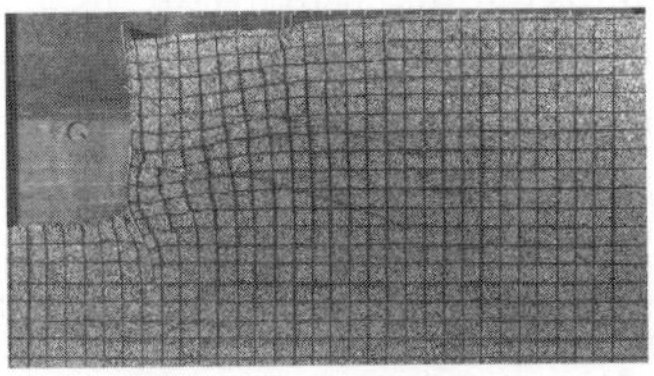

(a) after [BEL 98]

Mohr-Coulomb criterion with non-associated flow rule

ECP multimechanism elastoplastic model

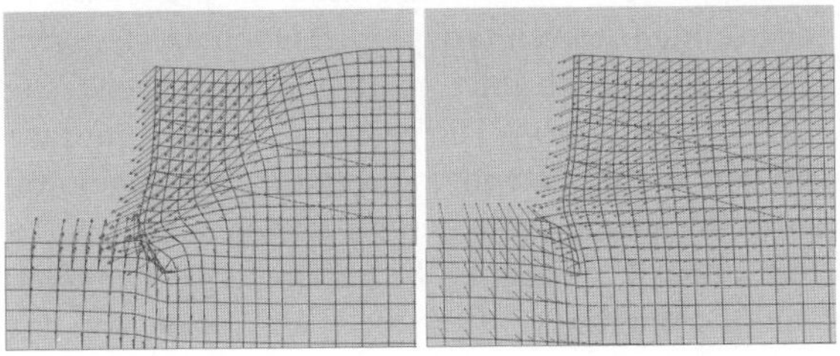

(b) (c)

(d) (e)

Figure 9.24. *Comparison between observations (a) on a flexible wall physical model [BEL 98] and results obtained by numerical modeling [VOS 01] with two different constitutive modes; (b) and (c): increments of displacement; (d) and (e): directions verifying Rice's condition and that of the second-order work* is observed on the face of the retaining wall. With the Mohr-Coulomb criterion, this area seems to bear the characteristics of an active zone. A more complete analysis of other geometries, configurations, and meshes is available in the PhD dissertation of K.C. Vossoughi [VOS 01].

9.5.2. *Vertically loaded piles*

The analysis of load transfers in vertically loaded piles necessitates a good comprehension not only of the soil behavior, but also of the soil-pile interaction. In this problem, the parameters such as the initial state before and after the installation of the pile, the loading history and stress path, as well as the roughness of the interface, play an important role. Moreover, it is important to analyze the cyclic behavior of the pile and the interface, especially for studying the effect of installation.

In the framework of Sofia D'Aguiar's PhD thesis [DAG 08], several series of numerical computations simulating non-displacement piles have been performed in order to simulate and to obtain the results of physical model tests performed in centrifuge by Fioravante [FIO 02]. Both rough and smooth piles have been tested in Toyoura sand with a relative density of $D_r = 0.93$. The properties of the interface are based on the behavior of the sand and should take into account the roughness of the interface. As these characteristics influence the friction mobilization at the interface, it is important to develop a coherent strategy to identify the parameters of the interface model based on the soil model and the roughness of the interface. Fioravante has performed shear tests on sand-iron interfaces with two different normalized roughnesses ($R_n = 0.01$ and 0.13; $R_n = \frac{R_t}{d_{50}}$, where R_t is the depth of surface asperities). This test is similar to a simple shear test where the normal stiffness is kept constant. Furthermore, the Toyoura sand behavior has been studied by many researchers [ISH 93, IWA 78, KOK 82, KAB 91].

The comparison of experimental results with those simulated numerically is given in Figure 9.25 for the Toyoura sand and in Figure 9.26 for the interfaces. Tables A.2 and A.3 in the annex in section 9.8 summarize the parameters used in the numerical models. In order to have a global view of the multimechanism model's response and the coherence of the chosen set of parameters for the Toyoura sand, interested readers are invited to consult S. D'Aguiar's PhD thesis [DAG 08].

The roughness of the interface is taken into account by reducing the friction angle ϕ'_{pp} from 30 to 16 degrees, by increasing the plastic compressibility β to reduce dilatancy and by changing the form of the yield surface (parameter b). Consequently, for $R_n = 0.01$ and smooth interface, contractancy/dilatancy is very low in comparison to the values for rough interfaces: $R_n = 0.13$.

The numerical modeling of load tests on piles conducted in the centrifuge is performed by applying a pile head displacement, starting from a geo-statistical stress

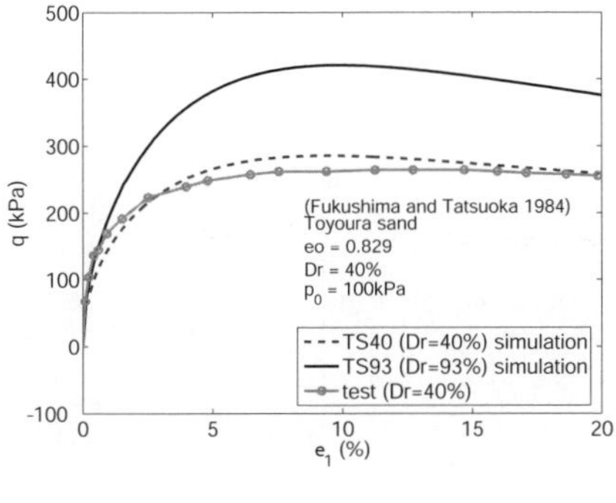

(a)

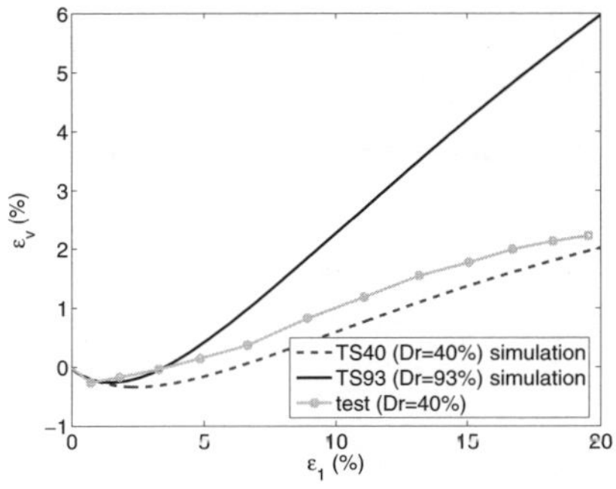

(b)

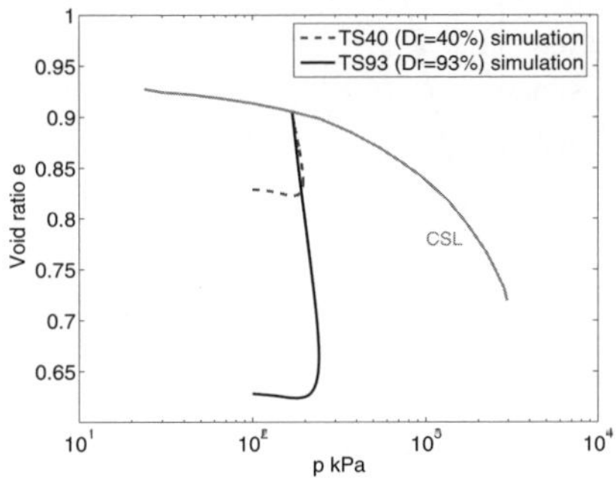

(c)

Figure 9.25. *Comparison of triaxial test results of Toyoura sand obtained in the laboratory ($D_r = 40\%$ and $D_r = 93\%$ for p'_0=100 kPa) [FUK 84] and those simulated numerically with the ECP model: a) $q - \varepsilon_1$ plane; b) $\varepsilon_v - \varepsilon_1$ plane; and c) void ratio $e - p$ plane*

field obtained by a previous calculation step, where the balance of forces between the pile, soil and interface due to different weights is checked. The strength mobilization in the shaft and at the base obtained by the numerical model compares very satisfactorily to those obtained by Fioravante [FIO 02] in the centrifuge testing two different interface roughnesses (Figure 9.27). The effect of roughness of the interface is very clearly highlighted. This roughness also affects the displacement of the pile head for which the maximum shaft strength is reached. This is obtained for s/d of about $12 - 15\%$ when $R_n = 0.45$, whereas it decreases to 2% for $R_n = 0.01$ (see Figure 9.27). Since measuring tools were placed on the piles, radial and tangential stresses were also measured at certain points along the pile ($\beta = \tau_s/\sigma'_{v0}$). These measures help to clarify

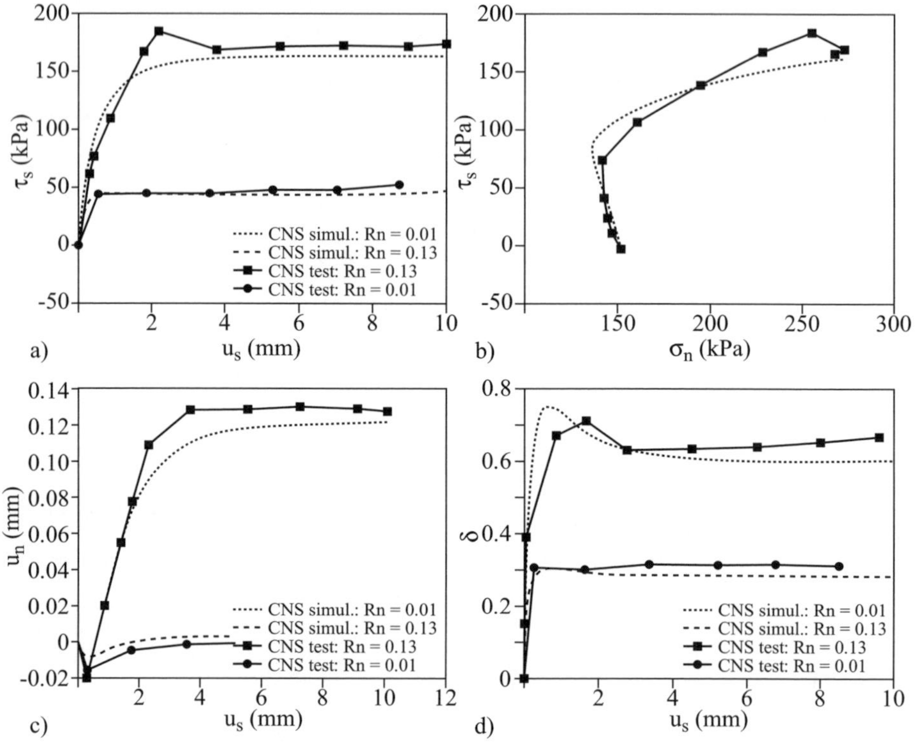

Figure 9.26. *Comparison between experimental results (data from Fioravante [FIO 02]) and the model's prediction [DAG 08] of a constant normal stiffness direct shear test (CNS), on Toyoura sand ($D_r = 90\%$), rough and smooth aluminum plates with $R_n = 0.13$ and $R_n = 0.01$, $K =$ 1,000 kPa/mm, $\delta = \tau_s/\sigma_n$*

the mechanisms of load transfer along the pile and can be compared with results of the numerical modeling (see Figure 9.28).

We note that the non-linear behavior of the sand is well represented by the soil and the interface models. The amplitude of $\beta = \tau_s/\sigma'_{v0}$ decreases with depth. A series of parametric studies can highlight the parameters that influence the mechanisms of load transfer from the pile to the granular media. Besides the interface roughness, the soil state parameters play a very important role in the response of the pile. Thus, the influence of the density of the surrounding soil and the length and the roughness of the pile were studied. The influence of a soil state (density, initial stress) can be grouped into a single state parameter, corresponding to the position of the point relative to the critical state, which can be characterized by the ratio p'_0/p_{c_0}. This state parameter is useful for quantifying the distance to the critical state line and the level of dilatancy and, consequently, the shear mobilization. This notion of state parameter is also used in the literature [KLO 01].

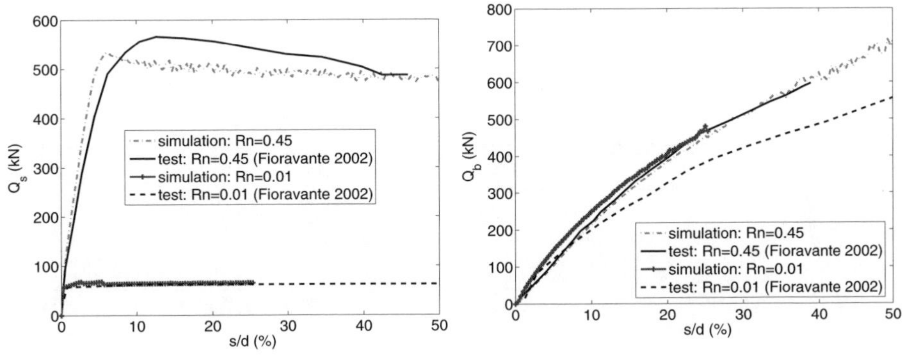

(a) (b)

Figure 9.27. *Force-displacement curves for the Toyoura sand ($I_D = 0.07$) ([FIO 02] and [DAG 08]). Results are obtained for two interface rugosities $R_n = 0.45$ and 0.01. (a) shaft; and (b) base*

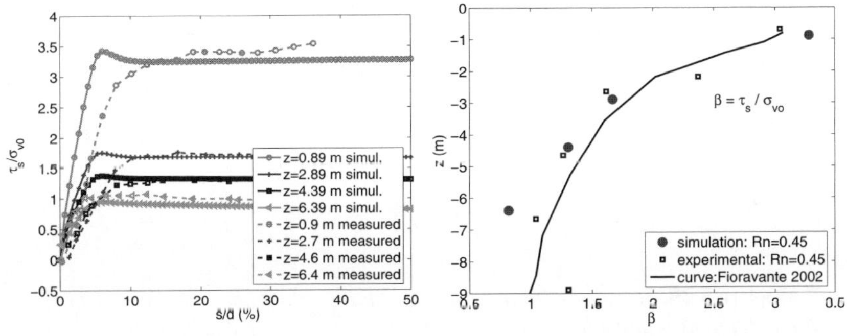

(a) (b)

Figure 9.28. *Comparison of computed and measured mobilized shear for a rugous interface $R_n = 0.45$. (a) Evolution of $\beta = \tau_s/\sigma'_{v0}$ with relative settlement at the head of the pile (s/d); and (b) Evolution of β with depth for $s/d = 25\%$*

The importance of this state parameter can also be seen in the study of the loading-unloading behavior, which highlights notions such as residual stress and cyclic fatigue.

If we unload an already loaded pile up to a high level of mobilized friction, we will find that, after the total force on the pile head has been suppressed, the residual stresses can be trapped inside the soil at the top and around the pile. The amplitude and distribution of these residual stresses depend on the level of the relative settlement

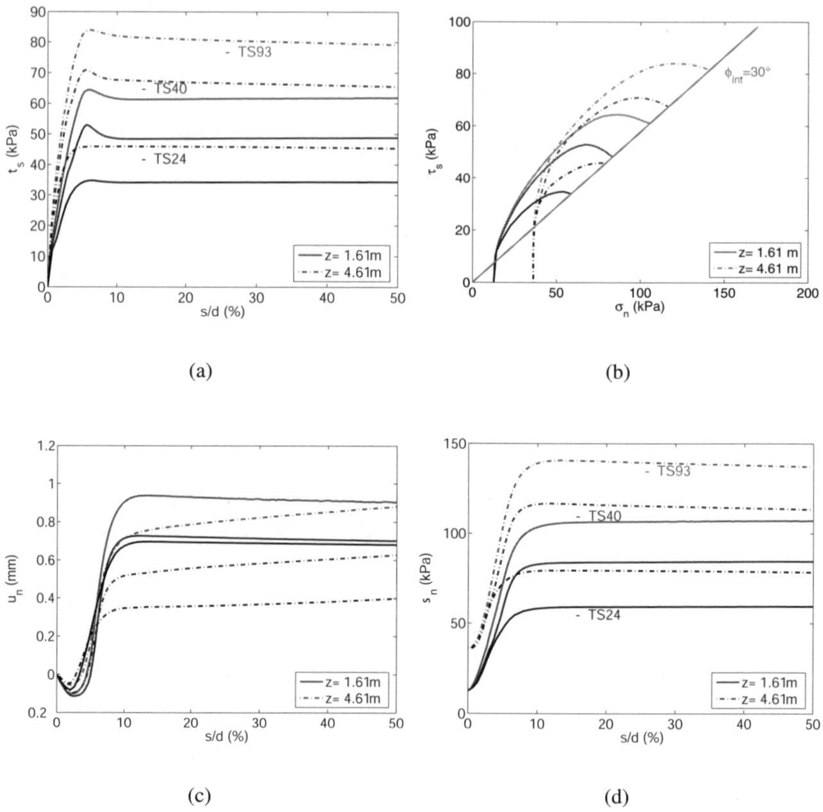

Figure 9.29. *Stress path in the interface at different depths, for soils with three different relative densities (D_r= 93, 40 and 24%): TS93, TS40 and TS24: effect of the initial state of the soil clump*

or on the total load applied. When the load on the pile head is removed, the vertical movement of the pile may be sufficient to generate a negative skin friction along the pile (Figure 9.30). If the pile is reloaded again and the force-displacement curve at the pile head is compared for the first and second cycles, the effect of the residual friction can be observed.

When applying several cycles of loading-unloading, the observed behavior depends on the load level. Thus, three mechanisms of friction degradation can be identified, depending on whether the maximum friction is reached during the first cycle, only the phase transformation (characteristic state) is reached, or neither of the two limits is reached. These scenarios correspond, by analogy, to the process of pile installation (jacked or driven piles).

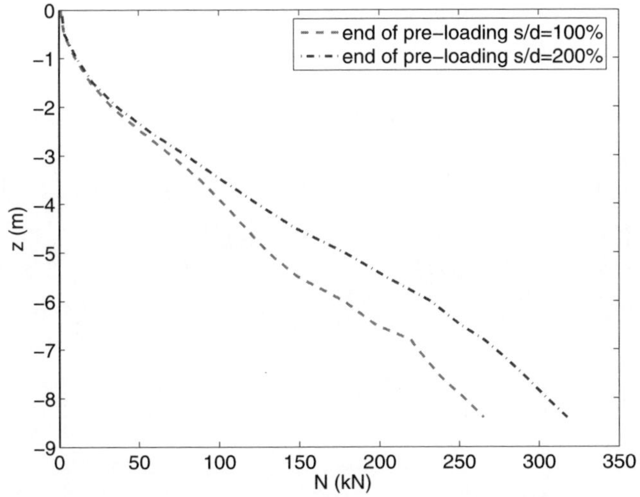

Figure 9.30. *Profile of the residual shear after the preloading-unloading cycle*

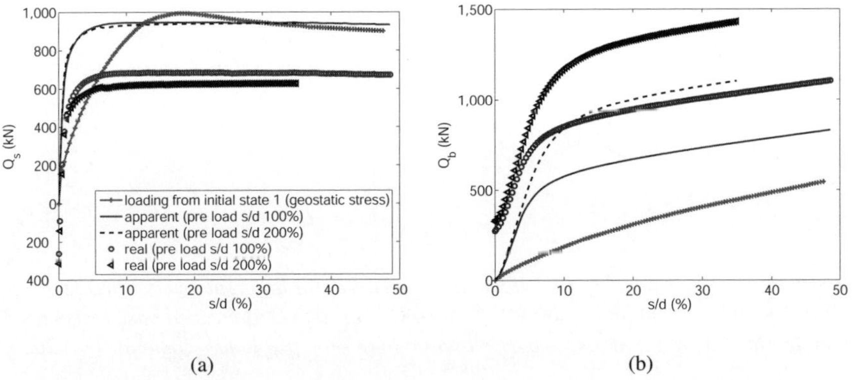

Figure 9.31. *Comparison of the "real" (taking into account the residual shear) and "apparent" (not taking into account the residual shear) force-displacement curves [DAG 08]: a) shaft; and b) base*

Three observations may be underlined in Figure 9.32:

– the level of loading plays an important role on the residual displacement obtained during cycles;

– this displacement increases with the number of cycles;

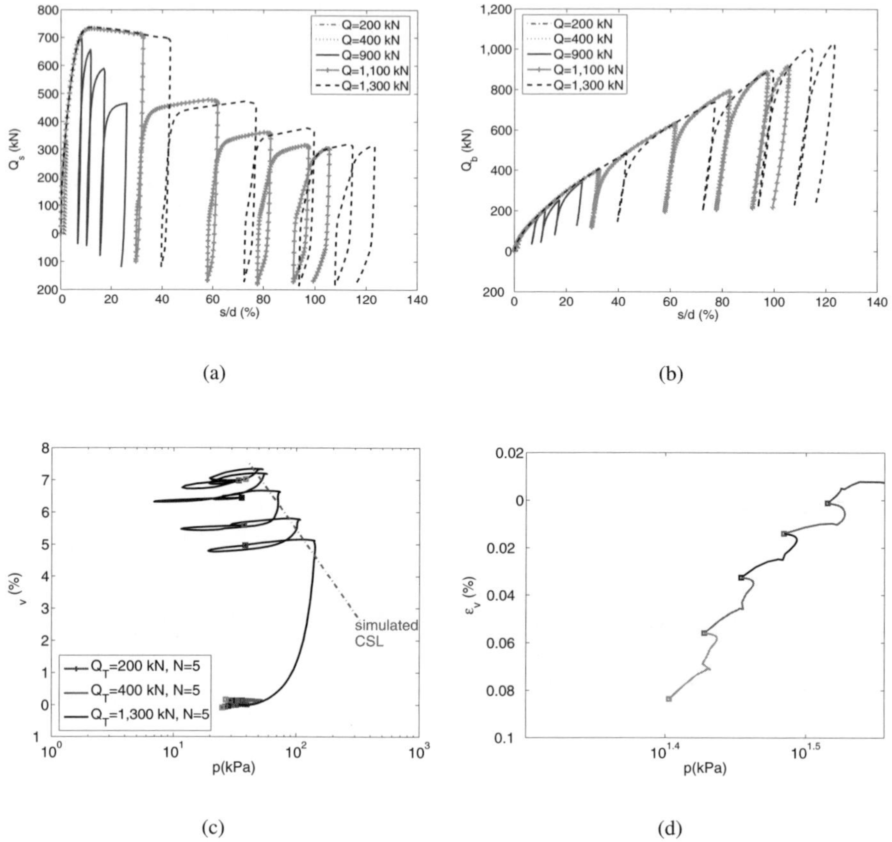

Figure 9.32. *Comparison of cyclic load-settlement curves for different levels of load (Q_T equal to* 200, 400, 900, 1,100 *and* 1,300 *kN) [DAG 08]: a) shaft resistance; b) base resistance; c) evolution in (e - log p') plane; and d) evolution in (e - log p')plane – zoom for Q = 200 kN*

– for high loads (900 1,100 and 1,300 kN), a reduction of friction peak (fatigue) is obtained.

On the other hand, the basic strength does not seem to be affected from one cycle to another.

We can conclude that changes in lateral stresses applied on the pile are governed by the volume changes in the shear zone. Therefore, if the mechanisms that control the volume variation are clarified, the mechanisms of load transfer will be better understood, enabling us to understand the mechanisms governing the friction fatigue phenomenon as well.

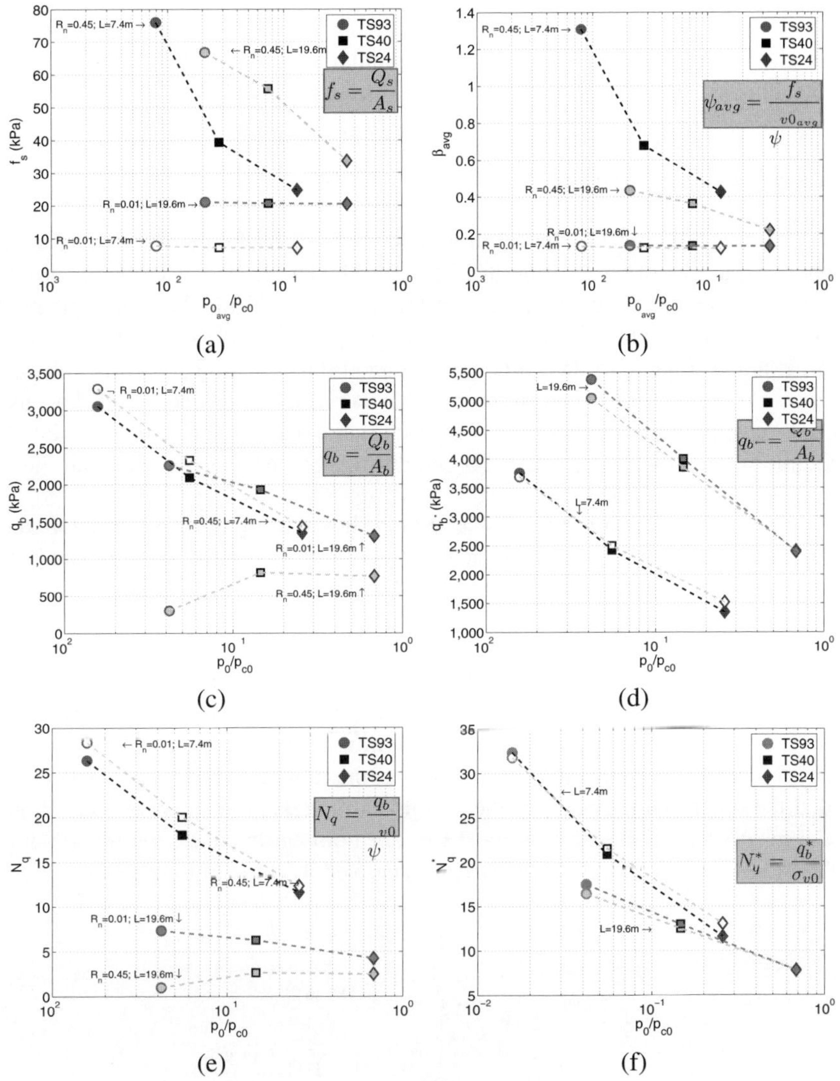

Figure 9.33. *Compilation of the results for Toyoura sand [DAG 08]:*
(a) unit shaft resistance – f_s*;*
(b) average load transfer coefficient in the shaft – β_{avg}*;*
(c) base resistance for $s/d = 10\%$ *–* q_b*;*
(d) base resistance for $s_b/d = 10\%$ *–* q_{b*}*;*
(e) limit load factor for $s/d = 10\%$ *–* N_q*;*
(f) limit load factor $s_b/d = 10\%$ *–* N_q^*

Even if these results require validation relative to observations on real structures, these numerical studies have helped to reveal the mechanisms and parameters that govern the behavior of piles installed in granular soil and loaded axially at a low velocity, considering the permeability of the surrounding soil.

9.5.3. *Earth and rockfill dams*

To illustrate the contribution of numerical methods in the field of earth dams, we present the example of the El Infiernillo dam in Mexico. In this example, we focus on two aspects. The first concerns the methodology of identifying the model parameters for a real structure containing rocks, which is particularly difficult to characterize.The second example relates to the hardening induced during loading and its subsequent influence on the response of the structure. This dam is of particular interest because its materials have been the subject of several studies and because it has undergone regular auscultations from the time it was built and during the earthquakes which hit the area.

The dam, whose maximum cross section is shown in Figure 9.34, is a zoned dam with a height of about 140 m. It consists of a core of low plasticity silty sand ($I_p = 26\% = 45\ w_L\%$), separated from the rockfill by well graded materials of the lateral transition zones. It is located on an area of high seismicity of the Pacific plate.

Ever since its construction, it has suffered from several earthquakes. The strongest ones were those of March 14, 1979 with a magnitude of 7.6 on the Richter scale, and those of September 19, 1985 with a magnitude of 8.1. For the former, with the epicenter located at a distance of 134 km, a maximum acceleration of 0.1 g was noted at the outcropping rock. For the second earthquake, the epicenter was at 68 km and the maximum acceleration was 0.13 g. For both earthquakes, the record showed accelerations on the crest, the downstream face, and the foundation rock.

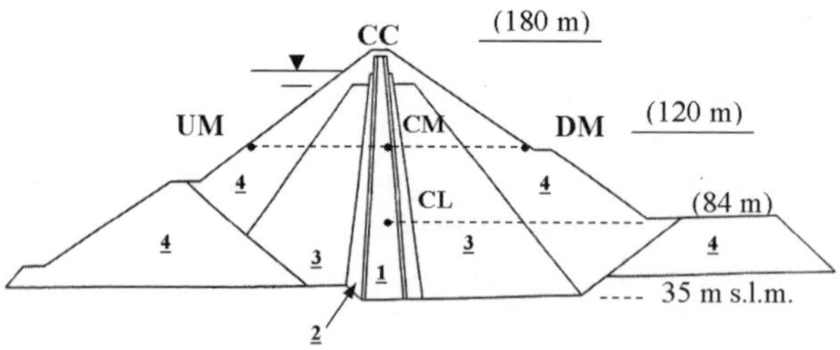

Figure 9.34. *Maximum cross-section of the El Infiernillo dam including different zones and monitoring points: zone 1: core of silty sand; zones 2 and 3: transition zones; zone 4: rockfill; U: downstream, D: upstream, C: center, L: bottom, M: middle, C: crest*

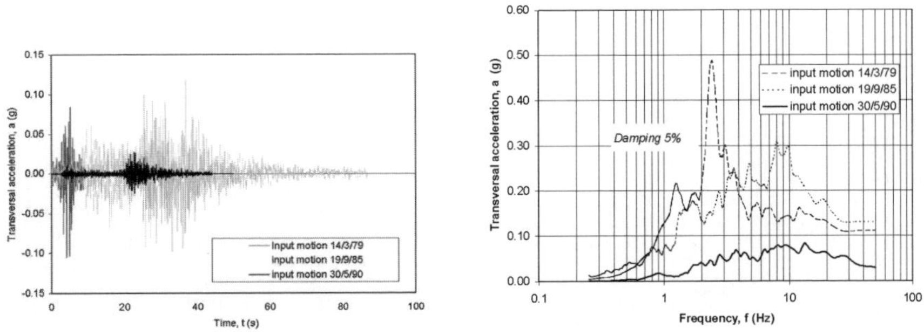

(a) (b)

Figure 9.35. *The registered accelerations (a) on the bedrock at the foundation of the dam and (b) their Fourier transformations*

Measurements of settlements and pore water pressure were also made during the periods of the construction and the earthquakes. These points are shown in Figure 9.34. The recordings on the cofferdam, placed directly on the rock of the foundation, and their Fourier transform are shown in Figure 9.35. Given the high permeability of the rocks, they were supposed to function under drained conditions. Thus, a single-phase model has been used for the rockfill, which takes into account their submerged mass and works under effective stress. However, in transition zones and in the core, a biphasic model taking into account the coupling of soil deformation and pore pressure and transient flow was considered.

A first set of parameters is chosen for the dam material using some experimental results (only monotonic tests) available and the existing correlations. The first calculation of the dam represents the staged construction, impoundment, followed by consolidation and then the earthquake of March 1979. This set of data is then refined by optimizing the settlements and pore pressures during construction and the peak accelerations and displacements induced during the first earthquake. The parameters β, a_1 and a_2 of the constitutive model are calibrated to obtain the best estimation of the settlement profile and pore pressure in the core after construction. In Figure 9.36, we see a comparison of the patterns of settlement in the core and the evolution of pore pressure in a piezometer located in the core.

The analysis of the calculated response of the dam during the earthquake of March 14, 1979 showed that, although the settlement profiles and pore pressure after construction (Figure 9.36) and settlements after the earthquake (Figure 9.37) were calculated in a satisfactory manner, the rigidity of the dam during the earthquake was overestimated. This is illustrated in Figure 9.38(a), where we see the shift towards

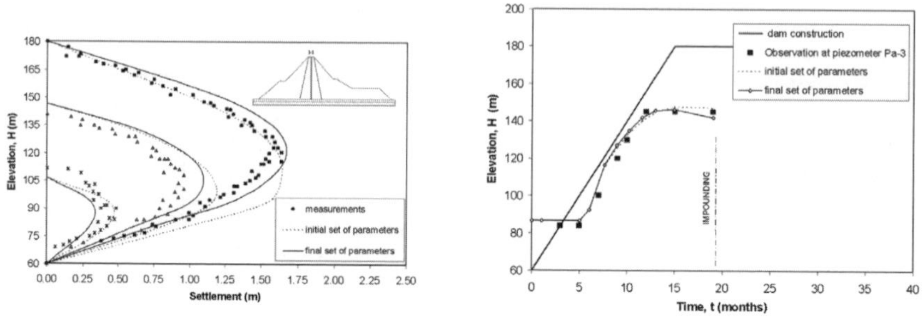

(a) (b)

Figure 9.36. *Comparison of settlements (a) and pore pressures (b) computed and measured at the El Infiernillo dam after construction [SIC 00]*

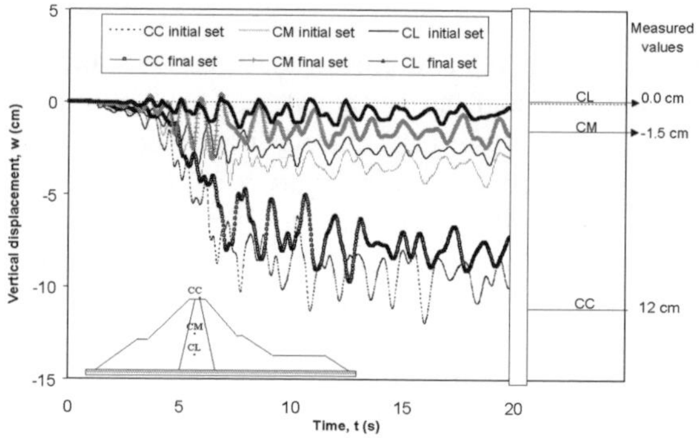

(a)

Figure 9.37. *Comparison of computed and measured vertical displacements at the El Infiernillo dam during the March 14, 1979 earthquake [SIC 00]*

higher frequencies of the calculated response compared to that of the measured signal. To reduce the dominant frequency of the dam, the shear stiffness had to decrease. Thus, to maintain settlement profiles and pore pressure in the core, the shear modulus of the rock was reduced. Moreover, this decrease in shear stiffness is compensated for by an increase in the volumetric stiffness through the elastic and plastic compressibility moduli. Both sets of parameters are summarized in Table A.4 (section 9.8). For lack of information, the mechanical parameters of the transition zone were chosen to be identical to that of the core. However, their permeability has been adjusted based on the materials'function in the dam.

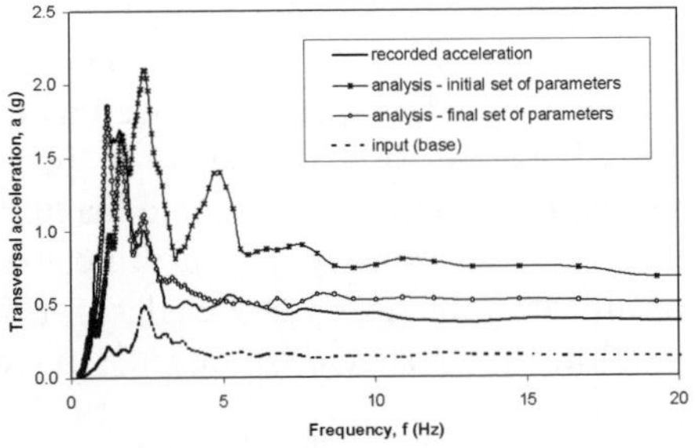

(a)

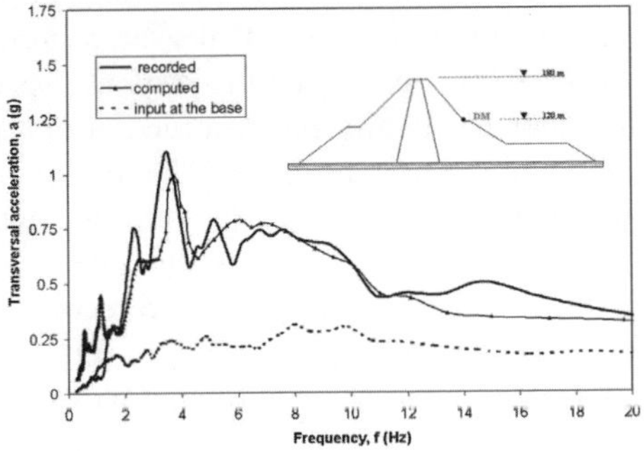

(b)

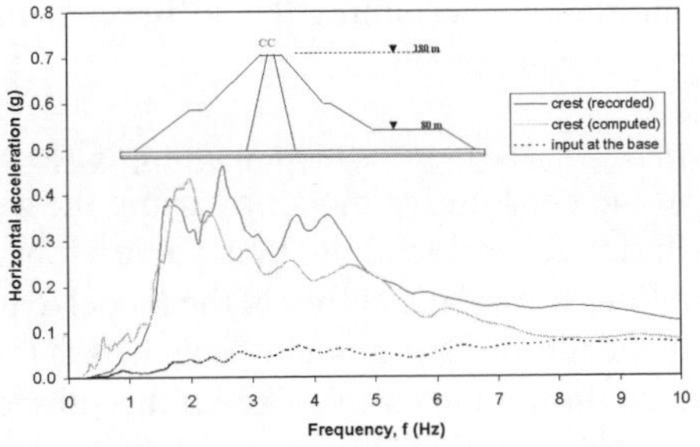

(c)

Figure 9.38. *Horizontal acceleration response spectra of El Infiernillo dam for different earthquakes a) March 14, 1979 at the crest; b) September 19, 1985 at point DM; c) May 30, 1990 [SIC 00]*

As can be seen in Figure 9.38(a), adjusting the stiffness of rocks yielded a better dynamic response. This highlights the important role played by rocks on the behavior of the entire dam and the sensitivity of the dynamic response to a slight modification of the shear modulus of these materials, especially since this modulus is difficult to determine *in situ* or in the laboratory. This analysis also shows that several sets of parameters can lead to the same response in monotonic loading conditions, but only one can provide the seismic response correctly. The identification of the first set could be more accurate if the results of cyclic tests were available. However, given the size of the rock fragments, it is difficult to characterize this material in the laboratory.

To evaluate the predictive quality of the model, the dam response from the earthquake of September 19, 1985 has been analyzed by this second set of data. The acceleration in DM was recorded during this earthquake. In Figure 9.38(b), we present the comparison between the recorded and calculated response spectra at point DM. We can see that the numerical simulations predicts the structure's response at point DM rather well. However, we note the acceleration at the crest measured during the earthquake of May 30, 1990. In Figure 9.38 (c), we can see the response spectra of the dam crest derived from the accelerations calculated and recorded at that point. There is, once again, a good prediction obtained using a calibrated set of parameters obtained using both static and dynamic behavior of the dam. This indicates the effectiveness of the parameter calibration procedure from a static-dynamic analysis in the absence of a suitable series of cyclic tests. This procedure may be recommended for estimating the seismic vulnerability of existing structures because, very often, there is a lack of experimental data and information concerning the material in place and the construction procedure.

Having highlighted the importance of stiffness on the model response, we must try to characterize this parameter as accurately as possible. This might be envisaged by a vibrational analysis of the structure. Thus, the vibration response and the dominant modes can be used for indirectly determining the stiffness of rockfills. This analysis obviously requires the proper equipment.

Through this example, we have presented a comprehensive methodology for identifying parameters of the constitutive model by using the overall response of the structure and especially during seismic loading. Being able to obtain both the frequency content and peak accelerations proved the ability of the model to predict the behavior of the dam under loadings in the future. This model was then used to analyze the influence of strain hardening induced during the earthquake on the subsequent behavior of the dam. Therefore, the dam was submitted again, albeit with different amplitudes, to the same magnitude of earthquake it had previously experienced. The freeboard loss resulting from a second application of the earthquake was smaller than that obtained for the first one because of the hardening induced in the materials. These results are summarized in Figure 9.39.

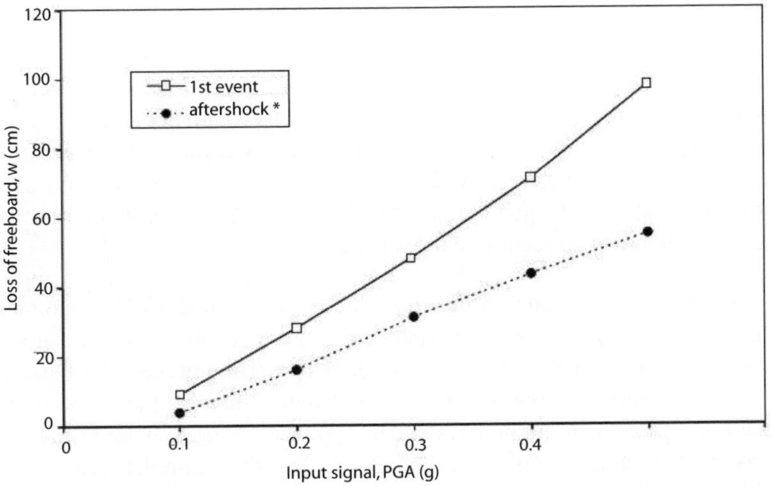

(a)

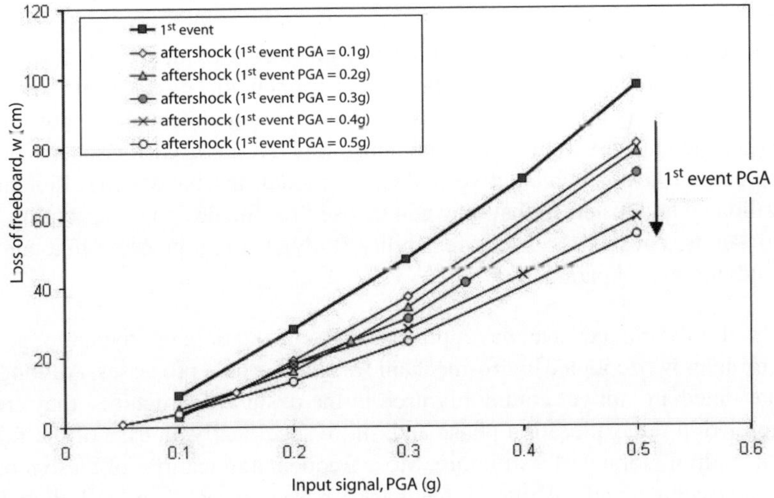

(b)

Figure 9.39. *Evolution of displacements with the maximum acceleration of the input signal; (a) for two consecutive earthquakes of identical amplitudes; (b) for two consecutive earthquakes of different amplitudes [SIC 08]*

9.6. Conclusions

The general principle for establishing a model that represents the behavior of a natural soil at a given spatial scale is to identify the essential mechanisms that govern its behavior. We have seen in this chapter that a relatively simple basic macroscopic model using the finite element method, when it contains the basic mechanisms of deformation, can be used to study diverse and complex problems. We have also shown how this model can be generalized; that is adapted to different geologies by taking into account the physical phenomena involved.

We have also discussed the importance of the model's key parameters and how they need to be identified. The latter presents two advantages: to ensure, on one hand, the consistency of model parameters, and to propose, on the other hand, an acceptable set of parameters in the absence of real data. Most of the work is carried out on disturbed, that is, remolded soils. The validity of the approach must, of course, be verified for natural soils. As obtaining undisturbed samples can be a very delicate task, it is also very important to extend the identification strategies to *in situ* tests such as the use of the pressuremeter, the penetrometer, etc. This can only be achieved after establishing a well-informed database. These various models respond to numerous loading conditions, but some unexplored ways to provide complete solutions to practical problems may still be found. We could take into account, for example, aging or the chemical interaction between soils in any given environment. Any development of this nature will require a rigorous theoretical framework, based on the thermodynamics of the irreversible processes taking place, complemented as always by careful experimental observation. The modeling tool can be used for several purposes. One objective is to obtain predictive models, i.e. models that have been sufficiently mastered, validated and tested so that they can be used for the design of structures. Another use for numerical models is as a feasibility study of a real problem that can be made by studying model parameters.

In this chapter we have presented some static and dynamic geomechanical problems with coupled hydro-mechanic or single-phase processes. Although the models presented are not yet commonly used in the design of structures, they are sometimes applied in the exploration phase and, more specifically, in case of the failure, that is, the malfunctioning of a structure. More frequent applications of these models to study geo-structures and calculation feedbacks, based on real data, will allow better model validation and will help establish, in a particular model, validity domains.

Let us not forget that these models do need real data; hence the need to develop measurement techniques and structure monitoring campaigns. A database containing the characteristics of these materials will help establish a correlation between the properties measured *in situ* and in the laboratory, on the one hand, and the physical and model parameters, on the other hand. Hence, we can limit the number of trials required for a structure to ensure data consistency. It is necessary to take into account uncertainties in the analysis of structures. Non-linear finite element models can be used progressively with this in mind. These methods can be used to assign probabilities of occurrence or to measure the margin of safety with respect to a set criteria. The models which represent physical phenomena remain important in these circumstances.

9.7. Acknowledgments

This document is based on several research projects and PhD theses conducted in the laboratory of soil, structures and materials' mechanics at *Ecole Centrale de Paris* in the framework of the *GEFDyn* project. The author wishes to thank all those who contributed to this project since the 1980s, especially Nabil Abu Bekr, Stefania Sica, Kamran Charles Vossoughi, Fernando Lopez-Caballero, Farhad Elmi (RIP), Kamel Hamadi, Dien Nguyen, Sofia D'Aguiar and Esteban Saezs, in order of chronology.

9.8. Appendix

Test	e_o	D_r	p'_0	β	p_{c0}		A=1,300		A=70	
					J1	J2	K_{ref}	G_{ref}	K_{ref}	G_{ref}
		[%]	(kPa)		(kPa)		(MPa)	(MPa)	(MPa)	(MPa)
146CD050	0.385	48	50	46	20	100	794	476	43	26
146CD100	0.373	51	100	45	70	100	810	486	44	26
146CD400	0.320	67	400	43	250	300	895	537	48	29
189CD200	0.310	70	200	43	600	600	911	547	49	29
200CD050	0.319	67	50	43	700	700	896	538	48	29
200CD200	0.275	80	200	43	1,800	1,200	911	547	49	29
200CD400	0.265	83	400	42	1,400	1,100	990	594	53	32
146CU050	0.380	49	50	45	7	10	801	481	43	26
146CU100	0.363	54	100	45	14	25	827	496	45	27
189CU200	0.296	74	200	43	400	300	935	561	50	30
$p_{ref} = 1$ [MPa]. $n_e = 0.6$ $\phi'_{pp} = 36$ degrees, $\psi = 36$ degrees $a_2 = 0.01$ $r_{ela} = 0.001$. $r_{hys} = 0.01$ $d = 10$. $c_m = 0.2$, $c_{cyc} = 0.1$ $r_{elaiso} = 0.0001$					$\alpha_\psi = 1.5$ $b = 0.2$ $a_1 = 0.005$	$\alpha_\psi = 0.5$ $b = 0.23$ $a_1 = 10^{-4}$	$r_{mbl} = 0.9$ $m = 1$		$r_{mbl} = 0.3$ $m = 3$	

Table A.1. *ECP model's parameters for the numerical simulation of the coarse gravel's behavior*

Sol: ECP multimechanism model's parameters ($TS40$-$TS93$)								
Elasticity		Critical state plasticity			Flow rule hardening		Limits of domains	
			D_r = 40%	D_r = 93%				
$K_{ref}(MPa)$	296	ϕ'_{pp} (degree)	30	30	$\psi(^o)$	30	r_{ela}	0.005
$G_{ref}(MPa)$	222	β	43	17	α_ψ	1	r_{hys}	0.003
n_e	0.4	b	0.22	0.22	a_1	0.0001	r_{mob}	0.8
$p_{ref}(MPa)$	1	$p_{c0}(kPa)$	1,400	4,900	a_2	0.15	$r_{elasiso}$	0.0001
					c_1	0.06		
					c_1	0.03		
					m	1		

Table A.2. *ECP model's parameters for the numerical simulation of Toyoura sand at relative densities of $D_r = 40\%$ and $D_r = 93\%$*

Interface: ECP model's parameters										
Elasticity		Critical state and plasticity			Flow rule			Limit of domains	
			R_n			R_n			
			0.45	0.01		0.45	0.01		
$E_{ref}(MPa)$	532.8	ϕ'_{pp} (degree)	30	14.5	ψ (degree)	30	14.5	r_{ela}	0.005
$G_{ref}(MPa)$	222.0	β	43	43	α_ψ	1	1	r_{hys}	0.003
n_e	0.4	b	0.22	0.0	a_1	0.0001	$0.5 \cdot 10^{-8}$	r_{mob}	0.8
$\sigma_{ref}(MPa)$	1	σ_{co} (kPa)	300	350	a_2	0.15	0.15		
					m	1	1		

Table A.3. *3D ECP model's parameters for the numerical simulation of Toyoura sand-aluminum interface at different rugosities ($R_n = 0.45$ and $R_n = 0.01$). Data used for the simulation of centrifuge tests*

	Core		Rockfill	
	Initial set	Final set	Initial set	Final set
Elasticity				
$K_{ref}(MPa)$	275	275	**470**	**484**
$G_{ref}(MPa)$	128	128	**200**	**165**
n	0.6	0.6	0.3	0.3
$p_{ref}(MPa)$	1	1	1	1
Initial state				
$p_{c0}(MPa)$	0.53	0.53	1.2	1.2
Critical state				
ϕ'_{pp} (degree)	28	28	43	43
β	**29**	**33**	**40**	**52**
d	2	2	2	2
b	0.9	0.9	0.12	0.12
Dilatancy				
ψ (degree)	28	28	40	40
α	1	1	1	1
Deviatoric hardening				
a_1	0.001	0.001	**0.010**	**0.012**
a_2	0.0005	0.0005	**0.005**	**0.006**
Isotropic hardening				
c_1	0.1	0.1	0.004	0.004
c_2	0.05	0.05	0.002	0.002
Behavior domain				
r_{ela}	0.0001	0.0001	0.02	0.02
r_{hys}	0.002	0.002	0.04	0.04
r_{mob}	0.05	0.05	0.8	0.8
r_{iso}	0.0003	0.0003	0.001	0.001
m_p	1	1	1	1
Hydraulic parameters				
k (m/s)	$2\ 10^{-10}$	$2\ 10^{-10}$	-	-
filters	$8\ 10^{-5}$	$8\ 10^{-5}$	-	-

Table A.4. *Parameters used for the material of El Infiernillo dam [SIC 00]; only the parameters in bold have been modified during calibration*

9.9. Bibliography

[ABO 92] ABOU-BEKR N., *Concept de Contraintes Capillaire, Bilan Annuel, Groupe 4: Interactions*, GRECO géomatériaux, 1992.

[ABO 95] ABOU-BEKR N., Modélisation du comportement mécanique et hydraulique des sols partiellement saturés, Doctoral thesis, Ecole Centrale de Paris, July 1995.

[ALO 90] ALONSO E., GENS A., JOSA A., "A constitutive model for partially saturated soils", *Géotechnique*, vol. 3, no. 40, p. 405-430, 1990.

[ANA 96] ANASTASIADIS J., PITILAKIS K., "Shear modulus G_o and damping of typical Greek soils at low strain amplitudes", *Technika Chronika*, vol. 16, no. 3, p. 9-18, 1996.

[AUB 82] AUBRY D., HUJEUX J., LASSOUDIÈRE F., MEIMON Y., *A Double Memory Model with Multiple Mechanisms for Cyclic Soil Behaviour*, Rotterdam, p. 3-13, Balkema, 1982.

[AUB 86] AUBRY D., CHOUVET D., MODARESSI A., MODARESSI H., *GEFDyn: Logiciel d'Analyse de Comportement Mécanique des Sols par Eléments Finis avec Prise en Compte du Couplage Sol-Eau-Air, Manuel Scientifique*, Ecole Centrale de Paris, LMSS-Mat, 1986.

[AUB 87] AUBRY D., BARD E., MODARESSI A., "A mixture approach to the mechanical behaviour of bituminous concrete", *Second International Conference and Short Course on Constitutive Laws for Engineering Materials 87*, Tucson Arizona, 1987.

[AUB 90] AUBRY D., MODARESSI A., MODARESSI H., "A constitutive model for cyclic behaviour of interfaces with variable dilatancy", *Computers and Geotechnics*, no. 9, p. 47-58, 1990.

[AUB 91] AUBRY D., BENZENATTI I., MODARESSI A., MODARESSI H., Exercice de Calcul de Stabilité des Ouvrages de Stockage des Déchets de Type B et C en Couche d'Argile, Rapport du Contrat crsa-andra/brgm, Ecole Centrale de Paris, LMSS-Mat, March 1991.

[AUB 96] AUBRY D., MODARESSI A., *GEFDyn, Manuel scientifique*, Ecole Centrale de Paris, LMSS-Mat, 1996.

[BAR 97] BARDET J., *Experimental Soil Mechanics*, Prentice Hall, Upper Saddle River, NJ, 1997.

[BEE 85] BEEN K., JEFFERIES M., "A state parameter for sands", *Géotechnique*, vol. 35, no. 2, p. 99-112, 1985.

[BEL 98] BELKACEMI H., Expériences sur modèles réduits bidimensionnels et comparaison à des calculs aux éléments finis, Doctoral thesis, Ecole Centrale de Paris, 1998.

[BIA 72] BIAREZ J., FAVRE J., *Corrélations de Paramètres en Mécanique des sols, Formation Continue*, Ecole Centrale de Paris, 1972.

[BIA 93] BIAREZ J., FLEUREAU J., TAIBI S., "Mechanical constitutive model for unsaturated granular media", THORNTON C. (ed.), *The Second International Conference on Micromechanics of granular Media*, Birmingham, Balkema, p. 51-58, Rotterdam,1993.

[BIA 94] BIAREZ J., HICHER P.Y., *Elementary Mechanics of Soil Behaviour, Saturated and Remolded Soils*, A.A. Balkema, Amsterdam, 1994.

[BIA 08] BIAREZ J., HICHER P.Y., "Mechanisms of soil deformation",*Constitutive Modeling of Soils and Rocks*, ISTE, London, John Wiley & sons, New York, 2008.

[BIG 91] BIGONI D., HUECKEL T., "Uniqueness and localization-I. Associative and non associative elastoplasticity", *International Journal of Solids and Structures*, vol. 28, p. 197-213, 1991.

[BIO 41] BIOT M., "General theory of three-dimensional consolidation", *Journal of Applied Physics*, vol. 12, p. 155-164, 1941.

[BIO 57] BIOT M., WILLIS P., "The elastic coefficients of the theory of consolidation", *Journal of Applied Mechanics*, vol. 24, p. 594-601, 1957.

[BIS 54] BISHOP A., "The use of pore pressure coefficients in practice", *Géotechnique*, vol. 4, no. 4, p. 148-152, 1954.

[BIS 63] BISHOP A., BLIGHT G., "Some aspects of effective stress in saturated and partly saturated soils", *Géotechnique*, vol. 13, no. 3, p. 177-197, 1963.

[BOU 82] BOUVARD D., Rhéologie des milieux pulvérulants: étude expérimentale et identification d'une loi de comportement, Doctoral thesis, Université Scientifique et Médicale et Institut Polytechnique de Grenoble, 1982.

[BRU 80] BRULL A., "Caractéristiques mécaniques des sols de fondations de chaussées en fonction de leur état d'humidité et de compacité", *Proceedings of the International Conference on Compaction*, p. 19-37, Paris, 1980.

[CAM 00] CAMBOU B., DI PRISCO C., "Constitutive modeling of geomaterials", *Revue Française de Génie Civil*, vol. 4, no. 5, 2000.

[CAS 38] CASAGRANDE A., *The Shearing Resistance of Soils and its Relation to the Stability of Earth Dams*, Proceedings, Soils and Foundation Conference of the U.S. Engineering Department, 1938.

[CAS 58] CASAGRANDE A., "Notes on the design of the liquid limit device", *Géotechnique*, vol. 8, p. 84-91, 1958.

[CAS 82] CASTRO G., ENOS J., FRANCE J., POULOS S., Liquefaction Induced by Cyclic Loading, Nsf/cee-82018, Report to the National Science Foundation, Washington, DC, 1982.

[CHA 99] CHAMBON R., CAILLERIE D., "Existence and uniqueness theorems for boundary value problems involving incrementally non linear models", *Journal de Mécanique*, vol. 12, no. 2, p. 235-274, 1999.

[CHA 77] CHANG C., DUNCAN J., Analysis of Consolidation of Earth and Rockfill Dams, Report no. TE 77-3, Department of Civil Engineering, University of California, Berkely, 1977.

[COR 05] CORONADO G.O., Etude du comportement mécanique de matériaux granulaires compactés non saturés sous chargements cycliques, Doctoral thesis, Ecole Centrale de Paris, September 2005.

[COR 07] CORONADO O., FLEUREAU J.M., GOMES CORREIA A., CAICEDO B., "Cyclic behavior of Bogotá unbound aggregates", *13th Panamerican Conference on Soil Mechanics and Geotechnical Engineering*, Isla de Margarita, 2007.

[COU 02a] COUSSY O., FLEUREAU J., *Mécanique des Sols non Saturés*, MIM, Hermès, 2002.

[COU 02b] COUSSY O., DANGLA P., *Mécanique des Sols non Saturés*, p. 137-173, MIM Hermès, 2002.

[CYB 98] CYBERQUAKE, A computer-aided design tool for evaluating seismic soil response, BRGM, 1998.

[DAF 82] DAFALIAS Y., HERRMANN L., Bounding surface formulation of soil plasticity *Soil Mechanics- Transient and Cyclic Loads*, p. 253-282, Wiley, USA, 1982.

[DAF 86] DAFALIAS Y., "Bounding surface plasticity. I: Mathematical foundation and hypoplasticity", *ASCE Journal of Engineering Mechanics*, vol. 112, no. GT8, p. 966-987, 1986.

[DAG 08] D'AGUIAR S., Numerical modeling of soil-pile axial load transfer mechanics in granular soils, Doctoral thesis, Ecole Centrale de Paris, Instituto Superior Técnico de Lisbonne, 2008.

[DAR 82] DARVE F., LABANIEH S., "Incremental constitutive law for sands and clays: simulation of monotonic and cyclic tests", *International Journal of Numerical and Analytical Methods in Geomechanics*, vol. 6, p. 243-275, 1982.

[DAR 99] DARVE F., LAOUAFA F., "Plane strain instabilities in soil. Appliction to slope stability", *Proceedings of the International Symposium on Numerical Models in Geomechanics*, Graz, 1999.

[DAR 00] DARVE F., LAOUAFA F., "Instabilities in granular materials and applications to landslides", *Mechanics of Cohesive Frictional Materials*, vol. 5, no. 8, p. 627-652, 2000.

[DOB 82] DOBRY R., LADD R., YOKEL F., CHUNG R., POWELL D., Prediction of Pore Water Pressure Buildup and Liquefaction of Sands During Earthquakes by the Cyclic Strain Method, Report no. 138, National Bureau of Standards, 1982.

[DRU 49] DRUCKER D., "Relations of experiments to mathematical theories of plasticity", *Journal of Applied Mechanics*, vol. 16, p. 349-357, 1949.

[DUN 70] DUNCAN J., CHANG C., "Non linear analysis of stress and strains in soil", *Journal of Soil Mechanics and Foundation Division*, vol. no. 5, p. 1629-1653, 1970.

[ELM 03] ELMI F., Détermination des propriétés des sols marins par reconnaissance géophysique et géotechnique: Intégration totale des données, Doctoral thesis, Ecole Centrale de paris, 2003.

[FAV 80] FAVRE J., Milieu continu et milieu discontinu: mesure statistique indirecte des paramètres rhéologiques et approche probabiliste de la sécurité, Doctoral thesis, UPMC, Paris VI, 1980.

[FIO 02] FIORAVANTE V., "On the shaft friction modeling of non-displacement piles in sand", *Soils and Foundations*, vol. 42, no. 2, p. 23-33, 2002.

[FLE 93] FLEUREAU J.M., KHEIRBEK-SAOUD S., SOEMITRO R. TAIBI S.,"Behaviour of clayey soils on drying-wetting paths", *Canadian Geotechnical Journal*, vol. 30, no. 2, p. 287-296, 1993.

[FLE 02] FLEUREAU J.M., Caractérisation du Matériau de la Digue, Rapport de Recherche, Ecole Centrale de Paris, 2002.

[FOU 10] FOUCAULT A., Modélisation du comportement cyclique des ouvrages en terre intégrant des techniques de régularisation, doctoral thesis, Ecole Centrale de Paris, 2010.

[FRE 77] FREDLUND D., MORGENSTERN N., "Stress state variable for unsaturated soils", *ASCE Journal of Geotechnical Engineering Division*, vol. 103, no. GT5, p. 447-466, 1977.

[FUK 84] FUKUSHIMA S., TATSUOKA F., "Strength and deformation characteristics of saturated sand at extremely low pressures", *Soils and Foundations*, vol. 24, no. 4, p. 30-48, 1984.

[GEU 48] GEUZE E., "Critical density of some Dutch sands", *Proceedings, Second International Conference on Soil Mechanics and Foundation Engineering*, Rotterdam, 1948.

[GHA 82] GHABOUSSI J., KIM K., "Analysis of saturated and partially saturated soils", *International Symposium on Numerical Models in Geomechanics*, p. 377-390, Zurich, September 1982.

[HAJ 84] HAJAL T., Modélisation élasto-plastique des sols, Doctoral thesis, Ecole Centrale de Paris, 1984.

[HAM 06] HAMADI K., Modélisation des bifurcations et des instabilités dans les géomatériaux, Doctoral thesis, Ecole Centrale de Paris, 2006.

[HAM 08a] HAMADI K., MODARESSI-FARAHMAND-RAZAVI A., DARVE F., "Bifurcation and instability modeling by a multimechanism elasto-plastic model", *International Journal for Numerical and Analytical Methods in Geomechanics*, vol. 32, no. 5, p. 461-492, 2008.

[HAM 08b] HAMADI K., MODARESSI-FARAHMAND-RAZAVI A., DARVE F., "Caractérisation numérique du phénomène de localisation des déformations dans des essais biaxiaux sur sable", *European Journal of Environmental and Civil Engineering*, vol. 12, no. 6, p. 651-671, 2008.

[HAR 63] HARDIN B., RICHART JR. F., "Elastic waves velocities in granular soils", *Journal of Soil Mechanics and Foundation Division*, vol. 89, no. 1, p. 33-65, 1963.

[HAR 68] HARDIN B., BLACK W., "Vibration modulus of normally consolidated clays", *Journal of Soil Mechanics and Foundation Division*, vol. 94, no. 2, p. 353-369, 1968.

[HAR 72] HARDIN B., DRNEVICH V., "Shear modulus and damping in soils: design equations and curves", *Journal of Soil Mechanics and Foundation Division*, vol. 98, no. 7, p. 667-692, 1972.

[HAR 78] HARDIN B., "The nature of stress-strain behavior for soils", *Proceedings ASCE Geotechnical Engineering Division, Specially Conference On Earthquake Engineering and Soil Dynamics*, vol. I, p. 3-89, Pasadena, CA, 1978.

[HAR 02] HARIWARDOYO S., Caractérisation des matériaux granulaires des très petites aux grandes déformations, Doctoral thesis, Ecole Centrale de Paris, 2002.

[HIC 94] HICHER P., RAHMA A., "Micro-macro correlations for granular media. Application to the modeling of sands", *European Journal of Mechanics. A/Solids*, vol. 13, no. 6, p. 763-781, 1994.

[HIC 02] HICHER P., SHAO J., *Elastoplasticité des Sols et des Roches*, MIM, Hermès, Paris, 2002.

[HUJ 79] HUJEUX J., Calcul numérique de problème de consolidation élastoplastique, Doctoral thesis, Ecole Centrale de Paris, June 1979.

[HUJ 85] HUJEUX J., *Une Loi de Comportement pour le Chargement Cyclique des Sols, Génie Prarasismique*, p. 287-302, Presses ENPC, 1985.

[IDR 73] IDRISS I., LYSMER J., HWANG R., RICHARD N. , SEED H.B., QUAD-4: a Computer Program for Evaluating the Seismic Response of Soil Structures by Variable Damping Finite Element Procedures, Pacific Earthquake Engineering Research Center, UCB/EERC-73/16, 1973.

[IND 91] INDARTO, Comportement mécanique et compactage des matériaux de barrages, Doctoral thesis, Ecole Centrale de Paris, 1991.

[ISH 75] ISHIHARA K., TATSUOKA F., YASUDA S., "Undrained deformation and liquefaction of sand under cyclic stresses", *Soils and Foundations*, vol. 15, no. 1, p. 29-44, 1975.

[ISH 93] ISHIHARA K., "Liquefaction and flow failure during earthquakes. 33rd Rankine lecture", *Géotechnique*, vol. 43, no. 3, p. 351-415, 1993.

[IWA 78] IWASAKI T., TATSUOKA F., TAKAGI Y., "Shear moduli of sands under cyclic torsional shear loading", *Soils and Foundations*, vol. 18, no. 1, p. 39-56, 1978.

[JAM 85] JAMIOLKOWSKI M., LADD C., GERMAINE J., LANCELLOTTA R., "New developments in field and laboratory testing of soils", *Proceedings of 11th International Conference Soil Mechanics And Foundation Engineering*, vol. 1, p. 57-154, San Francisco,1985.

[KAB 91] KABILAMANY K., ISHIHARA K., "Cyclic behaviour of sand by the multiple shear mechanism model", *Soil Dynamics and Earthquake Engineering*, vol. 10, no. 2, p. 74-83, 1991.

[KAL 99] KALLIOGLOU P., TIKA T., PITILAKIS K., "Dynamic characteristics of natural cohesive soils", *Proceedings 2nd International Conference On Earthquake Geotechnical Engineering*, Lisbon, 1999.

[KHA 98] KHALILI N., KHABBAZ M., "A unique relationship for the determination of shear strength of unsaturated soils", *Géotechnique*, vol. 48, no. 5, p. 681-688, 1998.

[KLO 01] KLOTZ E. U., COOP M. R., "An investigation of the effect of soil on the capacity of driven piles in sands", *Géotechnique*, vol. 51, no. 9, p. 733-751, 2001.

[KOD 83] KODAISSI E., Un modèle de loi de comportement viscoplastique appliqué à la mécanique des sols, doctoral thesis, UMPC Paris VI, 1983.

[KOK 82] KOKUSHO T., YOSHIDA Y., ESASHI Y., "Dynamic properties of soft clays for wide strain range", *Soils and Foundations*, vol. 22, no. 4, p. 1-18, 1982.

[KOL 00] KOLYMBAS D., "Elements of hypoplasticity", *Revue Française de Génie Civil*, vol. 4, no. 5, p. 85-108, 2000.

[KOP 48] KOPPEJAN A., VAN WAMELEN B., WEINBERG L., "Coastal landslides in the Dutch province of Zealand", *Proceedings, Second International Conference on Soil Mechanics and Foundation Engineering*, p. 89-96, Rotterdam, 1948.

[KOR 95] KORDJANI M., Caractérisation et modélisation du comportement en petites déformations des sols sur des chemins de sollicitations monotone et cyclique, doctoral thesis, Ecole Centrale de Paris, 1995.

[LAD 77] LADE P., "Elasto-plastic stress-strain theory for cohesionless soil with curved yield surfaces", *International Journal of solids and Structures*, vol. 13, p. 1019-1035, Pergamon Press Inc., New York, 1977.

[LAL 95] LALOUI L., Modélisation du comportement thermo-hydro-mécanique des milieux poreux anélastiques, doctoral thesis, Ecole Centrale de Paris, 1995.

[LAM 79] LAMBE T., WHITMAN R., *Soil Mechanics*, John Wiley and Sons, New York, 1979.

[LEM 09] LEMAITRE J., CHABOCHE J.L., BENALLA A., DESMORAT R., *Mécanique des Matériaux Solides*, Dunod, Paris, 2009.

[LLO 85] LLORET A., ALONSO E., "State surface for partially saturated soils", *Proceedings of 11th International Conference on Soil Mechanics and Foundation Engineering*, p. 557-562, 1985.

[LOP 03a] LOPEZ-CABALLERO F., Influence du comportement non linéaire du sol sur les mouvements sismiques induits dans des géo-structures, doctoral thesis, Ecole Centrale de Paris, 2003.

[LOP 03b] LOPEZ-CABALLERO F., MODARESSI A., ELMI F., "Use of laboratory and in-situ testing to identify an elastoplastic model parameters", *Symposium on the Deformation Characteristics of Geomaterials-ISLYON*, 2003.

[LOP 07] LOPEZ-CABALLERO F., MODARESSI FARAHMAND-RAZAVI A., MODARESSI H., "Non-Linear numerical method for earthquake site response analysis I- elastoplastic cyclic model & parameter identification strategy", *Bulletin of Earthquake Engineering*, vol. 5, no. 3, p. 303-323, 2007.

[LUO 80] LUONG M., "Phénomènes cycliques dans les sols pulvérulents", *Revue Française de Géotechnique*, vol. 10, p. 39-53, 1980.

[LYS 75] LYSMER J., UDAKA T., TSAI C., SEED H., *A Computer Program for Approximate 3-D Analysis of Soil-structure Interaction Problems*, Pacific Earthquake Engineering Research Center, UCB/EERC-75/30, 1975.

[MAG 91] MAGNAN J.P., MESTAT P., "Lois de comportement et modélisation des sols", *Techniques de l'Ingénieur, Traité Construction Volume C21*, Paris, 1991.

[MAN 87] *Manuel de Rhéologie des Géomatériaux*, Presse des Ponts, 1987.

[MAR 72] MARCUSON W., WAHLS H., "Time effects on dynamic shear modulus of clays", *Journal of Geotechnical Engineering Division*, vol. 98, no. 12, p. 1359-1373, 1972.

[MAS 26] MASING G., "Eigenspannungen und verfestigung beim Messing (Self stretching and hardening for brass)", *Proceedings of the Second International Congress for Applied Mechanics*, Zurich, p. 332-335, 1926.

[MAT 68] MATYAS E., RADHAKRISHNA H., "Volume change characteristics of partially saturated soils", *Géotechnique*, vol. 18, no. 4, p. 432-448, 1968.

[MES 93] MESTAT P., Lois de Comportement des Géomatériaux et Modélisation par la Méthode des Éléments Finis, Report no. GT52, LCPC, 1993.

[MIC 94] MICHALI A., Méthode pour l'identification des paramètres d'une loi élastoplastique à partir d'essais de laboratoire et in-situ, doctoral thesis, Ecole Centrale de Paris, 1994.

[MOD 89a] MODARESSI H., AUBRY D., "Numerical modeling for the flow of compressible fluids in systems of deformable rocks", *Proceedings of Numerical Models in Geomechanics, NUMOG III*, Niagara Falls, p. 391-398, Elsevier, 1989.

[MOD 89b] MODARESSI-FARAHMAND RAZAVI A., Modelisation mécanique des bétons bitumineux; applications aux barrages en Remblais, doctoral thesis, Ecole Centrale de Paris, 1989.

[MOD 94a] MODARESSI A., ABOU-BEKR N., "Constitutive model for unsaturated soils; Validation on a silty material", *Proceedings of Numerical Methods in Geotechnical Engineering, Manchester*, Balkema, Rotterdam, vol. 1 p. 91-96, 1994.

[MOD 94b] MODARESSI A., ABOU-BEKR N., "A unified approach to model the behavior of saturated and unsaturated soils", *Computer Methods and Advances in Geomechanics, Proceedings of the Eighth International Conference on Computer Methods and Advances in Geomechanics, Morgentown*, Balkema, Rotterdam, p. 1507-1513, 1994.

[MOD 95] MODARESSI A., MODARESSI H., "Thermoplastic constitutive model for unsaturated soils: A prospective appraoch", *Proceedings of Numerical Models in Geomechanics, NUMOG V*, Balkema, Rotterdam, p. 45-50, 1995.

[MOD 96] MODARESSI A., ABOU-BEKR N., FRY J., "Unified approach to model partially saturated soil unsaturated soils", *Proceedings of Numerical Methods in Geotechnical Engineering, Manchester*, Balkema, p. 1495-1502, Rotterdam, 1996.

[MOD 97] MODARESSI H., LALOUI L., "A thermoviscoplastic constitutive model for clays", *International Journal for Numerical and Analytical Methods in Geomechanics*, vol. 21, p. 313-335, 1997.

[MOD 03] MODARESSI-FARAHMAND-RAZAVI A., Modélisation des Milieux Poreux Sous Chargements Complexes, thesis, Institut National Polytechnique de Grenoble, 2003.

[MRO 67] MROZ Z., "On the description of anisotropic work-hardening", *Journal of Mechanics and Physics of Solids*, vol. 15, p. 163-175, 1967.

[MRO 81] MROZ Z., V.A. NORRIS, O.C. ZIENKIEWICZ, "An anisotropic critical state model for soils subjected to cyclic loading", *Géotechnique*, vol. 31, p. 451-469, 1981.

[NGU 05] NGUYEN T., FLEUREAU J., MODARESSI A., "Identification of parameters and validation of Hujeux model for a coarse soil", *Proc. 11th Int. Conf. of Int. Ass. of Computer Methods and Advances in Geomechanics*, Torino, 2005.

[NGU 06] NGUYEN T. D., Comportement des matériaux et des structures de barrages, Doctoral thesis, Ecole Centrale de Paris, 2006.

[NOV 04] NOVA R., "The role of non-normality in soil mechanics and some of its mathematical consequences", *Computers and Geotechnics*, vol. 31, p. 185-191, 2004.

[PAS 85] PASTOR M., ZIENKIEWICZ O., LEUNG K., "Simple model for transient soil loading in earthquake analysis. II Non-associative models for sands", *International Journal of Numerical and Analytical Methods in Geomechanics*, vol. 9, p. 477-498, 1985.

[PAS 95] PASTOR M., QUECEDO M., "A patch test for mesh alignment effects in localized failure", *Communications in Numerical Methods in Engineering*, vol. 11, no. 12, Dec 1995.

[PER 66] PERZYNA P., "Fundamental problems in viscoplasticity", *Advances in Applied Mechanics*, vol. 9, p. 243-377, 1966.

[PIC 91] PICCUEZZU E., Lois de comportement en géomécanique; Modélisation, mise en oeuvre, identification, Doctoral thesis, Ecole Centrale de Paris, 1991.

[POU 81] POULOS S., "The steady state of deformation", *ASCE Journal of the Geotechnical Engineering Division*, vol. 107, no. GT5, p. 553-562, 1981.

[POU 95] POULAIN D., ALONSO E., GIRARD H., "Pore pressure development in compacted wet clayey soils: an approach using simple laboratory tests", *Proceedings of Conference on Unsaturated Soils, Paris*, Balkema, Rotterdam, p. 541-546, 1995.

[PRE 90] PREVOST J., KEANE C., "Multimechanism elasto-plastic model for soils", *Journal of Engineering Mechanics*, vol. 116, no. 9, p. 1924-1944, 1990.

[RIC 76] RICE J., "The Localization of plastic deformation", KOITER W. T., (ed.), *Theoretical and Applied Mechanics, 14th IUTAM Congress*, Amsterdam, North-Holland Publishing Co., p. 207-220, 1976.

[ROM 95] ROMO M., "Clay behavior, ground response and soil-structure interaction studies in Mexico city. State of the Art paper", *3rd International Conference on recent Advances in Geotechnical Earthquake Engineering and Soil Dynamics*, vol. II, Prakash, p. 1039-1051, St. Louis, Missouri, 1995.

[ROS 58] ROSCOE K., SCHOfiELD A.N., WROTH C.P., "On the yielding of soils", *Géotechnique*, vol. 8, p. 22-53, 1958.

[SAI 97] SAÏM R., Des comportements repères des grains sans colle à un exemple de sol réel, Doctorate thesis, Ecole Centrale de Paris, 1997.

[SAN 99] SANTOS J., Caracterização de solos através de ensaios dinâmicos e cíclicos de torção; Aplicação ao estudo do comportamento de estacas sob acções horizontais estáticas e dinâmicas, Doctoral dissertation, Universidade Técnica de Lisboa, Instituto Superior Técnico, Portugal, 1999.

[SCH 02] SCHWEIGER H., *Benchmarking in Geotechnics, CGG-IR006-2002*, 2002.

[SEE 70] SEED H., IDRISS I., Soil Moduli and Damping Factors for Dynamic Response Analyses, Report no. EERC-70-10, University of California, Berkeley, CA, 1970.

[SEE 72] SEED H., LYSMER J., SCHNABEL P., *Shake: a Computer Program for the 1-D Analysis of Geotechnical Earthquake Engineering Problems*, University of California, Berkeley, UCB/EERC-72/12, 1972.

[SIC 00] SICA S., Analisi del comportamento dinamico di dighe in terra, doctorat, Università di Roma & Napoli 'Federico II', 2000.

[SIC 08] SICA S., PAGANO L., MODARESSI A., "Influence of past loading history on the seismic response of earth dams", *Computers and Geotechnics*, vol. 35, no. 1, p. 61-85, 2008.

[SKE 53] SKEMPTON A., "The colloidal activity of clays", *Proceedings of 3rd International Conference in Soil Mechanics and Foundation Engineering*, Switzerland, vol. I, p. 57, 1953.

[SKE 61] SKEMPTON A., *Effective Stress in Soils, Concrete and Rocks; Pore Pressure and Suction in Soils*, Butterworths, 1961.

[SLA 85] SLADEN J., D'HOLLANDER R., KRAHN J., "The liquefaction of sands, a collapse surface approach", *Canadian Geotechnical Journal*, no. 22, p. 564-578, 1985.

[TAI 94] TAIBI S., Comportement mécanique et hydraulique des sols soumis à une pression interstitielle négative, Etude expérimentale et modélisation, doctoral thesis, Ecole Centrale de Paris, 1994.

[TAT 01] TATSUOKA F., Impacts on Geotechnical Engineering of Several Recent Findings from Laboratory Stress-strain Tests on Geomaterials, Report, University of Tokyo, Geotechnical Engineering Laboratory, 2001.

[TAT 08] TATSUOKA F., DI BENEDETTO H., ENOMOTO E., KAWABE S., KONGKITKUL W., "Various viscosity of geomaterials in shear and their mathematical expression", *Soils and Foundations*, vol. 40, 2008.

[TER 67] TERZAGHI K., PECK R., *Soil Mechanics in Engineering Practice*, John Wiley & Sons, New York, 1967.

[TOL 90] TOLL D., "A framework for unsaturated soil behaviour", *Géotechnique*, vol. 40, no. 1, p. 35-38, 1990.

[VOS 01] VOSSOUGHI K., Etude numérique du comportement des ouvrages de soutènement à la rupture, doctoral thesis, Ecole Centrale de Paris, 2001.

[VUC 91] VUCETIC M., DOBRY R., "Effect of soil plasticity on cyclic response", *Journal of Geotechnical Engineering*, vol. 117, no. 1, 1991.

[VUC 94] VUCETIC M., "Cyclic threshold shear strains in soils", *Journal of Geotechnical Engineering*, vol. 120, no. 12, 1994.

[WRO 78] WROTH C., WOOD D., "The correlation of index properties with some basic engineering properties of soils", *Canadian Geotechnical Journal*, vol. 15, no. 2, p. 137-145, 1978.

[ZER 91] ZERHOUNI M., Rôle de la pression interstitielle négative dans le comportement des sols - Application aux routes, doctoral thesis, Ecole Centrale de Paris, 1991.

[ZIE 99] ZIENKIEWICZ O., CHAN A., PASTOR M., SCHREFLER B., SHIOMI T., *Computational Geomechanics*, John Wiley & Sons 1999.

Chapter 10

Evaluating Seismic Stability of Embankment Dams

10.1. Introduction

10.1.1. *A tribute to Jean Biarez*

Sometimes Jean Biarez began teaching with a newspaper. He would read and then comment upon the latest engineering incident. With these news items, that were likely to capture our attention, he would lead us to think critically about the overall behavior of the ground and the embedded structure. He pointed out that, well beyond formulae and regulations, the first rule in geotechnical engineering was observation. In this respect, his own work was exemplary: he contributed to soil mechanics by setting up a framework to solve the problems posed by the contradictions and enigmas between site observation and laboratory testing.

At a time when dams were hardly analyzed, dam failures provided Biarez with many lessons that he transformed into teaching material. As a member of the panel of technical experts following the double dam failures at Malpasset and Vajont, he contributed to developing the critical expertise required by the Permanent Technical Committee on Large Dams in France, which considered that each site or each dam was unique, hence requiring unique expertise. For decades, Biarez has been a model to engineers educated in France, who through experience and critical thinking try to obtain the particular solution suited to each site. Rather than apply the many existing standards and recommendations, he encouraged us to collect many observations, to

---

Chapter written by Jean-Jacques FRY.

sort and classify them in order to understand the links between them and, only then, to model their influence on the projected structure. In keeping with his approach, this chapter will analyze the seismic behavior of dams through an initial appraisal of observed accidents and assessed failures before launching any modeling.

Jean Biarez was among the first soil mechanics in France to focus on the specific non-linearities of the ground. Full of admiration for the pioneering work of Duncan and Chang, he promoted the development and use of constitutive laws to fit the observed performance of earthen structures. He showed that they were essential to predict the movements of dams. With the collaboration of Al Issa, he showed the way to model the construction and first filling of the Mont-Cenis dam (1968). He never lost sight of the main pedagogical and practical purpose, however, which was to identify, among all the possible reactions of the soil, the key phenomenon that governed the behavior of the structure under the studied loading.

Therefore, over the years, his disappointment with the inflation of increasingly complex numerical models became such that, on behalf of the practitioner, he demanded simpler models that were less ambitious but more dependable. Towards the end of his career, in order to help the "numerical analysts" simplify their models, he combined the acquired experimental soils to extract their "key behaviors", a set of behaviors unique to granular or cohesive soils ("soils with glue", as he liked to say).

Biarez never ceased to repeat that any numerical approach had to be validated by its ability to reproduce the "collapse" of any earthen structure, the ultimate phase of the behavior studied. The underlying idea was to find the main phenomenon, which triggered the failure of dams under seismic loading, and to have it implemented into the analysis. In accordance with this line of thought, a seismic modeling of an embankment dam is described in these pages by a deliberately simplified approach. Again, the basic idea is to provide the engineer with a rapid method with which to reinforce his/ her judgment on the dam performance, allowing him/her to improve his/her design of such works as quickly as possible. The most critical phenomenon leading to failure under seismic hazard is liquefaction. A coupled analysis will be presented in the last part of this chapter, illustrated by an application to embankments and their foundation in gravelly alluviums.

10.1.2. *Definitions*

In the field of dam engineering, failure is the word used in accordance with the practices of the International Commission on Large Dams (ICOLD). It means the loss of water storage capacity. This definition is adopted in this text. As a corollary, a sliding with no impact on the level of the reservoir is not considered a failure. In the following text, attention will be paid to the conditions of sliding that could cause collapse and release of water.

A distinction is made between large dams and small dams. If the consequences of the failure of a large dam are considered catastrophic, those of small dams – even though not negligible – are often socially acceptable to the extent that it is appropriate to distinguish them, as is the case in Japan. The only valid distinction between these two categories of dams is the nature of the consequences of a possible failure. Without data for assessing the importance of such consequences, the ICOLD distinction between small and large dams will be adopted: all dams with a height greater than 15 m are considered large dams.

10.2. Observed seismic performance

The pathology of the seismic behavior of dams is very limited in France. The strong earthquakes with an MSK intensity greater than VIII hit France at a time before books or reports existed. However, the international literature every year gathers the testimonies of more comprehensive data on the impact of earthquakes. The ICOLD regularly reports the damage that the strongest earthquakes cause on large dams [ICO 01]. American publications [SEE 73] and, more recently, Japanese or Chinese publications contain the post-seismic feedback from major earthquakes such as San Fernando (USA, 1971), Loma Prieta (USA, 1989), Kobe (Japan, 1995), Chi-Chi (Taiwan, 1999) and Wenchuan (China, 2008) and detailed observations on the seismic performance of dams.

10.2.1. *Earthquake performance of gravity dams*

On April 18, 1906, the Lower Crystal Springs Dam, a 38.5 m high curved concrete gravity dam, built of interlocking concrete blocks, withstood the San Francisco earthquake, M = 8.3 (estimated), without a single crack. The active fault rupture was located less than 180 m from the dam [ICO 01].

The Koyna Dam [CHO 72], built in 1963, is a 102 m high concrete gravity buttress dam in Iran. Its section is unusual: the 36 m upper part has a very steep slope H/V = 0.153, whereas the lower part has a more traditional slope of H/V = 0.72. The dam was calculated to withstand a pseudo-static acceleration of 0.05 g. On December 11, 1967, the dam was shaken by an earthquake suspected of being caused by reservoir-triggered seismicity [ICO 01] at a distance of 8–13 km and a depth of 13 km with a magnitude of M = 6.5. An accelerograph in a gallery on the right bank recorded the maximum values of the components of acceleration as 0.63 g perpendicular to the valley, 0.49 g in the direction of the valley and 0.34 g vertically with a strong phase of shaking of 6 s. Traces of relative movements between blocks were observed. Cracks were also observed in the gallery operating at half height and in the gallery foundation. Horizontal cracks of 35 to 45 m were visible at the downstream slope change. That damage was attributed to the high inertia of this quite unusual section, a phenomenon avoidable in modern structures.

The Kasho Dam with a height of 46 m at its base was subjected to an acceleration of 0.53 g during the Tottori-Ken Seibu earthquake (M = 7.3) in 2000. The peak of acceleration at the crest reached 2.00 g, with only a few cracks on the walls and slabs of the monitoring center!

The only large dam failure is a recent one. It concerns the 25 m gate-structure Shih Kang Dam built in 1977 in Taiwan, which broke during the Chi-Chi earthquake on September 21, 1999. The dam has two bottom valves and 18 discharge gates. The impressive picture of the damage is that of the concrete wall cut by the differential movements of 9 m following the rejection of the active fault at Chelungpu, which was sheared along a length of 105 km and shortened the valley by 6 m. The dam, built was calculated with a seismic coefficient of 0.15. The maximum acceleration measured nearest the site was 0.51 g horizontal and 0.53 g vertical. The reason for failure was the vertical rejection of the reverse fault at Chelungpu; previous studies and excavations had not revealed the position of the fault. The entire structure was severely cracked. Disorderly displacements of the foundation layers of argillite, siltstone and sandstone disassociated the foundations of the dam blocks. Such a construction on an active fault is not recommended in international practice and is prohibited in France. The difficulty here is that the fault branches in a complex system cannot be detected by investigations. The displacement of one of its ramifications sheared the tunnel. Despite these enormous irrecoverable displacements, the dam was repaired and continues to be operated with a reservoir of 0.4 instead of 2.7 hm^3 [WIE 03].

10.2.2. *Earthquake performance of buttress dams*

Hsinfengkiang Dam [SHE 74] in China has a height of 104 m and is supported by 19 buttresses. In 1962, it was shaken by a 6.1 magnitude earthquake, whose epicenter was only 1 km away. This dam developed a 82 m longitudinal crack near the top towards the right bank and smaller cracking at the same elevation on the left bank.

Sefi Rud Dam [AHM 92] is a 105 m high concrete gravity buttress dam in Iran with 23 counters. It was cracked in the central part by the Rudbar earthquake on June 21, 1990 (M = 7.6) at a time when the reservoir was 6 m below the level downstream. The maximum acceleration was estimated to be 0.7 g at the foundation and 2.8 g at the crest. It was calculated to have no traction on the upstream face under a pseudo-static coefficient of 0.1 and allowable stresses, less than 7 MPa in compression and 0.7 MPa in tension. The dam performed well during and after the earthquake, despite several cracks in the upper part [ZAK 03]. These cracks were concentrated in three locations: at the junction of the central buttresses and the wall; in the central buttresses at the construction joints; and at the joints between the buttresses. A horizontal crack opening a construction joint spanned 24 buttresses. Although a triangular wedge was formed under the buttress head, the maximum displacement did not exceed 2.5 cm. (see Figure 10.4). The separate parts of the buttresses underwent only small rotations without slipping. Vertical expansion joints were damaged and some leaked, but the leaks did not increase with time. A sharp drop in uplift was recorded at the base of the dam; on the other hand, leaks occurred downstream of the horizontal cracks, running through the plot. Despite the existence of horizontal cracks, the large accelerations, and the fact that the cracked buttresses could move freely in an upstream–downstream direction, no significant horizontal motion was observed. In the meantime, some indirect effects, which may lead to breakage in other circumstances, have been reported. The leak of the drainage gallery, from the opening of a crack at the location of a fault, increased over time. Numerous rock falls on the roads nearby, including the banks, have suspended traffic. Several power poles were cut down. The control room of a factory was badly damaged.

10.2.3. *Earthquake performance of arch dams*

On February 9, 1971, the Pacoima Arch Dam was subjected to the San Fernando, CA, earthquake (M = 6.5). The arch dam was located about 6 km from the epicenter of the earthquake. This arch, built in 1929, has a double curvature and a height of 113 m. An accelerograph, placed on the left bank at 40 m distance and 15 m above the crest, recorded an unprecedented peak acceleration of 1.25 g horizontal and 0.70 g vertical. The reservoir was 45 m below the crest. The arch was able to bear the shock, which indicates a margin of safety inherent in this type of structure. Yet the valley had narrowed by more than 2 cm. The greatest damage observed was the opening of a 6 mm to 1 cm contraction joint along a length of 15 m between the arch and the thrust block on the left bank and cracking of the block for a length of 2 m. During the Northridge earthquake, the only effects recorded on arch dams were a few openings in the vertical expansion joints after several cycles of opening and closing.

It is essential to analyze the movements of joints between blocks, for they dissipate energy and unacceptable tensions appear in a numerical model without joints [ARI 03]. After validating the 3D model attached to the Shintoyone Dam, where a maximum acceleration of 1.0 g was recorded on the crest in 1997 during an earthquake of M = 5.8, [ARI 03] showed that a major earthquake with a base acceleration of 0.5 g will lead to a maximum acceleration of 3.0 g at the crest without compromising safety.

10.2.4. *Earthquake performance of hydraulic fills*

The hydraulic fill method of construction was a cheap but dangerous method. Used in the first part of the 20th Century, it was abandoned in the 1940s, after the failure of the Fort Peck Dam in 1938 [JAN 76]. The process of transporting pumped fine soils, sands, silts, and their deposit by sedimentation had the advantage of being economical before earth-moving machines existed and the disadvantage of not compacting the soil. Thus, the soil was normally consolidated under its own weight or compacted after seismic loading. As a result, a number of hydraulic fills failed during construction or after an earthquake.

For instance, the Sheffield Dam is a 8 m high sandy hydraulic fill, built in 1917 on a sandy foundation. A layer of clay on the upstream face is extended by a cut-off trench in the foundation, which acts as a seal. A breach of 30 m was caused by the liquefaction of the sandy area under water during the 1925 Santa Barbara earthquake (M = 6.3).

The lower Van Norman Dam, a 42 m hydraulic fill, built between 1912 and 1930, experienced instabilities and leaks after construction that warranted extra downstream toe, although the top part was compacted. It was shaken by the San Fernando earthquake on February 9, 1971 (M = 6.6), with a maximum acceleration measured between 0.55 and 0.60 g. The downstream face moved downstream around 1 m without failure, while the upstream face disappeared in a major slide of 42m. During the earthquake, the crest was wavering, liquefaction softening the resistance of the lower part of the upstream shell by pore pressure rise, but the major sliding started 26 s after the earthquake and developed during 50 s with pore pressure dissipation of up to the residual strength. Overtopping of the crest and a major failure was miraculously avoided because the reservoir water level was relatively low when the earthquake occurred, leading to a final freeboard of 1.4 m.

The concerns regarding the vulnerability of hydraulic fills to liquefaction were so clearly confirmed after the Van Norman Dam sliding that the Department of Water Resources of the State of California [JAN 76] ordered a reassessment of all the 36 hydraulic fills in the state of California. Many of them had to be replaced or rehabilitated. According to this study, the seismic threshold of liquefaction appeared of the order of 0.2 g for a signal of long duration and low frequencies and 0.3 g for a signal of short duration with high frequencies.

10.2.5. *Earthquake performance of tailing dams*

Nine failures of tailing dams were reported by ICOLD (2001) in *Technical Bulletin 120*. In 1965, following the La Ligua earthquake (M = 7.1) in Chile, several tailing dams broke: El Soldado, La Patagua, and El Cobre. The mudslide caused by the failure of the El Soldado Dam (H = 35 m) caused 200 fatalities. In Peru, the Shorey Dam failed in 1969 and that of Huachocopolca in 1970. In 1971, three other failures occurred in Chile: Salamanca and Illapel under a M = 7.1 shake and Cerro Negro with a M = 7.5 shake. This last fill was broken again in 1985 under a stronger earthquake of M = 7.7. It is also interesting to note that the Izu-Ohshima earthquake (M = 7.0) in 1978 broke the first embankment of the Moshikoshi Tailing Dam, completed in Japan in 1964, during the main shock and the second embankment during an after-shock.

10.2.6. *Earthquake performance of road embankments and levees*

Tamotsu Matsui *et al.* [MAT 01] have analyzed the factors of harm to wooden houses and networks during the Kobe Hyogoken-Nambu earthquake. They show that the seismic intensity was particularly strong in a 6 km strip on either side of the fault. From statistics on the effects of 75 large earthquakes studied since 1885, Kazue Wakamatsu [WAK 01] identifies the areas most prone to liquefaction in geological and geomorphologic terms. She noted that it is mostly Holocene sediments that were liquefied. Current or recently abandoned river meanders are most prone to liquefaction. For very large earthquakes, Wakamatsu reported that the most sensitive areas are the recent floodplain deposits or *natural levees along a river*, because the water table is higher and the sediments are poorly consolidated, due to *the former river channels and the sand dunes between lowlands*.

The violent Chi-Chi M = 7.3 earthquake in Taiwan on September 21, 1999 caused 2,400 fatalities, injured over 10,000, damaged over 100,000 houses and led to more than 2,000 landslides. The lands most prone to liquefaction were polders, the alluvial deposits of ancient riverbeds and recent deposits along the banks. The soil in these zones is generally sandy, with a water table between 1 and 3 m deep and SPT (Standard Penetration Test) number counts less than 10. Settlement and displacement of many levees within these areas reached 1.5 and 2 m, respectively, with peak accelerations of up to 0.75 g.

10.2.7. *Earthquake performance of river hydroelectric embankments*

The embankments along the Danube are regularly shaken by earthquakes in the Vrancea region, for instance with a magnitude of M = 7.2 (1977), M = 6.9 (1986) and M = 6.7 (1990). Numerous longitudinal cracks appear, and sand boils indicate the presence of liquefaction, but the levees do not break. An interesting case of liquefaction has been reported [POP 93]. This is one of the embankments of the Draganesti hydropower plant on the most important river in Romania, the Olt. The length of the embankment is 28 km and its maximum height is 20 m. The typical section is a homogeneous gravelly embankment based on a drainage layer with upstream concrete slabs. The upper layer of the foundation is a 1.2 to 7.0 m thick continuous layer of silty sand resting on the alluvial gravel deposit (used for the body of the embankment). The silty sand was compacted by stone columns increasing the relative density from 10% to about 60% over the first 4 m and along 80% of the length. The earthquake of May 30, 1990, with an estimated maximum acceleration of 0.12 g, liquefied 60 m of the uncompacted foundation (silty sand with a relative density of 10%) in a leaking zone. The level of the reservoir was 4.8 m below the normal level. The hydraulic head above the ground surface was about 8.0 m when the earthquake occurred. Liquefaction caused a global movement to downstream, with contraction and lifting up of two slabs above the upstream toe and a sliding of the downstream toe, which filled the downstream drainage channel. Despite the liquefaction, the embankment did not fail (see Figure 10.1).

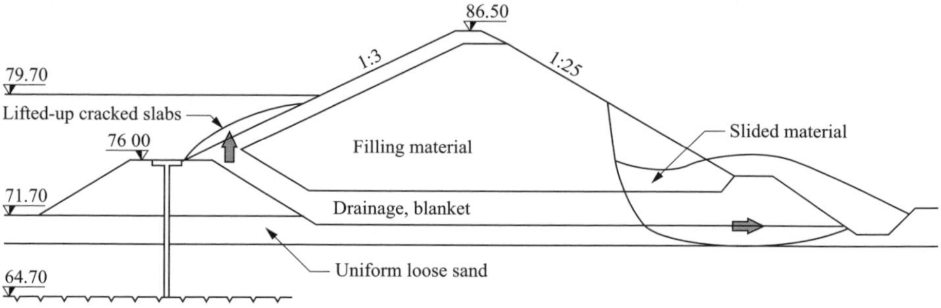

Figure 10.1. *Mean profile of the Romanian embankment damaged by the Vrancéa earthquake (1990)*

10.2.8. *Earthquake performance of small earth dams*

Failure of the 12 m Lo Ovalle Dam (1932) and the 18 m Lliu-Lliu Dam (1934) during the earthquake of March 3, 1985 in San Antonio is indicated by the ICOLD *Bulletin 99*, but not confirmed by *Bulletin 120* [ICO 01]. One explanation is that sliding occurred, but no water was released.

In Japan, the number of irrigation dams in 2000 was over 100,000, with 1,872 of them exceeding 15 m, out of which 1,688 are embankment dams. Several seismic failures were reported among embankments built before 1945 (the oldest is 1,600 years old). Tani [TAN 01] pointed out that the major cause of these failures is the liquefaction of the foundation or embankment.

The first seismic failure mentioned in the literature is the Manno-Ike Dam in Japan, occurring one month after the Ansei Nakai earthquake (1854). The failure was probably caused by the piping. The post-seismic re-assessment concludes that dams on sandy soil experienced the heaviest damage, suggesting that liquefaction is their cause.

Among the 1,200 small dams affected by the earthquake in Hyogo-Ken Nanbu (Kobe M = 7.2) on January 17, 1995, only nine were seriously damaged. On the island of Awaji, 950 small dams were damaged from among the 24,000, two of which broke. The failure of the 5.5 m Idenoshiri-Ike embankment, 155 m in length, occurred 7 s after the beginning of the earthquake when the site acceleration reached 0.45 g, according to Uchida Kazunori *et al.* [KAZ 01]. Liquefied areas in the foundation (SPT <5 in Figure 10.2) began at the toes and moved forward and backward, causing a downstream sliding when they met under the dam axis.

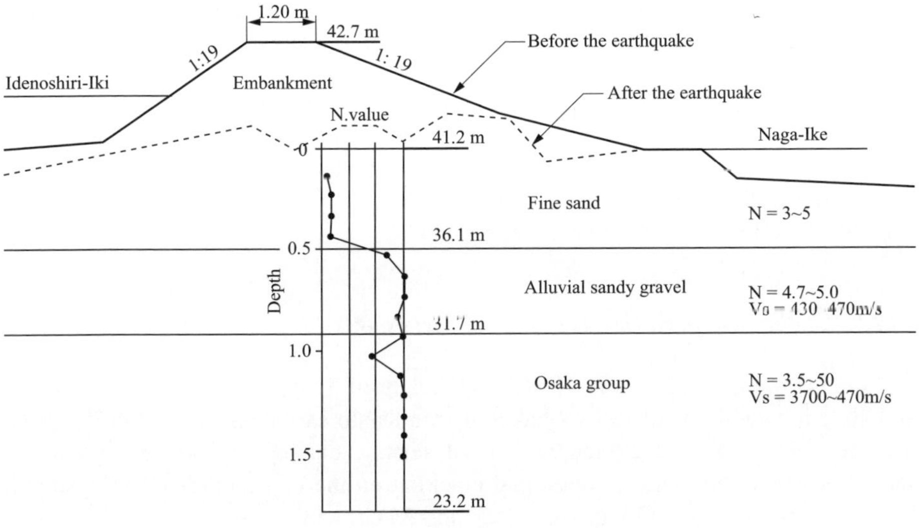

Figure 10.2. *Failure of the Idenoshiri-Iki Dam (Japan, 1995)*

On the island of Honshu, the majority of damaged dams are between 10 and 25 km away from the fault. Tamostsu Matsui *et al.* [MAT 01] consider that for an earthquake within a distance of 5 km, the dominant influence is maximum acceleration, and thereafter the resonant frequency of the soil governs the behavior. On the island of Honshu, the 12 m homogeneous Sugatadani-Ike Dam was built with a poorly compacted sandy-clayey material (4 <N <17, 30%, <80 μm, IP = 13). Located 26 km from the fault, it was shaken by a maximum site acceleration estimated at 0.23 g. The water reservoir level was half its normal height. Major sliding moved away the upstream face causing a step of 4 m, without overtopping the dam. The safety factor decreased from 2.8 to 1.0 during the earthquake, due to an increase in pore pressure of about 80%.

To ensure safety or to guarantee the availability of water, 245 small earth dams damaged in the Bhuj earthquake in India on January 26, 2001 (M = 6.9) required immediate repair [PAT 03]. The damage was mainly due to settlement of crests, transverse and longitudinal cracks at the crest and upstream face, cracks and steps in the upstream face of the highest section, openings of the contact with structures, floor structures and sand boils. No spillway, however, suffered damage. The soils of the Kachchh region are sandy and silty, poorly compacted, and vulnerable to liquefaction. Saturated at the bottom of the valley, they caused significant settlements and upstream landslides but, on account of the low water reservoir level, only one failure. For instance, the 18 m Shivlakha Dam (1954), with a core of masonry, was located 39 km from the epicenter. Traces of liquefaction of the foundation were observed upstream and downstream, causing sliding of the upstream slope and a ridge under the sliding of the downstream toe. The only reported failure was the 17 m homogeneous Chang Dam, built with a masonry core in 1959. This unusual design can barely sustain the probable seismic differential settlements. The earthquake liquefied the foundation 16 km from the epicenter, causing upstream and downstream sliding, accompanied by a collapse of the crest and the shear of the core. The reservoir was deliberately emptied through a breach to avoid damage following a probable high flood.

10.2.9. *Earthquake performance of large earth dams*

Ono Dam, built in Japan in 1914, has a height of 37 m above the ground surface and 40 m above its foundations. Shaken by the Kanto earthquake in 1923 (M = 8.3), it is perhaps the first documented case of seismic damage. It did not fail, but it settled nearly 30 cm, with longitudinal cracking on the upper third of the upstream face and crest up to 60 m long, 10 m deep and 25 cm wide.

The 28 m Russian Nizhni Svirskaya Dam, built in 1934, broke in 1935 after a mine fire [ICO 95], but this information has not been confirmed by ICOLD (2001).

The Hegben Dam, built in the USA in 1915, has a height of 34.5 m. Its reservoir was full when it was shaken by the Yellowstone earthquake in 1959 with a magnitude of 7.1. The earthquake caused a 5 m step along the active fault crossing the reservoir, 500 m from the dam. Four landslides downstream released

20 million m^3 of water into the Madison River, forming a natural dam of 70 m. A seiche swept the crest four times within a period of 10 to 20 minutes, causing scour and a 1.8 m upstream settlement and a 1.2 m downstream. The dam did not fail: the designer had planned a concrete core in this challenging seismic environment.

The 33.5 m Tokiwa Earth Dam was 700 m from the epicenter of the Kobe earthquake (1995), which had a magnitude of 7.3. It was cracked only at the crest, although the seismic center of Kobe recorded a maximum acceleration of 0.82 g.

Following the huge Tohoku underwater earthquake (M = 9.0) on March 11, 2011 almost all of the dams withstood minor to severe ground shaking and retained their reservoirs with generally minor to moderate damage (Matsumoto, 2011). The exception to this was the Fujinuma Dam, considered to be a pond-retaining structure under the ministry of agriculture regulation. Its construction began in 1937 and was halted during the Second World War. Following the war, the dam was completed in 1949. The failure of the dam resulted in the uncontrolled release of the entire reservoir, which flowed downstream into a small village and killed eight people (Matsumoto, 2011). This was the first time that fatalities were unfortunately caused by the seismic failure of a large dam. The causes are under investigation; however, once again poor compaction, poor construction design details, and liquefaction are reported.

During the Chi-Chi earthquake (M = 7.3) in 1999, most dams settled no more than 0.1% of their height. The conclusion can be drawn that a well-compacted fill may undergo a maximum acceleration of 0.6 g without major damage.

In conclusion, only two agricultural large dams – the Chang Dam in India, after the Bhuj earthquake (2001) and the Fujinuma Dam in Japan following the Tohoku earthquake (2011) – failed and resulted in the release of the entire reservoir. Note that the design of these two dams was poor and the lack of filter will, even without an earthquake, lead to a high risk of failure by internal erosion.

We should keep in mind, in accordance with the Japanese data, that no large dam with a height greater than 15 m built after 1959, according to modern design, has failed. Those built with poor design features have been seriously affected by earthquakes, however only two have failed. The two causes of the few seismic crest settlements over 1 m are:

– loose backfill material;

– liquefaction of sandy materials.

10.2.10. *Earthquake performance of large zoned dams with rockfill*

The Matahina Dam (1967) was shaken by the Edgecumbe earthquake (M = 6.3) on May 2, 1987. The maximum ground acceleration PGA = 0.33g triggered a sinkhole at the crest, revealing and boosting internal erosion.

Since 1975, the 60 m La Villita Zoned Dam with a thick core and the 148 m El Infiernillo Dam with a thin core have been shaken by a series of close earthquakes: five of them had a magnitude greater than 7.2 and one was 8.1. The damage can be summarized as cracks and settlements. It is interesting to note a trend of increasing settlements at La Villita, suggesting a progressive deterioration of the material strength that does not exist at El Infiernillo.

In 1984, the 105 m Makio Dam experienced the Nagano-Ken-Seibu earthquake (M = 6.8), being 5 km from the epicenter. Only 1.5 m deep cracks at the crest were observed, despite the ground acceleration being estimated at between 0.5 and 1.0 g.

All the modern zoned dams performed well, as has been generally predicted in prior evaluation studies.

10.2.11. *Earthquake performance of concrete face rockfill dams*

The 85 m Cogoti Dam was built in Chile in 1939 using dumped rockfill. It was shaken by several earthquakes, including that of Illapel in 1943, leading to the creation of a 40 cm crest settlement. The new settlement curve over time seems to have converged with the old one. It seems that the earthquake only accelerated the long-term settlement. Cracking slabs caused leaks that have increased over time and need repair.

The 146 m Zipingpu Dam, with a capacity of 1 billion m^3, was shaken on May 12, 2008 by the Wenchuan earthquake (M = 7.9), whose epicenter was 17 km away. A peak acceleration of 2.0 g was measured at the crest. Localized settlement could be seen on the downstream face near the spillway. The maximum settlement reached 73 cm and the maximum displacement reached 18 cm. Despite the many splits and cracks along the joints of the upstream slabs, the increase in leakage was very small. The main problem caused by this earthquake is not the performance of dams built by man, but the management of the 69 natural dams that formed after landslides in the valleys (see Figure 10.3). Their failures occurred without causing casualties. Some failures were deliberately initiated after evacuating the population.

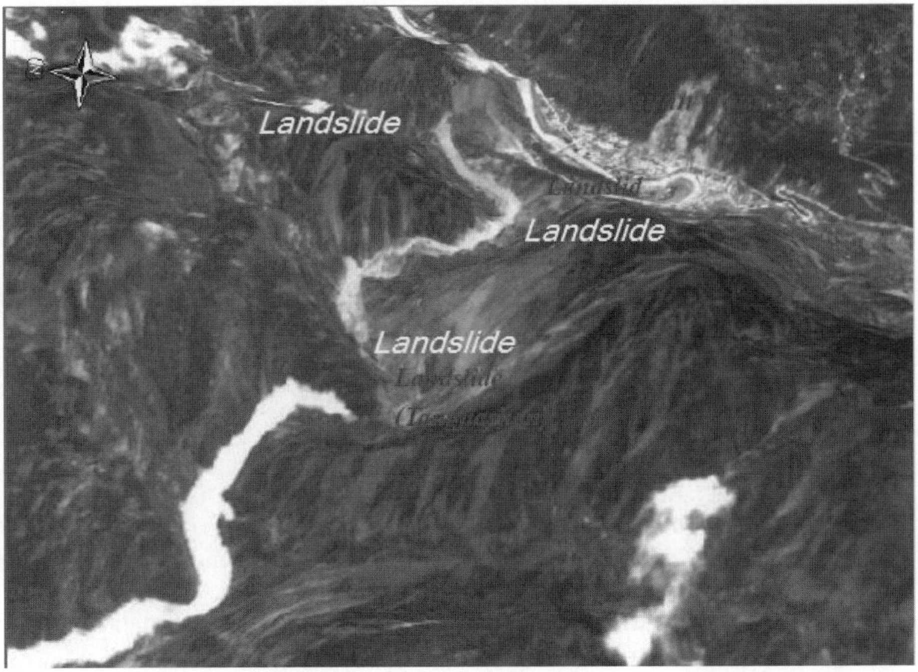

Figure 10.3. *Landslides blocking the rivers after the Wenchuan earthquake (2008)*

10.2.12. *Dynamic performance of physical models*

Physical models also provide information on modes of failure [MAS 03]. Thus, the vibrating table tests of an 80 cm dam with slope H/V = 1.5/1.0 in Toyoura sand with 5% moisture content and 50% relative density indicate the different phases until the collapse obtained at PGA = 0.6 g:

– longitudinal cracks, at the crest only if the shakes are horizontal and on the slopes if the vertical component is strong. The majority of cracks are produced by tension;

– onset of transverse cracks due to the extent of cracks;

– occurrence of landslides.

10.2.13. *Assessment of seismic damage on dams*

Even if earthquakes are still very destructive, to the point of causing thousands of human casualties per year, it is spectacular to note that no casualties have been caused by the failure of a large dam built in accordance with modern design features. Few dams were shaken by the maximum design earthquake (MDE), but some others were shaken by earthquakes larger than the MDE. This observation shows that there is less than one casualty for nearly one million dam-years and highlights the real resistance of these engineering works to seismic hazard, as long as the design respects the post-1959 state-of-the-art rules.

10.2.14. *Major seismic damage of large concrete dams*

Concrete dams performed very well under earthquakes. The performance of gravity dams and buttress dams is similar, as shown by the similarity in performance of the Koyna and the Sefirud dams: cracks are concentrated near the crest on the downstream face and around geometric changes in inertia. Apart from this cracking, their performance remained correct.

The most vulnerable areas are the locations where inertia changed at the crest and the interface with the foundation. The suspected common failure modes in gravity dams should be tilting after uplift in an open upstream crack and sliding along an open joint (see Figure 10.4), tilting along a joint, crushing of the downstream toe and downward tilting of the foot with crushing. These failure modes were identified on physical models.

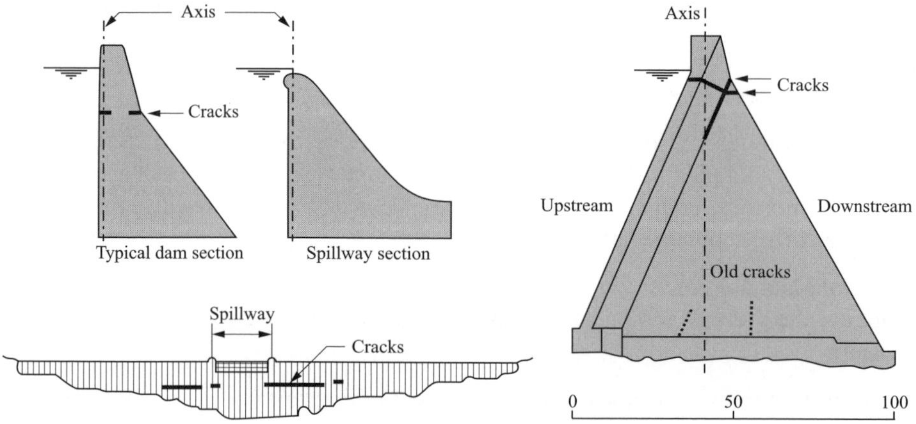

Figure 10.4. *Cracking in the Konya Dam (left) and Sefi Rud Dam (right)*

More generally, the main mode of rupture of concrete dams subjected to strong earthquakes is the formation of cracks in areas where the stresses are concentrated, such as the central part of the crest, and along planes of weaknesses: horizontal construction joint or vertical expansion joint. It is worth noting that leakages through these cracks lead to important uplift in these joints. Failure would be possible if, on the one hand, part of the crest or a block was initially isolated from the dam body by these discontinuities and, on the other hand, pushed to slide or tilt under the inertial and hydrodynamic forces. To our knowledge, this has not yet happened.

This reassuring feedback has to be taken with caution. The only failure, which occurred during the Chi-Chi earthquake (1999), concerns a failure mode not considered by physical models: the Shih Kang Dam moved on an active fault. This shows that previous or other failure modes are possible because the poorly designed structures have not been strongly shaken. Although the construction of a dam on an active fault is not acceptable to the majority of the scientific community, this failure nevertheless shows the difficulty of investigating and locating active faults.

10.2.15. *Seismic damage of large embankment dams*

Embankment dams are more vulnerable to earthquakes than concrete dams. This vulnerability varies greatly, however, with the type of design. It is important to remember that no major embankment dam anywhere in the world that has been constructed in accordance with modern practice or after 1960 has been destroyed by an earthquake. The only two large dams (H> 15m) that failed were the Chang Dam, built in 1959 in India, and the Fujinuma Dam, built in 1949 in Japan. Not only did they not comply with modern rules, but they suffered liquefaction due to a lack of compaction of their materials or foundations. This lack of compaction means that the tailing dams are by far the most likely to experience failure.

In conclusion, it is important to remember that liquefaction is the only cause of failure cited by the previous references. It is the main danger that requires a detailed investigation to be made on all existing dams. The evaluation of its effects on any embankment requires either a simplified risk analysis or a simplified coupled analysis developed later, depending on the issues involved.

10.2.16. *Delayed or indirect consequences of an earthquake*

Feedback shows that delayed or indirect effects can induce failure landslides in reservoirs, the development of internal erosion through cracks, loss of access by falling blocks, loss of control by destruction of technical control center and switching power poles. These are all very serious dangers that must be addressed in the analysis.

10.3. Method for analyzing seismic risk

A method for analyzing seismic risk is currently being re-evaluated in France by a panel of experts established by the Ministry of Ecology. The following section provides a general overview as well as my own contribution to this topic.

10.3.1. *Seismic classification of dams in France*

Seismic risk is defined by connecting seismic hazard, the vulnerability of a given dam shaken by a major seismic loading, and the potential risk to the downstream site. A panel of experts set up and led by the Technical Department of Electrical Energy, Large Dams and Hydraulics has already made a risk assessment and produced a method for classifying dams in relationship to seismic risk [DUB 03]. The purpose of this classification is to determine which dams need to undergo a safety re-evaluation. The three criteria mentioned above are given a coefficient whose value decreases whenever risk increases.

The coefficient *Cs* characterizes the risk of exposure to a major earthquake. It includes the intensity of a historically likely maximum earthquake (SMHV). Its values are presented in Table 10.1.

Coefficient *Cv* represents the vulnerability of a given dam subjected to a major earthquake. It has four components:

– the type of dam (C1);

– its age and behavior (C2);

– the type of foundation soil (C3);

– the slopes of the reservoir (C4).

Coefficient *Cr* characterizes the potential risk for the downstream site. It represents the energy released by the sudden failure of the reservoir. It is correlated to the number of victims after failure (not due to seismic loading) of large dams. It is calculated as a function of the parameter $H^2 \times \sqrt{V}$, H being the dam height in meters and V the volume of the reservoir in millions of cubic meters.

Values	Cs hazard	C1 conception	C2 age	C3 Foundation	C4 reservoir	Cr down-stream
1	intensity SMHV >= VIII MSK	Homogeneous embankment without upstream facing, vulnerable to seismic loading	Built before 1914	Soft soil likely to create problems	Unstable banks with dam prone to crest submersion	$H^2V^{1/2} >$ 100,000
2	intensity SMHV < VIII MSK	Homogeneous embankment without upstream facing, slightly vulnerable to seismic loading	Built between 1914 and 1945	Soft soil without a particular problem	Unstable banks with dam not prone to crest submersion	100,000 > $H^2V^{1/2} <$ 10,000
3		Embankment with internal water tightness	Built after 1945 and project having not been examined by the CTPB	Standard quality rock	Stable banks	10,000 > $H^2V^{1/2} <$ 1,000
4		Embankment by upstream facing and rockfill dams with clay core	Built after 1945 and project examined by the CTPB	Very high quality rock		1,000 > $H^2V^{1/2}$
5		Other rockfill dams				
6		Concrete or masonry dams other than arch dams				
7		Arch dams				
		Cv = min(C'v, C''v) C'v = min(C2,C1−1) + min(C3,C4)-1 C''v = Ent[1+($0.6 \times C2 \times \sqrt{C1} + 0.4 \times C3 \times C4 - 1$)/2]				

Table 10.1. *Values of the coefficients quantifying seismic risk (CTPB – Comité Technique Permanent des Barrages (Steady-state Technical Committee on Dams))*

Hence, the dams can be classified into three groups:

– dams located in moderate to significant seismic zones, including the zones with $Cs = 1$, which require a detailed safety verification;

– dams located in a zone of low seismicity with $Cs = 2$, but highly prone to seismic loading or to submersion due to landslide within the reservoir;

– other dams without any particular danger.

10.4. Evaluation of seismic hazard

A panel of experts was established in 2009 in France by the Ministry of Ecology in order to create a method for evaluating seismic hazard. Numerous discussions on how to meet the ever-growing social demand for better protection took place. As a result, the level of the probability of a maximum earthquake was raised. The rules are:

– to define the scenario for the different situations of dimensioning;

– to establish the seismic level for each situation;

– to characterize the seismic loading for each level;

– to define the accelerographs to be used.

10.4.1. *Scenarios for dimensioning a particular situation*

The method for dimensioning must include the verification of the durability of the geo-structure in accordance with its use (water stocking, control of the downstream flow and the upstream level) and its main functions (waterproofing, stability, filtration, drainage, evacuation and protection) during each life phase: construction and operation. A seismic loading will interfere with the reservoir.

Major earthquakes being a rare occurrence, we have to study two of the three types of scenarios proposed by the Euro Code: that corresponding to rare conditions (called transitory) and that corresponding to extreme conditions (called accidental). Indeed, we have to distinguish at least one level of the average seismic loading, cumulated with rare events, and one high level not cumulated with other loadings.

10.4.2. *Choice of seismic levels*

The ICOLD proposes four levels of seismic intensity, classified in decreasing order of occurrence or by increasing intensity:

– *Reservoir induced earthquake (RIE)*: whose impact on nearby engineering structures has to be examined during the project. The seismic level in France does not require this study.

– *Operating basis earthquake (OBE)* after which the dam and connected engineering works must stay in an operational state. This level assures the owner that his or her structure remains in operation after the earthquake. Even if the ICOLD proposes an average return period of 145 years, in accordance with the 50% probability of over 100 years, the final decision concerning the retained criterion belongs to the owner (for example, a 5% probability over the period of back investment, that is over 200 to 500 years).

– *Maximum design earthquake (MDE)* after which the dam must keep its stocking capacity and all its functions, and ensure population safety. The annual probability of such an event is between 1,000 and 10,000 years.

– *Maximum credible earthquake (MCE)*: treated on a case by case level, depending on the stakes and associated costs. For engineering works of significant importance, this level is the same as the one corresponding to the MDE. It is noteworthy that in Japan [JAP 02], it is recommended that two levels of MCE are estimated: level 1, applied to common engineering works, which must induce admissible seismic loads as computed by classical methods; and level 2, applied to the most important or most exposed structures, which induces irreversible but acceptable damages computed by more elaborate methods. Level 2 is similar to MCE (the largest seismic ground motions that ever occurred and may occur in the future at a given site).

10.4.3. *Choice of the seismic characteristics*

In France, the MCE is called *séisme maximal de dimensionnement* or *séisme de référence*. It is determined by intra-continental seismicity. The earthquakes induced by the encounter of the African and Eurasian plates are too far from France to be taken into consideration. Japan, in comparison, is shaken by earthquakes of magnitude 8 every 100 to 200 years at the bottom of the Pacific Ocean, and earthquakes of magnitude 6 to 7 along the active faults with a return period of 1,000 years. The convergence of the plates is 2 cm/year in France compared to 6 cm/year in Algeria. Furthermore, the successive and various orogeneses did not produce an important stress concentration within the French territory. These data as well as the seismic events, numbering 6,000, have been put into a database called *Sisfrance*. The epicentral depths are statistically located between 5 and 20 km.

The MCE estimation, realized through four steps, uses a deterministic method [DUB 03]:

– Identifying the seismic hazard zones: construction of a seismotectonic model that includes the active faults or, if not available, the precise location of homogeneous tectonic zones (a fault is active if some sliding took place during the Quaternary).

– Characterizing the source zones by their "maximum seismic potential", the historically known major earthquake, applied at any location along the fault.

– By seismotectonic synthesis: evaluating the seismic hazard level in the site from the different maximum seismic potentials. The maximum possible earthquake is the most destructive earthquake that could potentially occur at the site. This level is characterized by the epicentral intensity, the magnitude, the depth and the spectrum. If an active fault is found at the projected location of the dam, it is strongly recommended to change the building location. If an active fault crosses the reservoir, it is essential to define the induced rejection in order to determine the referential wave that could have an impact on the dam.

– Choosing the reference earthquake according to its maximum acceleration and its spectrum. These two factors can be accessed from regional correlations (earthquake type, distance, nature of the foundation soil). Correlations give two types of pseudo-acceleration spectrums issued from one degree of freedom oscillators for a 5% damping ratio: one corresponding to hard rock foundations with shear wave velocity higher than 800 m/s; the other for altered or soft soil foundations with shear wave velocity between 300 and 800 m/s. If the velocity is lower than 300 m/s, it is necessary to conduct a specific analysis (site effect). It should be noted that the most destructive earthquake for embankments is rarely the closest earthquake or the one with the highest acceleration on site [JAN 76]. Studies are underway to examine this aspect. In the next section, we suggest some alternative approaches.

10.4.4. *Choice of accelerographs*

Six methods are used, which will be discussed in the following sections.

10.4.4.1. *Empirical method*

The empirical method is the most frequently used method. It consists of choosing, in the databases of recorded accelerographs, those with a spectrum similar to the one corresponding to the reference earthquake. The maximum acceleration of a null period defined at the rocky substratum is then introduced in the selected accelerographs. A third parameter has to be added in order to reduce the scatter: the effective duration. The spectra for rocky foundations (*Vs* >800 m/s) should be compared to those coming from 293 Japanese recordings on 91 dam sites, from 63 earthquakes of magnitude higher than 5 [MAT 03]. Two types of events are distinguished: superficial, deep or interface failure. They take the general form:

$$\text{Lod } SA(T) = Cm(T)^{*}M + Ch(T)^{*}H - Cd(T)^{*}\log(R + 0.334\exp(0.653M)) + Co(T)$$

with *SA* being horizontal pseudo-acceleration, *T* period, *M* magnitude JMA (Japan Meteorological Agency), *H* focal depth limited to 100 km, and *R* minimum distance to the fault.

For $T = 0$, we obtain the acceleration of null period PGA (Peak Ground Acceleration).

The coefficients are given by bi-linear relations with the logarithm of the period connecting three particular points corresponding to the periods T1 = .01 s, T2 = .2 s, T3 = 4 s (Table 10.2).

Coefficient	**T1 = 0.01 s**	**T2 = 0.2 s**	**T3 = 4.0 s**
Cm	0.54	0.54	0.85
Ch	0.0054	0.0054	0.002
Co	1.25	1.25	-2.5
Cd	1.75	1.75	1.25

Table 10.2. *Spectrum coefficients [MAT 03]*

10.4.4.2. *Energetic method*

Experience has shown that if we apply the same acceleration to two earthquakes, one close by and one far away, the more destructive one will be the farther one. This result shows that it is not enough to characterize seismic hazard by the maximum acceleration factor. It also implies that there is an unknown law for seismic energy propagation. Knowing what this propagation law is would allow us to replace the couple spectrum and maximum acceleration by a sole parameter representing seismic hazard.

Nowadays, two characteristic factors are studied: Arias and Housner intensities. Recent studies [BRU 03] have shown that the Arias intensity is well correlated to the maximum displacements of the embankment dams, whereas the Housner intensity is well correlated to the maximum deformations at depth. The correlation is better if the earthquake duration is also taken into account. Current studies have tried to determine the connection of these two indexes with the epicentral distance on one hand, and the nature of the foundation soil, on the other.

10.4.4.3. *Semi-empirical method*

This method is used to generate strong seismic loads of level 2 and considers that there is a scale effect between local slides and the overall movement. The final loading is reconstituted from the summation of the local shocks using Green functions [HAR 78]. The fault is decomposed into several elements. Local slides are synchronized to the fault displacement by a probabilistic approach [BOO 83]. The application domain of this approach is limited to high frequencies.

10.4.4.4. *Theoretical method*

This method consists of calculating the transmission of the energy and the motion of the fault through a 3D mesh or a horizontal multilayer model [BOU 81]. However, the frequency content in the high-frequency domain depends very much on the path followed; the lack of geological knowledge of this path leads to an inaccuracy of the frequencies determined that are higher than 1 Hz.

10.4.4.5. *Hybrid method*

The method consists of adding the results obtained by the semi-empirical method in the high-frequency domain and the results obtained by the theoretical approach in the low-frequency domain.

10.4.4.6. *Probabilistic method*

This method benefits from a high amount of accumulated data from various seismic recordings. The US Geological Survey publishes maps of the spectral acceleration for a given probability and a given frequency. Therefore, it is possible to determine spectrums of a given probability, i.e. spectrums of pseudo-accelerations having the same probability of being exceeded for each frequency. By comparing the probabilistic and deterministic spectrums, we can estimate the return period of the latter. The probabilistic spectrum is obtained by combining three uncertain events [ATK 97; TOR 97]:

– the activation of each source;

– the activation of each magnitude given by a law of the type $\log N = A - bM$, with N being the number of earthquakes of magnitude equal or smaller than M;

– the activation of each distance by an attenuation law.

All these approaches have to be improved by knowing the effect of the mechanism of fault failure: the type of failure, the energy released, the length, etc.

For each method, it is necessary to test the coherence and precision of the results with:

– all the sources (magnitude and distance to the dam);

– local earthquakes;

– local geology;

– the first natural frequencies of the dam.

For this purpose, the AFPS (*Association Française de Génie Parasismique*) intends to develop a probabilistic approach at the national scale [AFP 07]. In the meantime, it is possible to use the probabilistic approach developed by GEOTER at EDF's request, which is based on a Bayesian analysis of the measurements given by the accelerograph network installed by the RAP as well as measurements made on nuclear power plant sites [GEO 08]. This study has produced maps of the maximum acceleration in France for a given return period.

10.5. Re-evaluation of seismic stability

10.5.1. *Maximum risk associated with seismic loading: liquefaction*

After having reevaluated the seismic hazard, we now have to reevaluate the seismic risk attached to the most exposed dams, that is to say the embankments prone to liquefaction. It is possible to distinguish four deformation modes: superficial sliding of the embankment slope, deep sliding, loss of foundation bearing capacity, and embankment settlement. For each of them, liquefaction appears to be the driving force of the deformation. Therefore, it has to be carefully studied.

Liquefaction develops in saturated undrained soils. Its well-known appearances are the quicksands in the zones of lake and shore sedimentation. Less well-known are the failures during construction, as for example during the construction the Calaveras Dam. Finally, liquefaction-induced seismic loading has been highlighted by the failure of the Van Norman Dam. The back analysis of this failure has given us the means to predict the most destructive effects of liquefaction [SEE 73].

What is liquefaction? A non-cohesive material subjected to cyclic shear stresses develops contractive volumetric strains. In undrained conditions (or if the loading is fast enough to prevent any drainage), the volume of a saturated material remains constant. The contractancy manifests itself by an increase in the pore water pressure, which induces a decrease in the mean effective stress, counterbalancing the tendency of pore volume decrease during cyclic shearing. When the pore water pressure reaches the value of the total stress, the soil is liquefied.

The term liquefaction includes several states and modes of deformation. These different modes have to be distinguished in order to have a better understanding of the liquefaction effects. Liquefaction is in theory a state of the soil corresponding to zero effective stress. In reality, the effective stress decreases to non null but small enough values to lead to, in the case of a loose soil, a liquefied sliding zone. In other cases, where the dam and foundation soils are more compact, the effective stress can increase again, leading to what is called "cyclic mobility", which stops at the end of the seismic loading. Therefore, two phenomena driven by liquefaction can occur:

– liquefied soil sliding, which is due to an abrupt decrease in the soil resistance until the geometry of the soil mass produces a new stress state compatible with the mobilized residual strength. This was the case for the San Fernando Dam whose upstream facing slid 42 m;

– cyclic mobility, which is the consequence of the pore pressure increase during seismic loading and leads to irreversible strains, but does not affect the post-seismic soil strength.

To simplify, geo-structure failure is quasi-certain when liquefied sliding occurs, whereas fissures and settlements are more probable consequences of cyclic mobility. We, therefore, have to select the engineering works constructed on non-cohesive loose soils. Then, we have to determine the seismic hazard for the site. Post-analyses of geo-structure behavior give way to a first evaluation of liquefaction occurrence:

– the failure criterion expressed in terms of maximum acceleration would be $PGA = 0.3$ g for a close-by earthquake (short duration and high frequencies of seismic loading) and $PGA = 0.2$ g for a distant earthquake (long duration and low frequencies of the seismic loading) [JAN 76];

– from the results of parametric calculations, the failure criterion expressed in terms of Arias intensity would be around 0.15 m/s for fine soils. For gravels, it would be 0.5 m/s for any initial density. These criteria remain to be confirmed;

– the sites located less than 6 km from an active fault and those located on a gravely foundation less than 4 km from an active fault are potentially prone to being liquified [MAT 01];

– the liquefiable sites are located at a distance smaller than the distance *Re* to the epicenter or *Rf* of the active fault, depending on the amplitude of the moment *Mw* or the surface shear waves *Ms*:

$$Mw = -0.31 + 2.65.10^{-8}\ \mathrm{Re} + 0.99\log_{10}(\mathrm{Re}) \text{ with Re in cm [AMB 88]}$$

$$Mw = 0.18 + 9.2.10^{-8}\ \mathrm{Re} + 0.99\log_{10}(\mathrm{Rf}) \text{ with Re in cm [AMB 88]}$$

$$Ms = 1.5 + 3.1\log_{10}(\mathrm{Re}) \text{ with Re in km [GAL 00]}$$

10.5.2. *A recommended step-by-step methodology*

We recommend a step-by-step methodology of seismic analysis in order to formulate well-founded arguments on geo-structures when considering public safety. The superposition of different simple approaches appears to be a good method because it is accessible to most of the consulting firms and because these approaches are able to justify the structural stability concerning distinct but complementary physical phenomena. The methodology is developed in seven steps (see Table 10.3).

Step	Method for the seismic analysis of dams
1	Identification
2	Pseudo-static approach of stability
3	Approach of liquefaction
4	Pseudo-dynamic approach to displacements
5	Non-linear dynamic approach to liquefaction
6	Post-seismic approach to residual stability
7	Assessment

Table 10.3. *The seven steps and associated methods for seismic evaluation*

10.5.3. *Identification*

The first step concerns the inventory and critical analysis of the data necessary for the study. These data are in connection with the three phases of the project: studies, construction and operation. The hypotheses concerning the soil characteristics are issued from the studies. During the phase of operation, fundamental information is gathered, which enriches the project data: incidents, visual observations of abnormal behavior, measurements on the geo-structure, and maintenance works. The absence of such information leads to the need for significant investigation programs based on *in situ* testing. The data needed are assembled within four site models:

– the geometrical model which fixes the actual dimensions of the dam and the different soil layers of the foundation. Often, the “as-built drawings” are not available;

– the geological model that defines the nature of the foundation, the position of the faults and the intensity of the seismic hazard;

– the geomechanical model that characterizes the soil's mechanical properties and the mechanical strength value required for the calculation;

– the hydraulic model that quantifies the hydraulic properties of the site, including the permeability tensors, Skempton's coefficient B, and the boundary conditions.

It would be dangerous to start the numerical modeling without any in-depth assessment of the construction of these four models. The reality is unknown; it has been replaced by the theoretical construction of the four models. It is therefore important to judge their pertinence in the light of some elements of reality. The strongest uncertainty concerns the seismic hazard that is part of the geological model. It is very difficult to obtain a good knowledge of the active faults that are hidden or remolded by the alluvial filling. For example, the geologists working on the Shih-Kang Dam project did not localize the fault that was passing under the dam which was responsible for its failure during the Chi-Chi earthquake.

10.5.4. *Pseudo-static analysis of stability*

Pseudo-static analysis of stability under seismic loading consists of calculating the safety factor of a dam subjected to a horizontal force considered representative of the earthquake action. This force is taken to be equal to the mass multiplied by a percentage K of the gravity acceleration g, called the pseudo-static coefficient. This coefficient is usually taken equal to 2/3 the ratio between the maximum acceleration at the base of the dam and g.

The analysis of seismic loading on pier retaining walls has led us to consider that the seismic load can be assimilated to a horizontal force with a pseudo-static coefficient $K = 0.33(PGA/g)^{1/3}$ [NOD 75]. The pseudo-static analysis is the first step of the stability verification. It gives access to the critical pseudo-static coefficient Kc for which the stability is lost.

An evaluation of the irreversible deformations of the geo-structure can be achieved by comparing PGA/g and Kc in the absence of liquefaction. It is the first step of the following method.

10.5.5. *Pseudo-static analysis of displacement*

During the earthquake, the geo-structure can have momentarily a safety factor smaller than 1, without the overall security being questioned. If the duration of this transient disequilibrium remains smaller than several milliseconds, the resulting displacement is negligible. It is therefore better to define criteria in terms of displacement amplitude in order to estimate the impact of a safety factor smaller than 1. This is the objective of the approaches presented here.

Newmark's approach was the first to calculate the displacement of an unstable mass when the factor of safety was smaller than 1 or $K > Kc$. $\Gamma_{mean}(t)$ is the acceleration applied by the seismic loading to the soil mass studied at time t:

– if $\Gamma_{mean}(t) > K_c g$, the sliding is calculated by integrating the acceleration $\Gamma_{mean}(t) - K_c g$;

– if $\Gamma_{mean}(t) < K_c g$, there is no irreversible movement (see Figure 10.5).

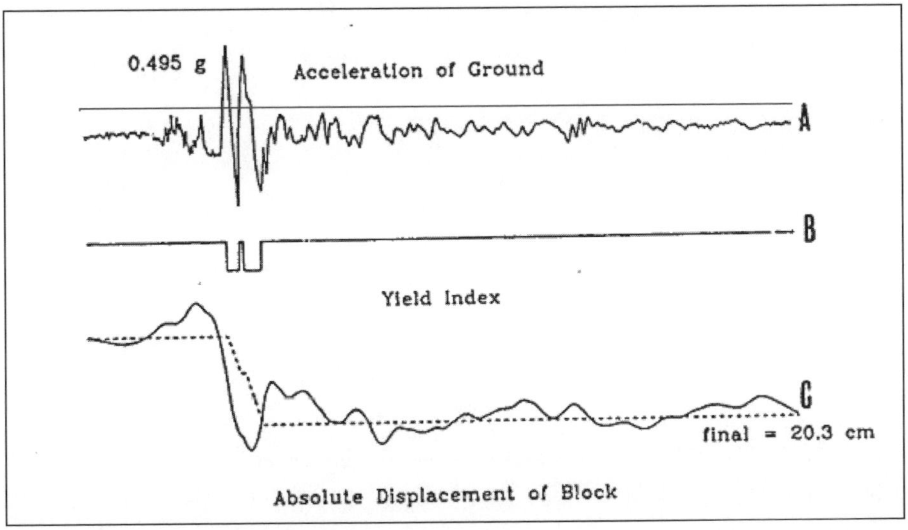

Figure 10.5. *Example of an accelerogram k_{MOY} (A), of Kc (B), and of integrated displacement (C) compared to measured displacement (C), after Ambraseys*

There are a lot of applications of this approach that distinguish themselves by the hypotheses on the accelerograph, the studied mass and the method for evaluating $\Gamma_{mean}(t)$. If we assume the absence of liquefaction, we can calculate the irreversible deformations within the geo-structure by comparing $K_c g$ to the maximum acceleration of the sliding mass $K_{max} g$ ($\max(\Gamma_{mean}(t))$) at any time t of the accelerograph duration.

Ambraseys *et al.* [AMB 67] calculated K_{max} as a function of the natural period of the geo-structure T_0 and the geometry of the unstable zone. They have applied the El Centro seismic loading on a triangular mass at the vertical distance $y = n.h$ from the crest, having a natural period T_0 and a damping factor ξ. To determine the displacements they used the abacus in Figure 10.6 proposed by Ambraseys and Menu [AMB 88b], which gives the displacements relative to the velocity *V* given by the accelerograph applied at the base of the dam function of Kc/K_{max}. The abacus is applicable to earthquakes of magnitudes between 6.55 and 8.25.

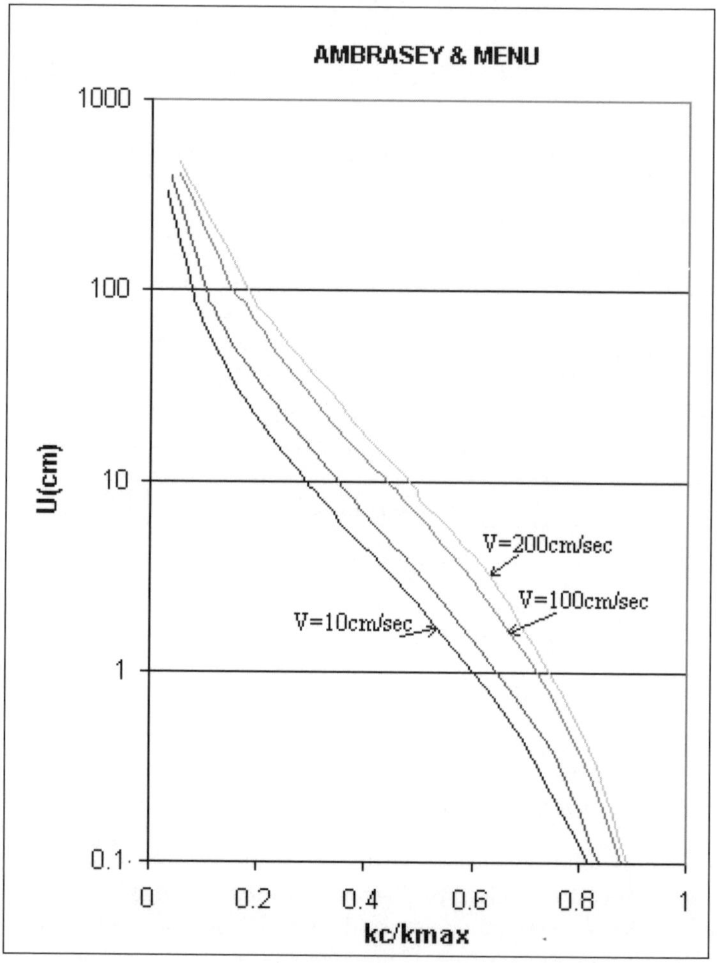

Figure 10.6. *Displacements versus accelerations [AMB 88b]*

Figure 10.7 gives measured mean values for dams with heights between 30 and 180 m, and shear wave velocities between 92 and 300 m/s.

K_{max} is the maximum acceleration of the sliding mass from the crest to the depth y. K_{max} is determined from the abacus in Figure 10.6 depending on the natural period and the maximum acceleration at the crest u_{max}.

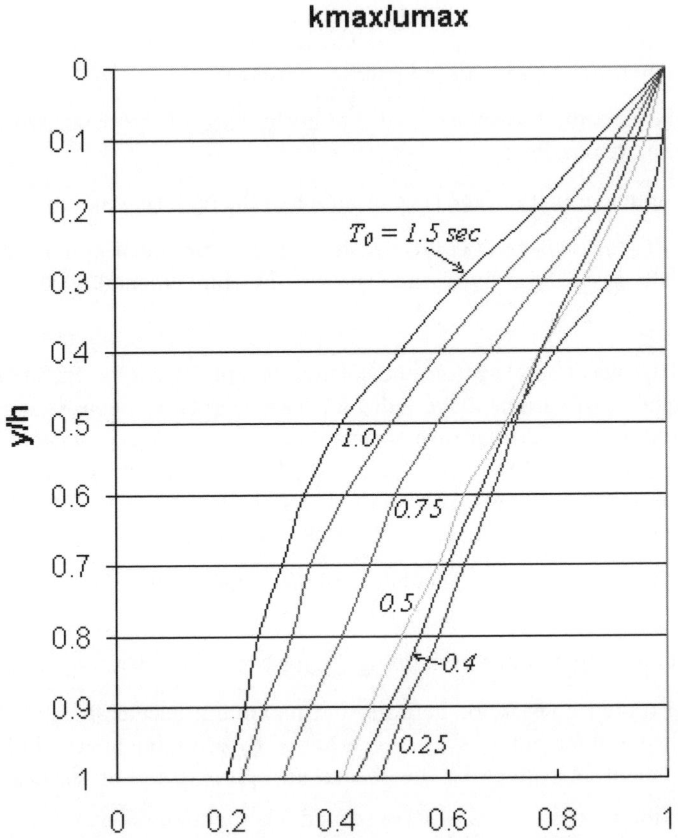

Figure 10.7. *k_{max}/u_{max} versus y/h (Ambraseys and Sarma)*

T_0 is the first natural period of the dam:

$T_0 = 2.61.h/Vs$

u_{max} is the maximum acceleration at the crest obtained from the spectral analysis, equal to 1.6 times the spectral acceleration for the natural period of the dam $Sa(T_0)$:

– $u_{max} = 1.6\ Sa(T_0)$;

– h is the height of the dam; and

– Vs is the velocity of the shear waves crossing the dam from bottom to top.

The maximum value of the displacement is given as function of K_{max}, Kc and V, where:

– Kc is determined by the pseudo-static approach;

– V is the maximum displacement rate at the base of the dam; two methods of determination are possible:

- maximum velocity of the accelerograph at the base (in cm.s^{-1});

- $V = PGA/(2;n_0)=4 \times PGA \times P$ with n_0 being the mean number of times the accelerograph passes through zero and P the fundamental frequency of the earthquake.

Figure 10.7 was obtained from the calculations of the displacements induced by three earthquakes of various magnitudes for dam heights between 50 and 65 m. The authors think that the displacements calculated for $Kc/K_{max} < 0.2$ are unreliable. However, a comparison between calculated and measured displacements on Chilean dams have shown that Ambraseys' method gave better results than the Makdisi and Seed's method [MAK 78]. The latter contains three steps:

– the first step consists of calculating the modulus and the damping ratio of the dam;

– the second step consists of defining the maximum acceleration at the crest;

– the third step consists of defining the maximum acceleration of the circular sliding surface at the depth y; K_{max} is determined by using the abacus in Figure 10.6. The displacement U is given by the abacus in Figure 10.8. For various earthquake magnitudes, the ratio $\dfrac{U}{k_{\max} \times T_0 \times g}$ (in seconds) is function of Kc/K_{max}.

10.5.6. *Analysis of the liquefaction risk*

The analysis of the risk of liquefaction is based on the evaluation of the resistance of the foundation and embankment by penetrometer testing. The *in situ* resistance is too scattered to be determined by laboratory testing. Youd has organized a workshop in order to establish norms for the prediction of *in situ* liquefaction. The general report of this colloquium remains the state of the art [YOU 01].

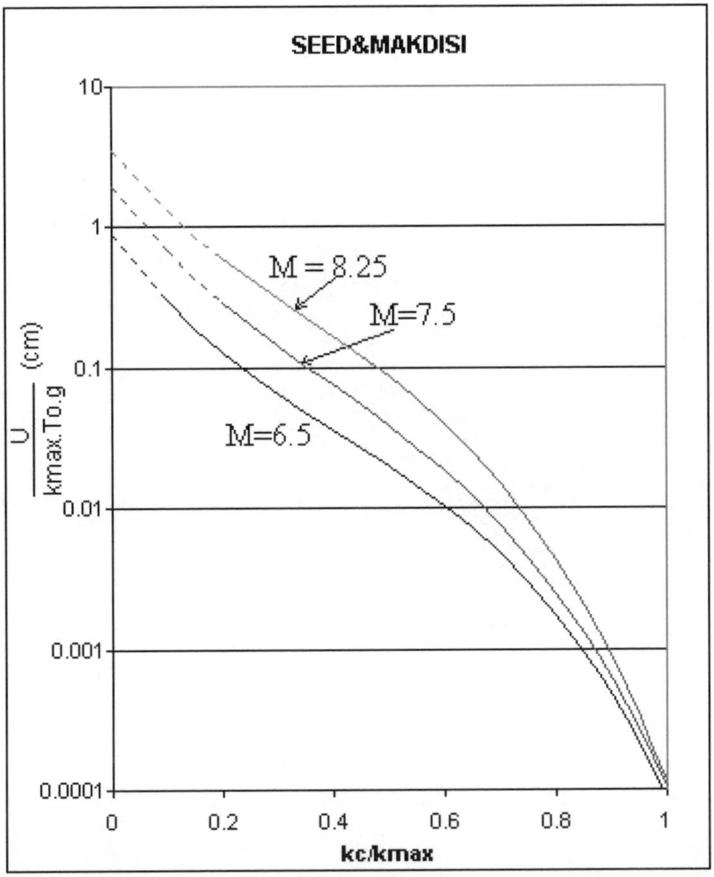

Figure 10.8. *Variation of acceleration with normalized mean displacement*

The seismic loading is represented by the cyclic stress ratio (CSR), which is the ratio between the shear stress and the initial vertical effective stress. CSR is calculated by the analytical expression of a shear wave propagating vertically to the free horizontal surface of the site function of the seismic maximum acceleration. The resistance is called the cyclic resistance ratio (CRR). It value is estimated empirically. CRR is the minimal CSR value that has led to a liquefaction for a given penetrometer resistance. The liquefaction criterion is then represented by all the measured CRR, which give a limit curve in the plane (seismic stress, *in situ* resistance) separating the points representative of the liquefied and non-liquefied sites (see Figure 10.9).

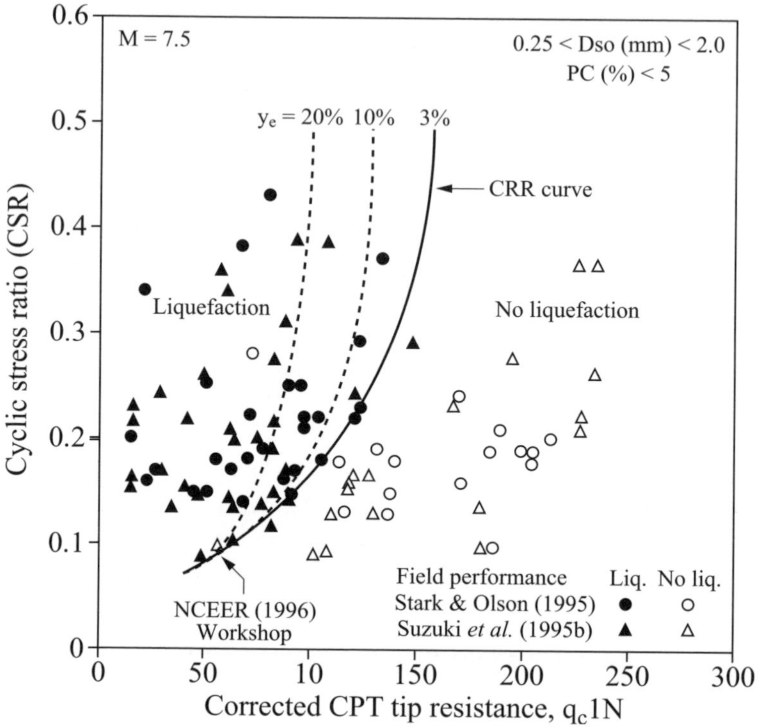

Figure 10.9. *Resistance to liquefaction measured by SPT or CPT (Cone Penetration Test) [YOU 01]*

The *in situ* resistance is measured by the SPT, which has the advantage of being not only a penetrometer test but also a boring machine. Since the development of this experimental procedure, new faster and more precise testing procedures have been studied. Among them we can cite destructive borings with recording of machine parameters, stato-dynamic penetrometers such as PANDA or the ones with an electrical cone. The most effective one appears to be the piezocone with an electrical cone. The penetrometer with an electrical cone permits a soil classification without boring and a fast investigation of a large zone, therefore being more economical. The piezocone is not so fast but gives a measurement of the soil permeability. It requires some pre-borings to bypass zones that are too strong and additional points to cover up the point breakage.

The safety factor is given by the ratio CRR/CSR; the contours of its value can be mapped into the foundation. The practice shows that potentially liquefiable zones are often found. The *in situ* investigation has to be sufficiently precise in order to quantify the spreading of these potentially liquefiable zones.

In some cases, the soil used in the foundation is not common enough to be present in the published abacus, such as the silts. Laboratory testing is then indispensable and a new abacus has to be set up. The evaluation of the seismic behavior of gravels and silts requires joint laboratory and *in situ* testing. The quality of the test results is an essential factor in preventing designs that are too conservative. For example, the strength of silts measured in triaxial apparatus can be lowered by more than 50%, if the samples are completely remolded. Moreover, it is not possible to obtain undisturbed gravelly samples. To overcome this problem, we have to determine the material's *in situ* state and link it to the liquefaction resistance. In the United States, the Becker penetrometer supplants the SPT. In France, EDF uses the CPT, whose results have been validated within a calibration chamber. We can also find correlations between shear wave velocity and liquefaction resistance, which allow us to overcome the difficulty of reliable measurements.

A potentially liquefiable zone is not likely to liquefy under certain given boundary conditions and more specifically under the influence of drainage. Therefore, if a potentially liquefiable zone is detected, the next step is to perform coupled or semi-coupled non-linear calculations in order to quantify the drainage influence.

10.5.7. *Coupled non-linear analysis*

Coupled non-linear dynamic calculation is based on the integration of a constitutive law for the continuous medium and is applied on the weakest embankment section for stability analysis. The finite element method or the finite difference method can both be used.

In the last part of this chapter, a non-linear constitutive model is presented. The results are studied in terms of liquefaction factor (ratio between pore pressure rise and initial effective stress), shear strains and final displacements.

The gravel permeability plays a fundamental role. Whereas the undrained triaxial tests show a liquefaction potential of gravels similar to the one obtained on sands at the same relative density, the calculations show that the pore pressure generation during cyclic loading is dissipated in few seconds in the gravels (see Figure 10.10) due to high permeability.

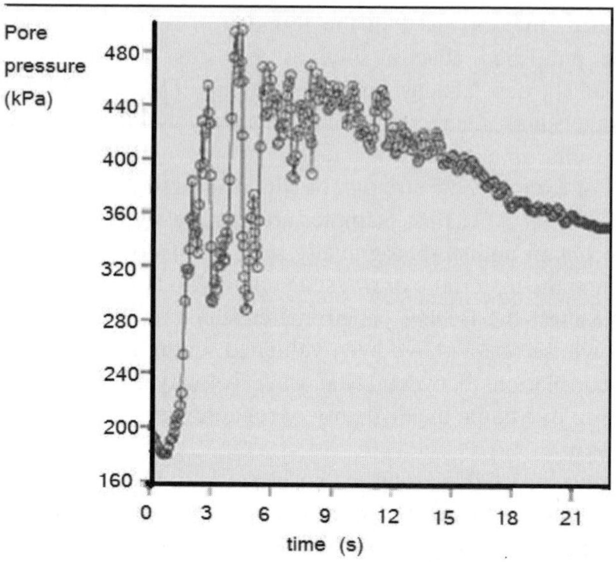

Figure 10.10. *Pore pressure increase within a gravelly foundation*

The failure evaluation is based on criteria taking into account the modeling errors [FRY 03]. The validation of these criteria on real failures, such as the failure of the San Fernando Dam, allows us to generalize their application to embankment and rockfill dam projects. The criteria express the consequences of a seismic loading at the end of the earthquake, considered as non-admissible in terms of strains, irreversible displacements and liquefaction factor obtained by the numerical modeling:

– The strains are analyzed by considering the evolution of the second invariant of the strain tensor. The interpretation is based on the visualization of the kinematics of failure or of highly sheared zones represented by the locus of the invariant's biggest values. The threshold for the failure criterion corresponds to a shear strain amplitude higher than 5%, following Seed's study [SEE 73]. A generalized failure corresponds to a zone of shear strain amplitude above this threshold, crossing the whole embankment from upstream to downstream.

– The threshold in displacement corresponds to a vertical settlement greater than 1/3 or 1/5 of the "freeboard" and to a horizontal displacement greater than 1/3 or 1/5 of the filter thickness or more than 1% of the dam height.

– The liquefaction factor is the ratio between the pore pressure increase and the initial effective stress. Liquefaction is considered to be reached when this factor is greater than 80%.

10.5.8. *Analysis of post-seismic stability*

A simplified approach to the post-seismic stability consists of calculating the final safety factor from the residual strength Su_r obtained from triaxial tests on samples prepared in the *in situ* state. This residual strength corresponds to the critical state. In practice, its minimum value is obtained by an extension triaxial test at the end of the undrained cyclic loading, using the two relations:

– the initial void ratio e_0 as a linear function of the log of the critical stress p'_c, giving what is called the critical state line;

– the constant value of the ratio Su_r/p'_c.

Another approach uses penetrometer tests by correlating the residual strength of a series of liquefied sites to the point resistance. This approach is more conservative because it gives the residual strength after failure by liquefaction.

The experimental discrepancy has to be carefully examined in order to determine a characteristic value for each layer of the foundation soil. Castro has proposed taking the mean value minus half the standard deviation as the characteristic value of the residual strength, according to the analysis of the San Fernando Dam failure [CAS 89].

Comparison of the results obtained by the above analysis, the simplified analysis of liquefaction and the non-linear calculations must show a global coherence of these different approaches. This coherence being verified, the *in situ* tests are calibrated for establishing the global diagnosis of safety under seismic loading.

10.5.9. *Assessment*

Study of the re-evaluation is not complete with the sole calculation of the stability under seismic loading. *In situ* results have shown that the failure is often delayed by several minutes to several weeks (Manno-Ike). Indeed, the failure can be induced by a mode different from sliding. Therefore, in order to undertake a complete study, we have to gather all the possible causes that initiate each type of failure and evaluate the likelihood of its occurrence for the case being studied [BRU 03]. Three modes of failure are likely to happen:

– internal erosion;

– external erosion;

– sliding.

The main causes of failure by external erosion are:

– settlement due to consolidation during seismic or post-seismic loading, leading to the submersion of the dam;

– landslide within the reservoir;

– loss of control of the sluice (blockage of the gate, falling rocks, blocked road).

The causes that can induce internal erosion are:

– movement of an active fault within the foundation;

– fissures and water leakage after the earthquake;

– shearing of the filter.

The cause of failure by sliding is due to the loss of stability induced by liquefaction.

To conclude, liquefaction is the main cause of embankment failure and therefore requires tools dedicated to the evaluation of its risks of occurrence, but the other possible causes should not be forgotten in a global safety assessment.

10.6. Semi-coupled modeling of liquefaction

10.6.1. *Objectives*

Updating of the seismic risk regulation in France in 2010 means a number of engineering works will need to be recalculated. For this purpose, consulting firms will require powerful numerical models adapted to cyclic and dynamic loadings that are easy to use and already implemented in current commercial codes. In this section, we present the work of Le [LE 06] who, during his PhD thesis at the hydro-engineering center of EDF and at 3S-R laboratory in Grenoble, attempted to develop a simple and reliable numerical approach to the stability of embankment dams under seismic loading, taking into account the liquefaction phenomenon.

10.6.2. *Constitutive model*

The Mohr-Coulomb model is the simplest and most common elastoplastic model. However, it is not adapted to reproduce soil behavior under cyclic loading. Here, we present a modified version of the Mohr-Coulomb model based on elastoplasticity and on the observation of sand behavior under monotonic and cyclic loading. It incorporates Byrne's model for volumetric strains and Ramberg-Osgood's model for shear strains, and is named the BRO (Byrne–Ramberg-Osgood) model.

10.6.3. *Failure criterion*

The failure criterion corresponds to the Mohr-Coulomb criterion. The material cohesion is equal to zero. The friction angles are determined from drained triaxial tests, based on stress–strain curves through the following expression:

$$\frac{\sigma'_1}{\sigma'_3} = \tan^2\left(\frac{\pi}{4} + \frac{\phi}{2}\right)$$

with:

- ϕ, friction angle;
- σ'_1, major effective stress;
- σ'_3, minor effective stress.

Three friction angles control the behavior of cohesionless materials: the characteristic angle ϕ_{CAR} that corresponds to the change from contractive to dilative behavior; the peak friction angle ϕ_{MAX} that corresponds to the maximum strength; and the critical friction angle ϕ_C that corresponds to the residual strength and is, therefore, extremely significant. In most cases, where liquefaction is analyzed, the material density is low and the friction angle considered is the critical friction angle. In the other cases, the peak friction angle is defined as a function of the initial e_0 and critical e_C void ratios, or as a function of the overconsolidation ratio p'_c/p'_o as follows:

$$\tan(\varphi_{MAX}) - \tan(\varphi_C) = A.\log\left[\frac{p'_C}{p'_0}\right] = \frac{A}{C_C}.[e_0 - e_C]$$

Biarez never stopped repeating that the critical state in soil mechanics is a unique phenomenon. During liquefaction, discontinuities are created, corresponding either to a thin water film at the base of an impervious layer or to shear strain localizations and fissures in the neighboring material that disrupt the undrained condition. For this reason, the relation between peak strength and the void ratio is measured from undrained compression tests, whereas the ultimate strength S_{us} at the end of the earthquake, called the post-seismic or undrained residual strength, is measured from extension tests following undrained cyclic loading. The ultimate strength S_{us} is a crucial parameter for the model because its value controls the post-seismic stability. The model limits the pore pressure value in order to correctly simulate this ultimate strength.

10.6.4. *Shear strain law*

During an earthquake, the soil is shaken by cyclic loads that induce strains of various amplitudes, from values below 10^{-5} corresponding to reversible strains to significant irreversible strains. The cyclic stress–strain relationship is non-linear; the shear modulus decreases and the soil dissipates a fraction of the seismic energy. In order to reproduce this non-reversible behavior, a non-linear constitutive law, Ramberg-Osgood's law, has been introduced [RAM 43].

In the elastic domain, the shear modulus G_{MAX} and the bulk modulus K_{MAX} are made dependent on the effective mean stress, following Hertz' law:

$$G_{MAX} = G_1 \left(\frac{p'}{p_1} \right)^n$$

$$K_{MAX} = K_1 \left(\frac{p'}{p_1} \right)^n$$

with:

– G_1: shear modulus for p' = 1MPa;

– K_1: bulk modulus for p' = 1MPa;

– n: non-linearity exponent (n = 0.5);

– p': mean effective stress (MPa);

– p_1: reference stress, p_1 = 1MPa.

Beyond the elastic domain, the hysteretic non-linear law represents the decrease of the moduli and the increase of the hysteretic damping due to energy dissipation during shearing and to grain rearrangement using Ramberg-Osgood's expression (see Figure 10.11):

$$\frac{\tau - \tau_C}{\gamma - \gamma_C} = G_{MAX} \cdot \frac{1}{\left[1 + \alpha \left(\frac{|\tau - \tau_C|}{n.\tau_Y} \right)^{r-1} \right]}$$

with:

– τ: shear stress;

– γ: shear strain;

– τ_c and γ_c: shear stress and strain, respectively, at the last change of loading direction;

– G_{MAX}: initial tangent shear modulus;

– $n = 1$ at first loading and n = 2 afterwards;

– r, α: material constants;

– τ_y: maximum stress corresponding to Coulomb's criterion;

– $\tau_y = G_{MAX}.\gamma_y$.

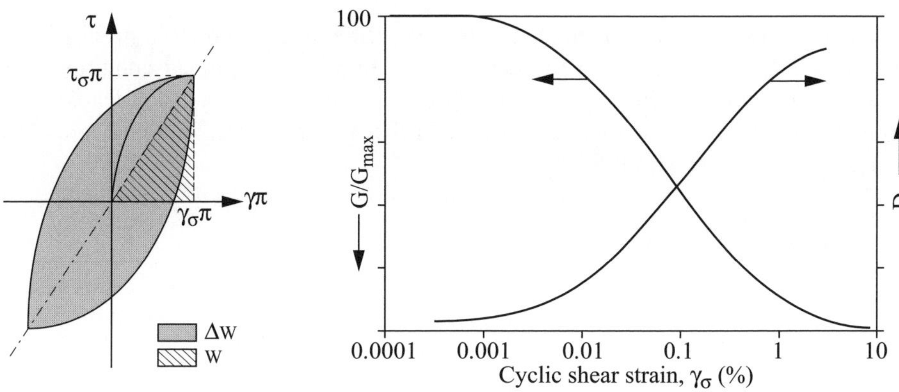

Figure 10.11. *Shear modulus and damping ratio versus shear strain*

During one cycle, the elastic energy $W = 1/2.\tau_C.\gamma_C$ is restored, whereas the plastic energy is dissipated during the material shearing. The damping ratio D measures the dissipated fraction of the energy:

$$D = \frac{\Delta W}{4\pi W}$$

By integrating Ramberg-Osgood's expression, the coefficient of the damping ratio D permits us to measure the parameter r and the ratio G/G_{MAX}:

$$D = \frac{2}{\pi}\frac{r-1}{r+1}\left(1-\frac{G}{G_{max}}\right)$$

Beyond the critical state, the dilatancy is activated and controlled by the following equation:

$$\Psi = 1.6.(\phi_{pic} - \phi_{crit})$$

In the conditions of potential liquefaction, the relative density is sufficiently small for the dilatancy angle Ψ to be between 0 and a few degrees.

10.6.5. *Volumetric strain law: liquefaction*

Byrne's law is an incremental law that describes the volumetric strain or the material contractancy under cyclic deviatoric loading. In undrained condition, the total volume remains constant and, therefore, the material contractancy leads to an isotropic unloading created by an increase in the pore pressure. The pore pressure change *du* is given by [BYR 94]:

$$du = \frac{1}{1 + \frac{nK}{K_w}} \cdot K \cdot \Delta\varepsilon_{vd}$$

with:

– du: pore pressure change;

– n: porosity;

– K: material bulk modulus;

– K_w: water bulk modulus;

– $\Delta\varepsilon_{vd}$: volumetric strain increment.

Byrne's law is expressed as follows:

$$\Delta\varepsilon_{vd\,1/2} = 0.5.C_1.\,\gamma \exp\left(-\frac{C_2 \cdot \varepsilon_{vd}}{\gamma}\right)$$

with:

– $\Delta\varepsilon_{vd\,1/2}$: volumetric strain increment created during a half-cycle of shear strain amplitude γ;

– ε_{vd}: sum of the increments due to the preceding cycles;

– C_1, C_2: material constants.

10.6.6. *Model implementation*

The model has been implemented in FLAC 2D (Fast Lagrangian Analysis of Continua), version 3.40. It is based on the finite difference method and has been developed in order to analyze coupled hydromechanical problems. The code has three advantages: it is easy to use, easy to implement constitutive models, and possible to calculate in large strains. However, the numerical resolution for cyclic loading requires us to distinguish between applied loads and numerical cycles used to guarantee the calculus convergence. Indeed, the material cyclic contractancy is taken into account at the end of each half-cycle, which requires the reactualization of the effective stress state and of the mechanical properties in Ramberg-Osgood's formulation. In the 1D case, the notion of the cycle is clear: a cycle corresponds to the maximum strain amplitude. For a 2D analysis, the strain tensor has three independent components, and for a 3D case, the seismic loading is even more complex. A half-cycle is determined by detecting the maximum values in the strain space. The change in the loading direction is defined by a negative value of the variable corresponding to the projection of the actual strain state on the strain state at the end of the last cycle. This variable can be expressed by the scalar product of the actual strain vector and of the reference unit vector, written in 2D as:

$$d = \frac{\left(\varepsilon - \varepsilon^{c}\right).\left(\varepsilon^{c} - \varepsilon^{cp}\right)}{\sqrt{\left(\varepsilon^{c} - \varepsilon^{cp}\right).\left(\varepsilon^{c} - \varepsilon^{cp}\right)}}$$

– ε, ε^{c}, ε^{cp}: strain vectors $(\varepsilon_{11} - \varepsilon_{22};\ 2\varepsilon_{12})$ of the actual state, of the last change (index *c*) and of the penultimate change (index *cp*) of loading direction, respectively.

– $(\varepsilon - \varepsilon^{c})$ corresponds to the strain evolution during the actual cycle since the last direction change.

– $(\varepsilon^{c} - \varepsilon^{cp})$ corresponds to the reference vector of the last half-cycle.

Once the extremum in memory has been confirmed as correct, the number of half-cycles is reactualized and the volumetric strain of the actual half-cycle is injected as the soil response to the loading.

10.6.7. *Model qualification in the case of the San Fernando dam failure*

Every design method has to be numerically validated and physically justified. This procedure is called qualification. A numerical model applied to a dam can be compared to a doctor examining a patient. Whereas anybody is capable of diagnosing a state of good health, only a competent practitioner can correctly diagnose an illness that can lead to near death early enough. In other words, each numerical model has to demonstrate its ability to reproduce the failure of real geo-structures. This exercise offers the opportunity to control the values of the instability criteria.

The phase of model validation showed that the number of cycles or the value of the extremum could be affected by the resolution scheme. To detect a change in the loading direction would require a capacity to accurately differentiate an effective change from a numerical change due to an oscillatory convergence after a real extremum. Tests showed that this detection was not very effective, which leads to the calculation of a number of cycles depending on the numerical choices during the model integration. An improved version for the determination of the number of cycles was developed [BAG 08]. Concerning the justification, the seismic behavior of the San Fernando Dam was retained as modeling case. The interesting feature of this site was that it had been subjected to several earthquakes and was very well documented. In 1952, an earthquake with $PGA = 0.12$ g did not produce any damage, whereas one in 1971 led to the failure of the upstream facing of the lower embankment, but not the one of the upper embankment.

10.6.7.1. *Instability criteria*

In the analysis carried out by the numerical model, the consequences of an earthquake are estimated during and after the seismic loading in terms of strains, irreversible displacements and the occurrence of liquefaction. The comparison with *in situ* observations allows us to evaluate and then generalize the criteria of seismic instability and failure to embankment and rockfill dams.

10.6.7.2. *Shear strains*

The evaluation in terms of strain amplitude is based on the analysis of the second invariant of the strain tensor. This term allows us to determine the position of the localized shear bands within the geo-structure. The interpretation is based on the visualization of the possible kinematics of failure or of zones of significant shearing. The EDF practice consists of connecting:

– a liquefied zone to a zone with shear strain greater than 5%, generalizing Seed's criterion to the whole earth structure obtained from laboratory experiments;

– a generalized failure to a concentration of strains greater than 5% across the whole dam from upstream to downstream.

It is essential to be able to visualize the shear zones in order to estimate the spreading, localization and concentration of these zones. The risk of failure becomes significant when these zones connect the upstream and downstream sides of the dam without discontinuity, mobilizing a large volume of soil.

10.6.7.3. *Displacements*

The failure threshold in displacement corresponds to:

– a vertical settlement greater than 1/3 of the “freeboard”;

– a horizontal displacement greater than 1/3 of the filter thickness or, in the absence of filter, no higher than 1% of the total height of the dam or to 50 cm.

10.6.7.4. *Liquefaction factor*

The liquefaction factor is a key parameter that is calculated by the following relation:

$$T_{liq} = \frac{u - u_0}{\sigma_{yy} - u_0}$$

with:

– u: pore pressure at the time of the study;

– u_o: pore pressure after impounding;

– σ_{yy}: vertical stress at the time of the study.

A value of T_{liq} higher than 0.8 is considered to represent the occurrence of liquefaction.

10.6.7.5. *Parameter determination*

The model validation was conducted by simulating the behavior of the two dams of the Van Norman developments in the San Fernando Valley. The determination of the BRO model parameters is based on laboratory test results, mainly monotonic and cyclic triaxial testing.

The common procedure consists of the step-by-step determination of the initial state and, specifically, first the relative density, second the parameters of the shear and volumetric strain laws, and third the residual strength. This procedure requires the following experimental data:

– the elastic moduli K_{max} and G_{max};

– the curve (G/G_{max}, γ) giving the shear modulus decrease with the shear strain amplitude;

– the curve (D, γ) giving the damping ratio increase with the shear strain amplitude;

– the curve giving the liquefaction potential.

The latter is taken from [SEE 73]. The residual strength was measured by Castro [CAS 89]. All the parameter values used in this study are assembled in Table 10.4.

Embankment	**Lower**					**Upper**			
Material	**Compacted embankment**	**core**	**Shale layer**	**Sandy embankment**	**Alluvial foundation**	**Compacted embankment**	**Core**	**Sandy embankment**	**Alluvial foundation**
e_{max}	**0.93**	**1.15**	**1.15**	**0.93**	**1**	**0.93**	**1.15**	**0.93**	**1**
$e_{in\ situ}$	**0.571**	**0.76**	**0.76**	**upstream: 0.663 and downstream: 0.652**	**0.29**	**0.571**	**0.76**	**0.652**	**0.5**
e_{min}	**0.42**	**0.51**	**0.51**	**0.42**	**0.29**	**0.42**	**0.51**	**0.42**	**0.29**
c (KPa)	**0**	**0**	**0**	**0**	**0**	**0**	**0**	**0**	**0**
ϕ_{crit} (°)	**34**	**34**	**34**	**34**	**35**	**34**	**34**	**34**	**35**
Su_C (kPa)	**104**	**33**	**33**	**upstream: 26 and downstream: 33**	**103**	**104**	**33**	**33**	**103**
G_l (MPa)	**381**	**382**	**360**	**298**	**727**	**360**	**418**	**208**	**761**
K_l (MPa)	**781**	**924**	**1,010**	**1,041**	**1,700**	**968**	**1,109**	**524**	**1,478**
C_c	**0.154**	**0.194**	**0.194**	**0.154**	**0.215**	**0.154**	**0.194**	**0.154**	**0.215**
r	**2.45**	**2.4**	**2.4**	**2.3**	**2.45**	**2.45**	**2.4**	**2.3**	**2.45**
α	**4**	**5.8**	**5.8**	**8**	**1.8**	**5.5**	**5.5**	**13**	**2**
C_1	**0.07**	**0.09**	**0.09**	**0.024**	**0.026**	**0.04**	**0.09**	**0.012**	**0.05**
C_2	**20**	**41**	**41**	**43**	**17**	**31**	**36**	**47**	**21**

Table 10.4. *Material parameters for the modeling of San Fernando embankment dams*

10.6.7.6. *Justification of the behavior of the lower dam during the 1952 earthquake*

The construction of the San Fernando lower dam started in 1912; the upper dam was built afterwards, in 1919. The "semi-hydraulic fill" technique was applied in order to build the upper dam, which appears to be an improvement over the classical hydraulic fill technique. In order to validate the constitutive model used in this study, and to ascertain that the chosen parameter values do not correspond to a pessimistic misrepresentation of the soil resistance leading to early failure, we have to verify that the earthquakes before 1971 could not have led to the dam failure. With no information available for the 1952 earthquake, the accelerograph of the 1971 earthquake was selected and only the maximum acceleration was modified to the measured value equal to 0.12 g.

Figure 10.12. *Horizontal (upper graph) and vertical (lower graph) displacements of the lower dam calculated by modeling the 1952 earthquake*

The calculated displacements are small and relatively close to those measured: calculated settlement of the top of the dam equal to 4.7 cm compared to the measured ones from 5.0 to 8.0 cm; maximum horizontal displacement equal to 16.0 cm. The shear strains are also small: the maximum value is around 0.8%. The pore pressure rise remains limited: the liquefaction factor remains below 22.5%.

Figure 10.13. *Maximum shear strain (upper graph) and liquefaction ratio (lower graph) in the lower dam calculated by modeling the 1952 earthquake*

The numerical model demonstrates that the dam remains stable. Only small deformations and displacements are generated by the earthquake for a maximum acceleration of 0.12 g. No failure threshold is reached during this seismic loading.

10.6.7.7. *Justification of the lower dam failure during the 1971 earthquake*

The results of the seismic loading calculation of the lower dam show a good agreement between the curves giving the isovalues of the shear strain and the observed shear bands (see Figure 10.14). Shear strains greater than 5% correspond to the slip zone, whereas the zones with calculated shear strains greater than 20% do not coincide exactly with the geometry of the observed failure zone. Therefore, the 5% shear strain criterion is representative of the failure zone. The liquefaction factor increased within the whole soil mass at the base of the embankment by up to 80%, which agrees with the analysis that liquefaction was the cause of the failure of the upstream facing and of the core. The calculated displacements are significant, although smaller than the real ones; they reach 7 m for horizontal displacement of the upstream facing and 5 m for the vertical settlement of the dam. All the stability criteria have been exceeded.

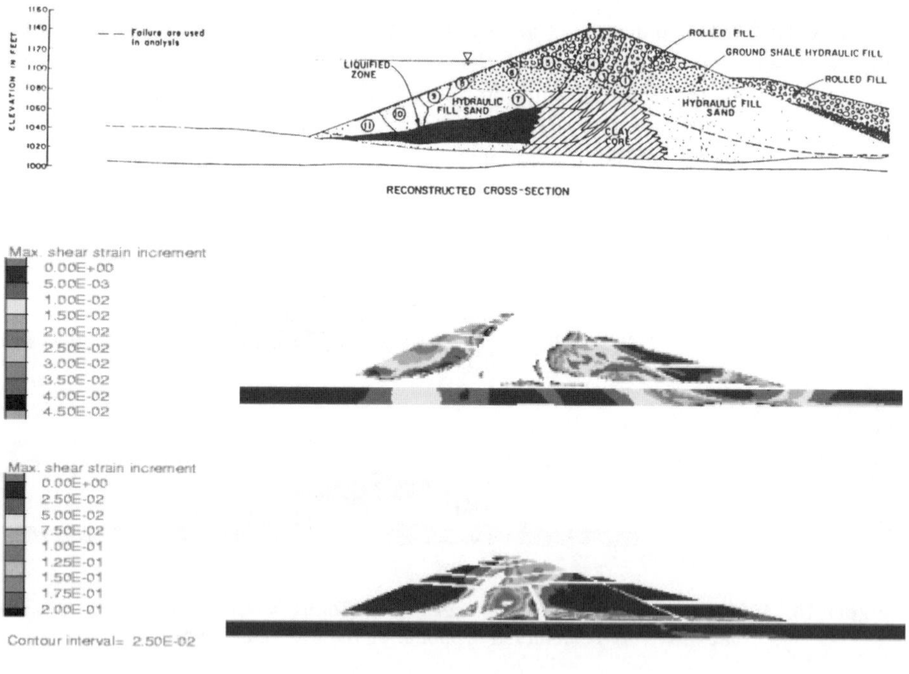

Figure 10.14. *Comparison between reconstituted in situ and calculated shear bands for the 1971 earthquake (from top to bottom: shear strains <5%, shear strains <20%, liquefaction factor, horizontal displacement and vertical settlement)*

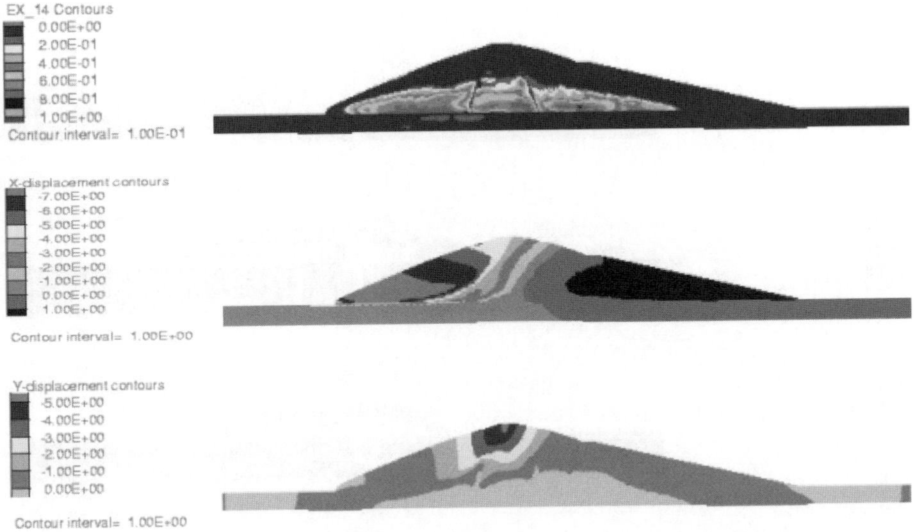

Figure 10.14. (continued) *Comparison between reconstituted in situ and calculated shear bands for the 1971 earthquake (from top to bottom: shear strains <5%, shear strains <20%, liquefaction factor, horizontal displacement and vertical settlement)*

10.6.7.8. *Justification of the stability of the upper embankment during the 1971 earthquake*

The numerical results for the upper dam are different from those obtained for the lower dam. First of all, the curves of 5% shear strain values are not continuous from upstream to downstream. We obtain zones of strain concentration, but they are not generalized from one side of the dam to the other. In reality, failure did not occur, which is in agreement with the model prediction showing a non-continuous 5% shear strain zone. The liquefaction factor reached only 50%, which indicates that the liquefaction did not spread within the embankment base. The calculated displacements remained small, reaching 0.75 m for the horizontal displacement due to the sliding of the upstream and downstream facings, compared to the 2.20 m displacement measured downstream. The calculated vertical settlement reached 0.70 m, in complete agreement with the measured value of 0.76 m. For this embankment, the stability criteria were not exceeded, which is confirmed by its real behavior during the earthquake (see Figure 10.15).

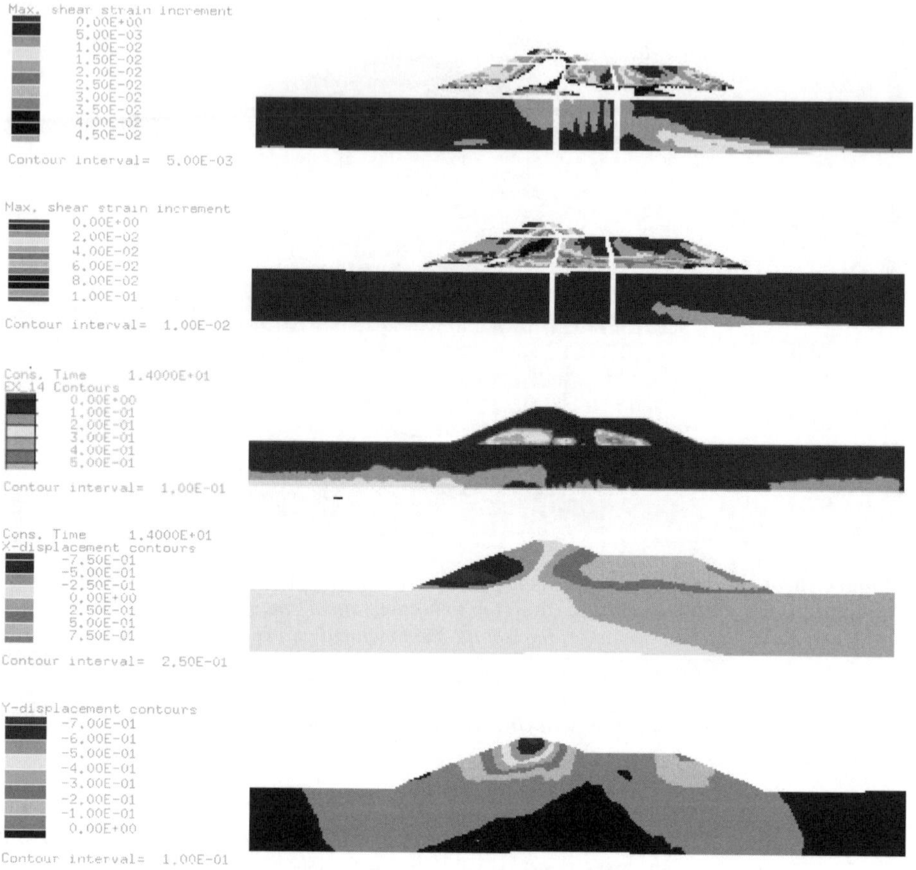

Figure 10.15. *Comparison between reconstituted in situ and calculated shear bands for the 1971 earthquake (from top to bottom: shear strains <5%, shear strains <20%, liquefaction factors, horizontal displacements and vertical settlements)*

The main assets of this model are that its easy use as well as its ability to reproduce the liquefaction potential from the pore pressure or shear strain increase. Its potential has also been demonstrated in the successful numerical analysis of the behavior of the San Fernando dams under seismic loadings.

10.6.8. *Model application to canal embankments*

The numerical model was applied to the case of a canal embankment. The main problem concerning the study of canal embankments is the variability of the nature and state of the foundation soil all along the reach. The objective of the study was to test the sensibility of the seismic resistance of the embankment to the state of the alluvium controlled by its relative density. This type of calculation should give us elements for defining the seismic risk in various zones of the embankment. Indeed, the relative density of the materials constituting the embankment and its foundation can easily be determined by an extensive series of penetrometer tests. According to the test results, it will be possible to characterize the seismic risk for a given embankment at a first attempt.

The numerical model studied is presented in Figure 10.16. The embankment is assumed to be homogeneous and constructed with soils coming from the foundation. The embankment is 8 m high and its slope is H/V = 3/1. The model is made of 3,102 quadrilateral elements and 3,254 nodes, and is 217 m long and 33 m high.

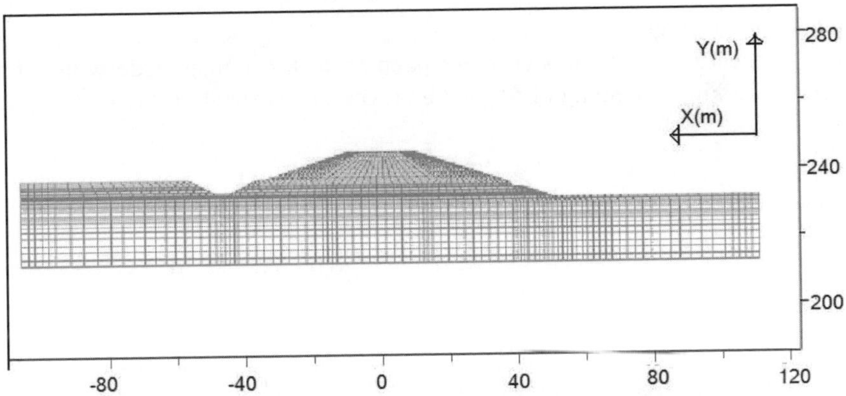

Figure 10.16. *Geometry and meshing of the numerical model*

The mechanical properties have been simplified on purpose. The relative density is assumed to be identical in the foundation and in the embankment. Four values were used: 30%, 50%, 70%, and 100%. The mechanical properties, dependent on the relative density, are provided in Table 10.5.

The failure criterion is expressed as follows:

$$tan\phi = 0.22*\log(p'c/p') + 0.726$$

The hydraulic characteristics are the following: for the alluviums a horizontal permeability of 2^{-3} m/s and a vertical permeability of 5^{-4} m/s, assumed to be the same for the embankment and the foundation and independent of the relative density.

Dr	G_1 (MPa)	G_{max}. (for $\sigma'_3 = 200$ kPa)	τ_y (kPa)	r	alpha	C_1	C_2
50	370	165	118	2.4	8	0.09	5.0
60	390	174	118	2.5	7	0.08	4.5
70	410	183	122	2.6	6	0.06	3.0
80	430	192	123	2.7	5	0.05	2.5
100	470	210	129	2.8	4	0.04	1.5

Table 10.5. *Parameter values*

Only one earthquake was considered, given by a horizontal accelerograph applied at the base of the model, corresponding to a 6.1 magnitude with a focal distance of 12 km, a damping of 5% and a maximum acceleration $PGA = 0.25$g (see Figure 10.17).

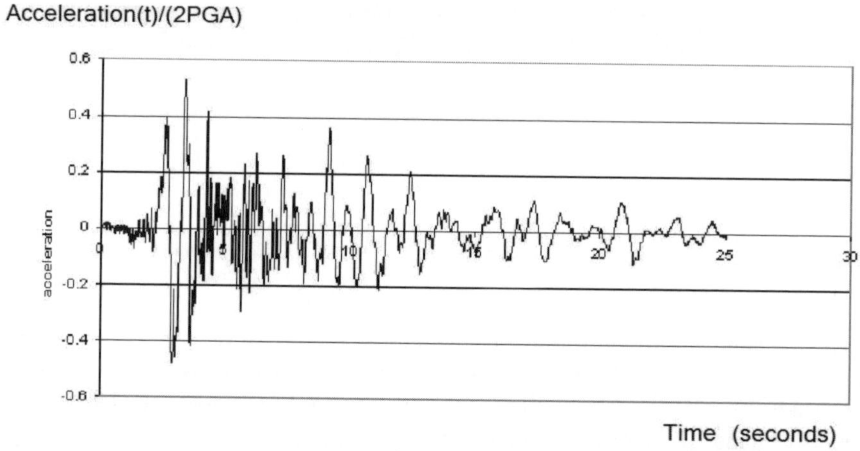

Figure 10.17. *Chosen accelerograph*

The results showing the influence of the relative density are quite interesting. They demonstrate that a relative density of 50% corresponds to a change in the embankment behavior.

Table 10.6 gives a comparison between the numerical results and the thresholds characterizing the seismic resistance, as discussed in the above section.

Criteria		Dr = 30%	Dr = 50%	Dr = 70%	Dr = 100%
Liquefaction	Maximum liquefaction factor	0.8	0.8	0.8	0.8
	Localization	Foundation and upstream embankment	Upstream embankment	Upstream embankment	Upstream embankment
Failure	Maximum shear strain	>20%	5%	2%	1%
	Local shear (1)	-	Track	-	-
	Shear band strain >5% (2)	Upstream and downstream embankment	-	-	-
Displacement	Maximum horizontal displacement (m)	1.5	0.4	0.2	0.1
	Security criterion (3)	Failure	No	No	No
	Maximum vertical displacement (m)	0.65	0.15	0.075	0.025
	Security criterion (4)	Failure	No	No	No

Table 10.6. *Results of the parametric study*

Figures 10.18 and 10.19 provide a comparison of the behavior indicators for the four initial states of the embankment. This comparison leads to the following conclusions.

Liquefaction factors: these increase during the earthquake but start to decrease before the end of the seismic loading. This is due to the permeability of the gravels, which is greater than for sands. If the increase in pore pressure during the seismic loading is the same for the two soils in undrained condition, this is not the case *in situ,* where the possible drainage in the gravels reduces the risk of liquefaction. The presence of slabs prevents the pore pressure dissipation, which induces higher liquefaction factors in their surroundings. If the slabs can move under the influence of the pore pressure increase, then the hypothesis of imperviousness retained for the calculation is probably too conservative.

Shear strains: important shear strains induce the sliding of the two facings for $Dr = 30\%$, which is no longer the case for $Dr = 50\%$, where only the track is sliding. No sliding occurs for $Dr = 70\%$. We can note, however, that the bank of the side drainage canal is still affected by significant straining, without any effect on the overall stability.

Horizontal displacement: for $Dr = 30\%$, the zones that have a horizontal displacement greater than 50 cm are only the two facings; for $Dr = 50\%$, this is only the case for the bank of the side drainage canal (the maximum horizontal displacement of the facings is equal to 0.4 m); for $Dr = 70\%$, no significant displacement can be noted.

Vertical displacement: for $Dr = 30\%$, the zones having a settlement greater than 20 cm are the two facings; for Dr equal to or greater than 50%, no significant settlement is obtained.

This parametric study gives very important results, confirmed by *in situ* observations:

– failure with large sliding is probable for $Dr = 30\%$;

– $Dr = 50\%$ corresponds to the limit between acceptable and pathological behaviors;

– no pathology is observed for $Dr > 70\%$.

Dr = 30%

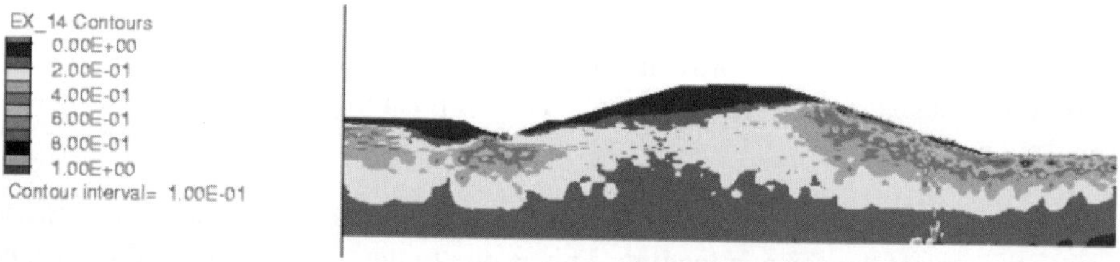

Dr = 50%

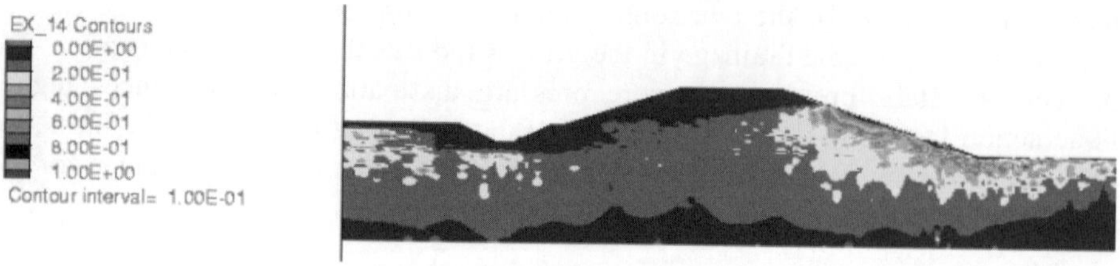

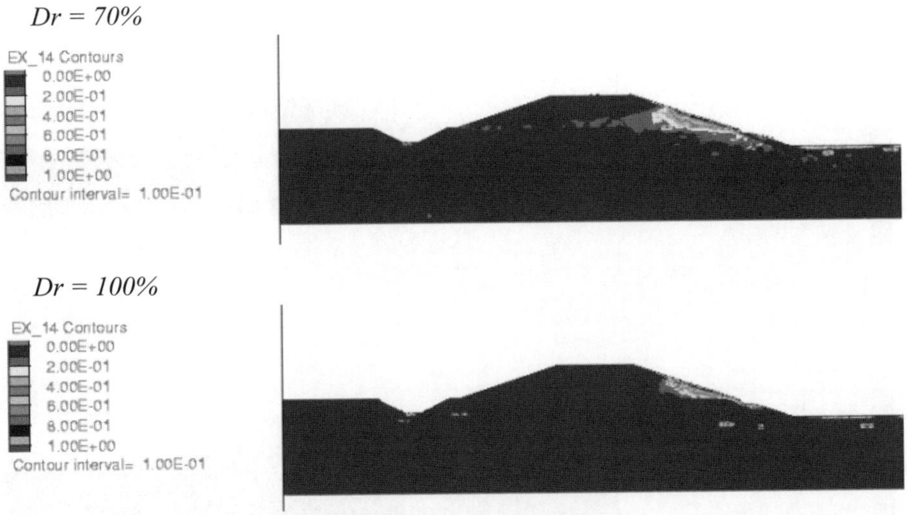

Figure 10.18. *Liquefaction factor at t = 24 s for different relative densities*

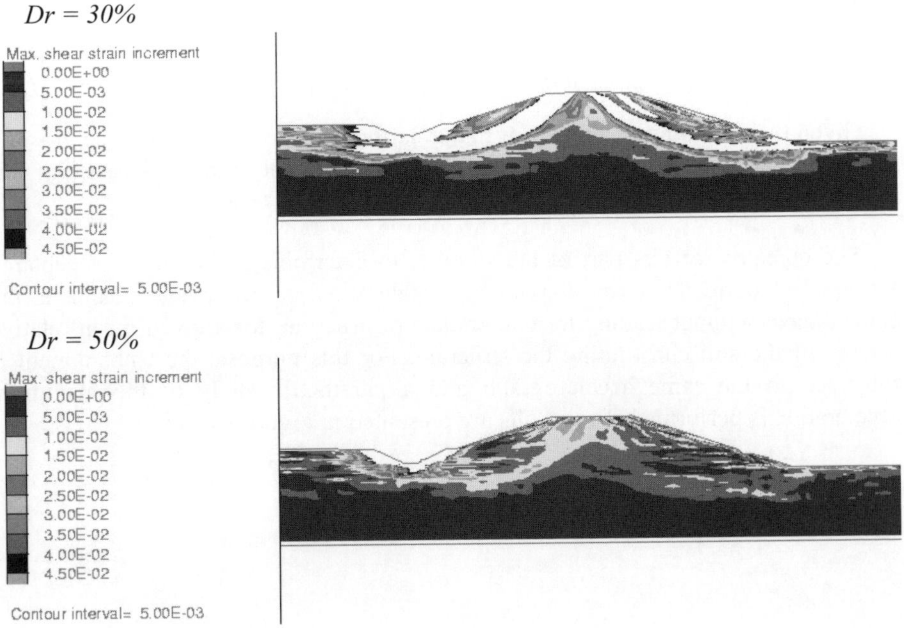

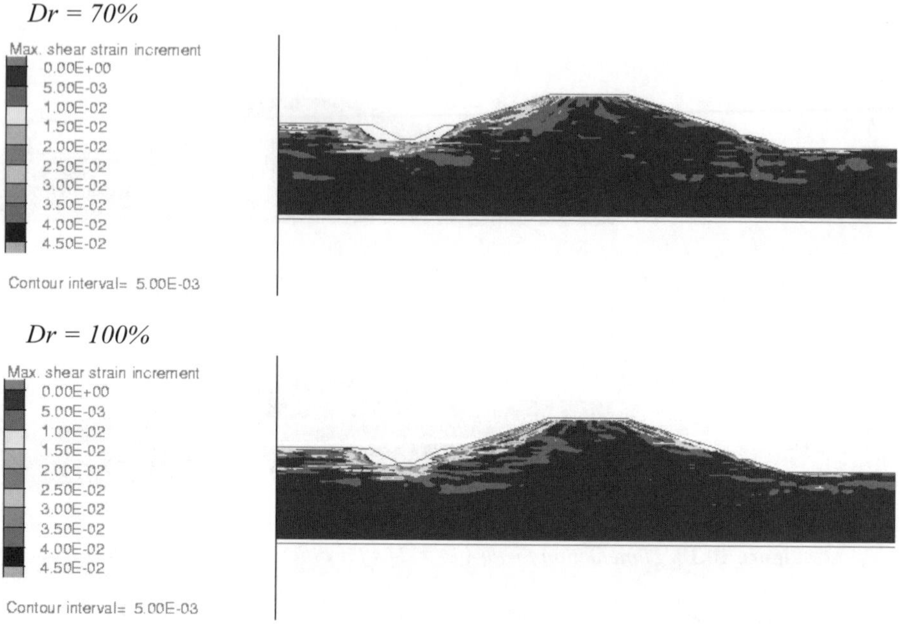

Figure 10.19. *Shear strains from 0 to 5% at t = 24 s for different relative densities*

10.6.8.1. *Liquefaction factor at the end of the earthquake*

Given the very limited number of major earthquakes recorded in France, it is not surprising that the question of uncertainty over the seismic signal is raised. For this reason, a parametric study on the seismic hazard effect was carried out.

The objective of this part of the study is to determine the strongest acceptable earthquake, defined by its maximum acceleration, which would damage the embankment without leading to its complete destruction, for a given mean relative density of the soil constituting the structure. For this purpose, the embankment is subjected to the same accelerograph and a parametric study of the maximum acceleration is performed. The results are presented in Figure 10.20.

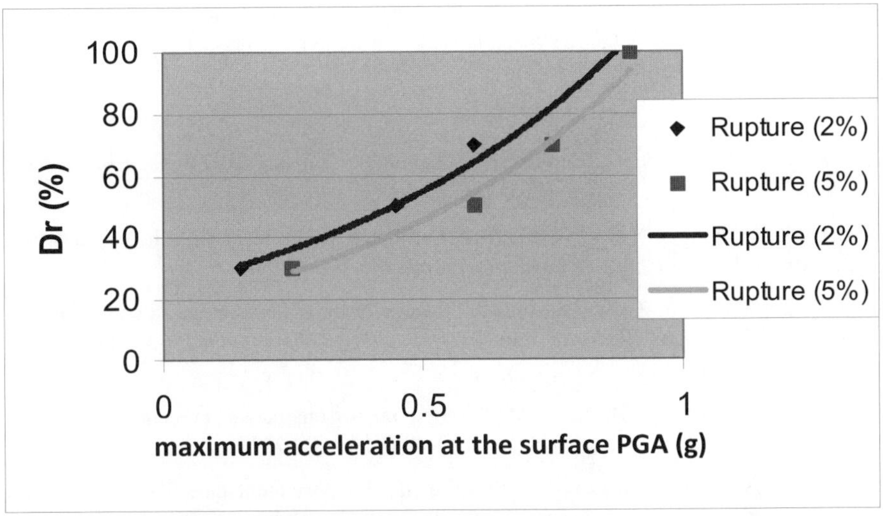

Figure 10.20. *Joint influence of the relative density* Dr *and of the maximum acceleration PGA*

The criteria adopted for characterizing the failure of the embankment are the following:

– shear strain >2% along a sliding surface upstream-downstream or downstream-upstream;

– maximum shear strain >5% in a zone within the embankment.

The conclusions of this study are:

– with *Dr* around 30%, the strongest earthquake acceptable has a PGA = 0.2g;

– with *Dr* around 50%, the strongest earthquake acceptable has a PGA = 0.5g;

– with *Dr* around 80%, the strongest earthquake acceptable has a PGA = 0.8g.

The model justification based on real failure and numerical study appears to be quite satisfactory. An improvement in the robustness of the numerical method remains to be achieved for assuring the model qualification.

10.7. Bibliography

[AFP 07] AFPS, Groupe de Travail Zonage 2005-2007, Résolution du Conseil, Avis du CST et Rapport, Paris, 2007.

[AHM 92] AHMADI M.T., KHOSHRANG G., "Sefi Rud dam's dynamic response to the large near-field earthquake of June 1990", *Dam engineering*, vol. 3, no. 2, pp. 85-115, 1992.

[AMB 67] AMBRASEYS N., SARMA S., "The response of earth dams to strong earthquakes", *Géotechnique*, vol. 17, pp. 181-213, 1967.

[AMB 88a] AMBRASEYS N.N., "Engineering seismology", *Earthquake Engineering Structural Dynamics*, vol. 17, no. 1, pp. 1-105, 1988.

[AMB 88b] AMBRASEYS N.N., MENU J.M., "Earthquake-induced ground displacements", *J. Earthquake Eng Structural Dynam*, vol. 16, pp. 985-1006, 1988.

[ARI 03] ARIGA Y., CAO Z., WATANABE H., "Seismic stability assessment of an existing arch dam considering the effects of joints", *21st CIGB Congress*, Q83 R33, pp. 553-575, Montréal, Canada, 2003.

[ATK 97] ATKINSON G.M., BOORE D.M., "Some comparisons between recent ground-motion relations", *Seismological Research Letters*, vol. 68, no. 1, 1997.

[BAG 08] BAGLALI Y., Amélioration d'une loi de comportement de sol sous sollicitation sismique et d'une loi de comportement d'interface sol structure sous chargement cyclique, TFE EDF, 2008.

[BAK 79] BAKHTIN B.M., DUMENKO V.I., "Seismic stability of concrete having a lightweight profile", *Hydrotechnical Construction,* vol. 5, 1979.

[BAR 96] BARTHURST C., "Deterministic sliding block methods for estimating seismic displacement of earth structures", *Journal of Soil Dynamics and Earthquake Engineering*, vol. 15, p. 255-268, 1996.

[BEL 96] BELISLE C., LINARD M., "Quelles nouvelles compétences des acteurs de la formation dans le contexte des TIC? ", *Education Permanente*, vol. 127, pp. 19-47, 1996.

[BET 03] BETCGB, *Classification des Barrages vis-à-vis des Séismes*, Ministère de l'Industrie, Paris, December 2003.

[BOO 83] BOORE D.M., "Stochastic simulation of high-frequency ground motion based on Seismological Models of the radiated Spectra", *Bulletin of the Seismological Society of America*, vol. 73, pp. 1865-1894, 1983.

[BOU 81] BOUCHON M., "A simple method to calculate Green's function for elastic layered media", *Bulletin of the Seismological Society of America*, vol. 71, pp. 957-971, 1981.

[BRU 03] BRUNET C., FRY J.J., HAMMOUD I., LAIGLE F., "Analyse du choix des critères de rupture des barrages en remblai en vue de la réévaluation de la stabilité sismique", *CIGB Congrès de Montréal*, vol. 3, Q83 R47, pp. 839-848, Montreal, Canada, 2003.

[BYR 87] BYRNE P.M, CHEUNG H., YAN L., "Soil parameters for deformation analysis of sand masses", *CGJ*, vol. 24, no. 3, pp. 366-376, 1987.

[CAS 89] CASTRO G., KELLER T.O., Re-evaluation of the San Fernando dams – Report 1, US Army Corps of Engineers 9/1989.

[CAS 03] CASTRO G, WALBERG F.C., PERLEA V., “Dynamic properties in foundation of an embankment dam in Kansas”, *CIGB Congrès de Montréal*, vol. 3, Q83 R30, pp. 497-518, Montreal, Canada, 2003.

[CHO 72] CHOPRA A.K., CHAKRABARTI P., “The earthquake experience at KOYNA dam and stresses in concrete gravity dams”, *Earthquake Engineering Structural Dynamics*, vol. 1, no. 2, pp. 151-164, 1972.

[CHO 88] CHOPRA A.K., “Earthquake response analysis of concrete dams”, in R.B. JANSEN (ed.), *Advanced Dam Engineering for design Construction and Rehabilitation*, Van Nostrand Reinhold, New York, 1988.

[CIG 89] CIGB, Bulletin 72, Choix des Paramètres Sismiques pour les Grands Barrages, CIGB, 1989.

[CIG 95] CIGB, Bulletin 99, Sécurité des Barrages, CIGB, 1995.

[CIG 01] CIGB, Bulletin 120, Aspects de la Conception Parasismique des barrages, CIGB, 2001.

[DON 91] DONLON W.P., HALL J.F., “Shake table study of concrete dam monoliths”, *Earthquake Engineering and Structural dynamics*, vol. 2, no. 20, 1991.

[DUB 03] DUBIE J.Y., LE DELLIOU P., “Méthodes de classement synthétique des barrages vis-à-vis du risque sismique”, *21° CIGB Congrès de Montréal*, vol. 3, Q83 R45, pp. 797-811, Montreal, Canada, 2003.

[FER 02] FERC, Engineering guidelines for the evaluation of hydropower projects, chap. III, Gravity dam Federal Energy Regulatory Commission Department of Energy Washington, DC, 2002.

[FRY 03] FRY J.J., “Analyse des incertitudes de la réévaluation de la stabilité sismique des barrages en remblai”, *Congrès CIGB de Montréal*, vol. 5, Montreal, Canada, 2003.

[GAL 00] GALLI P., “New empirical relationships between magnitude and distance for liquefaction”, *Tectonophysics*, vol. 324, pp. 244-265, 2000.

[GEO 08] GEOTER, Evaluation Probabiliste de l’Aléa Sismique sur le Territoire National Actualisée par Méthode Baysienne sur le REX: Atlas Cartographique, Rapport GTR/EDF/0508-458, 31 May 2008.

[HAR 78] HARTZELL S.H., “Earthquake aftershocks as Green’s function”, *Geophysical Research Letters*, vol. 5 pp. 1-4, 1978.

[JAN 76] JANSEN R.B., DUKLETH G.W., BARRETT K.G., “Problems of hydraulic fills dams”, *12th CIGB Congress, Mexico*, Q44 R16, pp. 337-364, Mexico, 1976.

[JAP 02] JAPAN DAM COMMISSION, Technical Committee on Seismic Safety Dams, Present Situations and Issues Regarding the Seismic Safety Evaluation of Existing Dams, Summary Japan Dam Commission on Large Dams, JDC, 2002.

[KAZ 01] KAZUNORI U., TORII T., TSUJINO S., ANDO S., HIGASIO K., YAMAMOTO T., "A study on failure mechanism of embankment dams for irrigation damaged by the Hyogoken-Nanbu Earthquake", *Fourth International Conference on Recent Advances in Geotechnical Earthquake Engineering and Soil Dynamics*, Paper no. 5.27, San Diego, CA,

[KOK 95] KOKUSKO T., TANAKA Y., KUDO K., KAWAI T., "Liquefaction case study of volcanic gravel layer during 1993 Hokkaido-Nansei-Oki earthquake", *Proc. 3rd Int. Conf. Recent Adv in Geotech. Earthquake Eng. and Soil Dynamics*, pp. 235-243, Saint-Louis, MO, USA, 1995.

[LE 06] LE T., Modélisation du Comportement des Barrages en Terre sous Séisme, Doctoral thesis, University of Grenoble, 2006.

[LEG 03] LEGER P., TINAWI R., JAVANMARDI F., LECLERC M., "Transient damping and uplift pressure responses of cracked concrete gravity dams subjected to earthquakes", *CIGB Congrès de Montréal*, vol. 3, Q83 R50, pp. 903-924, 2003.

[LIN 93] LIN G., ZHOU J., FAN C., "Dynamic model rupture test and safety evaluation of concrete gravity dams", *Dam Engineering*, vol. 4, no. 3, 1993.

[MAK 78] MAKDISI F.I., SEED H.B., "Simplified procedure of estimating dam and embankment earthquake-induced deformations", *Journal of the Geotechnical Engineering Division*, vol. 104, no. GT7, pp. 849-867, 1978.

[MAL 03] MALLA S., WOIELAND M., "Simple model for safety assessment of cracked concrete dams subjected to strong ground shaking", *CIGB Congrès de Montréal*, vol. 3, Q83 R11, pp. 151-167, Montreal, Canada, 2003.

[MAS 03] MASUKAWA S., YASUNAKA M., TANAKA T., KOHGO Y., "Dynamic progressive failure of earthfill dam models due to strong motions and their analyses", *21st CIGB Congrès de Montréal*, vol. 3, Q83 R36, pp. 613-629, Montreal, Canada, 2003.

[MAT 01] MATSUI T., MITAMURA M., TOSHIHIKO S., FUMIO S., TERUYUKI H., YASUO T., SEIJI S., HIROSHI T., AKIRA T., "Relevance of some damage factors to structures damage in the 1995 Kobe earthquake Paper no. 10.27 Fourth International", *Conference on Recent Advances in Geotechnical Earthquake Engineering and Soil Dynamics*, San Diego, CA, USA, March 26-31, 2001.

[MAT 03] MATSUMOTO N., YOSHIDA H., SASKI T., ANNAKA T., "Response spectra of earthquake motions at dam foundations", *21st CIGB Congrès de Montréal*, vol. 3, Q83 R35, pp. 595-611, Montreal, Canada, 2003.

[MON 25] MONONOBE C., "Characteristics of a water storage gravity dam and its rational design method", *Journal of the Japan Society of Civil Engineers*, vol. 11, no. 5, 1925.

[NEW 65] NEWMARK N.M., "Effect of earthquakes on dams and embankments", *Géotechnique*, vol. 15, no. 2, pp. 139-159, 1965.

[NOD 75] NODA S., UWABE T., CIBA T., "Relation between seismic coefficient and ground acceleration for gravity quay wall", *Report of the Port and Harbor Research Institute,* vol. 14, no. 4, 1975.

[PAT 03] PATEL M.S., BRAHMABHATT V.S., "Restoration and rehabilitation of embankment dams damaged due to BHUJ earthquake of January 2001", *21st CIGB Congrès de Montréal*, vol. 3, Q83 R22, pp. 339-355, Montreal, Canada, 2003.

[POP 93] POPOVICI A., ABDULAMIT A., TOMA I., PASCU D., "Comments upon an earth dam severely damaged by foundation", *Third International Conference on Case Histories in Geotechnical Engineering*, no. 3.35, Saint-Louis, MO, USA, June 1-4 1993.

[SAR 75] SARMA S.K., "Seismic stability of earth dams and embankments", *Géotechnique*, vol. 25, no. 4, pp. 743-761, 1975.

[SEE 73] SEED H.B., LEE K.L., IDRISS I.M., MAKDISI F., Analysis of the Slides in the San Fernando Dams During the Earthquake of February 9, 1971 Earthquake, Engineering Research Center, University of California, Report No E.E.R.C. pp. 73-2, 1973.

[SHE 74] SHEN H.C., CHANG C.H., HUANG L.S., LI T.C., YANG C.Y., WANG T.C., "Earthquakes induced by reservoir impounding and their effect on Hsinfengkiang dam", *Scientia Sinica*, vol. 17, no. 2, pp. 239-272, 1974.

[TAN 02] TANAKA T., SAITO Y., KOHGO Y., HARADA D., "Dynamic progressive failure of earth dams by finite element method and model experiment", *Proceedings of the 3rd International Conference on Dam Engineering*, pp. 323-330, 2002.

[TAN 01] TANI S., "Consideration of the earthquake resistance of large fill dams paper no. 10.03", *Fourth International Conference on Recent Advances in Geotechnical Earthquake Engineering and Soil Dynamics*, San Diego, CA, USA, March 26-31, 2001.

[TOR 97] TORO G.R., ABRAHAMSON N.A., SCHNEIDER J.F., "Model of strong ground motions from earthquakes in central and eastern North America Best estimates and uncertainties", *Seismological Research Letters*, vol. 68, no 1, 1997.

[UEN 01] UENG T.S., LIN M.L., CHEN M.H., "Some geotechnical aspects of 1999 Chi-Chi, Taiwan earthquake. SPL –10.1", *Fourth International Conference on Recent Advances in Geotechnical Earthquake Engineering and Soil Dynamics*, San Diego, CA, USA, March 26-31, 2001.

[WAK 01] WAKAMATSU K., YAMAMOTO A., TANAKA I., "Geomorphological criteria for evaluating liquefaction potential considering the level-2 ground motion in Japan", *Fourth International Conference on Recent Advances in Geotechnical Earthquake Engineering and Soil Dynamics*, Paper no. 4.56, San Diego, CA, USA, March 26-31, 2001.

[WIE 03] WIELAND M., BRENNER P.R., SOMMER P., "Earthquake resiliency of large concrete dams: damage, repair and strengthening", *21st CIGB Congrès de Montréal*, vol. 3, Q83 R10, pp. 131-150, Montreal, Canada, 2003.

[YAN 03] YAN L., KUM K.Y., "Liquefaction assessment of gravelly soils for dam safety evaluation", *21° CIGB Congrès de Montréal*, vol. 3, Q83 R44, pp. 765-796, 2003.

[YEG 91] YEGIAN M.K., MARCIANO E.A., GHAHRAMAN V.G., "Earthquake-induced permanent deformations: probabilistic approach", *Journal of the Geotechnical Engineering Division*, vol. 117, no. 1, pp. 1158-1167, 1991.

[YEG 94] YEGIAN M.K., MARCIANO E.A., GHAHRAMAN V.G., HARUTIUNYAN R.N., "Liquefaction and embankment failure case histories, 1988 Armenia earthquake", *Journal of the Geotechnical Engineering Division*, vol. 120, no. 3, pp. 581-596, 1994.

List of Authors

María Eugenia ANABALÓN
Arcadis Chile
Santiago
Chile

Edgar BARD
Arcadis Chile
Santiago
Chile

Jean-Louis BORDES
Ecole Centrale de Paris
Châtenay-Malabry
France

Bernard CAMBOU
Laboratoire de Tribologie et Dynamique des Systèmes
Ecole Centrale de Lyon
France

José CAMPAÑA
Arcadis Chile
Santiago
Chile

Félix DARVE
Laboratoire Sols Solides Structures Risques
Ecole Nationale d'Ingénieurs pour l'Énergie, l'Eau, l'Environnement
Grenoble
France

Jean-Louis FAVRE
LMSSMat
Ecole Centrale de Paris
Châtenay-Malabry
France

Etienne Flavigny
Laboratoire Sols Solides Structures Risques
Ecole Nationale d'Ingénieurs pour l'Énergie, l'Eau, l'Environnement
Grenoble
France

Jean-Marie FLEUREAU
Laboratoire de Mécanique des Sols, Structures, Matériaux
Ecole Centrale de Paris
Châtenay-Malabry
France

Etienne FROSSARD
COYNE & BELLIER
Gennevilliers
France

Jean-Jacques FRY
Laboratoire de Tribologie et
Dynamique des Systèmes
Ecole Centrale de Lyon
France
&
EDF-CIH
Le Bourget du Lac
France

António GOMES CORREIA
University of Minho/ DEC
Guimaraes
Portugal

Daniel GRIMM
Ecole Centrale de Paris
Châtenay-Malabry
France

Sigit HADIWARDOYO
Faculty of Engineering
University of Indonesia
Indonesia

Mahdia HATTAB
LEM3
Paul Verlaine University
Metz
France

Pierre-Yves HICHER
Institut de Recherche en Génie Civil
et Mécanique
Ecole Centrale de Nantes
France

Farid LAOUAFA
Institut National de l'Environnement
industriel et des Risques
Verneuil en Halatte
France

Sylvain LIGNON
Groupe de Recherche en Sciences
pour l'Ingénieur
University of Reims
Champagne-Ardenne
France

Arezou MODARESSI-FARAHMAND-RAZAVI
Laboratoire de Mécanique des Sols,
Structures, Matériaux
Ecole Centrale de Paris
Châtenay-Malabry
France

Cécile NOUGUIER-LEHON
Laboratoire de Tribologie et
Dynamique des Systèmes
Ecole Centrale de Lyon
France

Florent PRUNIER
Laboratoire de Génie Civil et
Ingénierie Environnementale
INSA de Lyon
France

Hanène SOULI
Laboratoire de Tribologie et
Dynamique des Systèmes
Ecole Nationale d'Ingénieurs de
Saint-Etienne
France

Said TAIBI
Laboratoire d'Ondes & Milieux
Complexes
University of Le Havre
France

Index

C, D, E

civil engineering, 1, 2, 4-7, 11, 15, 16, 35, 36, 43, 53, 78, 79
clay, 8, 12, 35, 113, 115, 117, 126, 128, 130, 131, 138, 144, 145, 164, 196, 198, 199, 201-204, 215, 231, 235, 250-253, 255, 258, 260, 270, 279-283, 285-288, 290, 291, 293, 295, 296, 338, 349
critical void ratio, 118-121, 131, 188, 215, 251, 252, 262, 287
deformability, 73, 77, 146, 197, 198
deviatoric compression, 172
drainage, 106-110, 169, 253, 254, 274, 281, 291, 337, 340, 350, 355, 365, 383
ECP, 2, 8, 9, 13, 165, 166, 178, 249, 250, 269, 279, 280, 295, 297, 305, 309, 323, 324
effective stress, 96, 117, 118, 132, 144, 153-155, 159, 160, 174, 176-178, 191, 193, 195, 196, 202, 203, 205, 212, 215, 216, 234, 246, 251, 257, 260, 268, 270-274, 276, 283, 287, 288, 291, 316, 355, 356, 363, 365, 366, 369, 370, 373
embankment, 63, 65-70, 72, 73, 79, 333, 334, 340, 341, 347, 349, 353, 355, 362, 365, 366, 368, 374, 376, 378, 379, 381, 383
energy dissipation by friction, 36, 37, 39, 45, 46, 48, 79

G, I, J

geometrical anisotropy, 19, 24
grain
 breakage, 16, 28, 36, 46, 53-57, 59-61, 63, 65, 68, 73, 74, 78, 79
 size distribution, 53, 55-58, 61, 65, 68, 69, 71, 73, 86-89, 91-97, 99-101, 109, 164, 166, 177, 183, 192, 206, 231, 279, 281, 283, 298
granular materials, 4, 8, 15, 19-21, 24, 32, 35-37, 53, 56-59, 61, 65, 73, 78, 84, 90, 94, 97, 100, 109, 121, 122, 137, 145, 146, 177, 183, 184, 196, 204, 206, 215
IDIEM, 85, 110
isotropic compression, 29, 118, 157, 171-173, 198, 283, 285
Jean Biarez, 1-5, 7-13, 15-21, 23-25, 27, 29, 31, 32, 113-117, 119, 123-126, 128, 132-134, 136-138, 140-147, 183, 191, 194, 199, 214, 249, 250, 255, 271, 333, 334

L, M, N, O, P, R

landslide, 221, 222, 234, 350, 368
liquefaction, 8, 83, 107, 108, 135, 195, 221, 252, 253, 297, 302, 334, 338-343, 347, 355-359, 362-369, 372, 374-380, 383, 385, 386
material instability criterion, 221, 304
microstructural model, 160, 271
numerical modeling, 17, 25, 245, 249, 307-309, 358, 366
oedometer test, 28, 29, 30, 115, 285
plasticity, 3, 5, 8, 11, 118, 129, 131, 133, 134, 137, 175, 177, 201, 221, 225, 231, 237, 239, 245, 246, 248, 252, 255-257, 259, 263, 265, 266, 273-275, 279, 281-283, 285, 288, 293
Roscoe, 123, 129, 135, 136, 252, 261

S, T, W

sand, 8, 12, 19, 28, 46, 106, 113, 120, 123, 126, 128, 130, 134, 136, 153, 160, 164-168, 172-175, 177, 186, 190-195, 208, 209, 213, 215, 250-253, 258, 260, 270, 279-283, 287-289, 291-293, 295-297, 308, 309, 316, 339, 342, 345, 368
scale effects, 36, 58, 63, 65, 66, 70, 71, 73-75, 78, 79, 110, 158
second-order work, 222-224, 227, 228, 236-239, 304, 305, 307
seismic
  behavior, 334, 335, 365, 374
  classification, 348
  damage, 342, 345-347
  hazard, 334, 346, 348, 350, 352, 353, 355-358, 386
  level, 350, 351
  stability, 355, 367, 369
semi-coupled modeling, 368
soils, 4, 6, 8, 10, 12, 15, 84, 107, 113-115, 117, 123, 129, 146, 153-155, 158, 169, 174, 177, 178, 183, 200, 202, 214, 222, 232-234, 239, 249-253, 255-257, 259, 265, 268, 270-273, 276, 279-281, 312, 322, 334, 338, 342, 355, 356, 381, 383
stability, 9, 24, 36, 53, 63, 65-70, 79, 86, 104-107, 110, 222, 224, 227, 234, 237, 238, 248, 250, 251, 299, 301, 333, 350, 357, 358, 365, 367, 368, 378, 379, 384
strain, 4, 5, 8, 19, 24, 26, 27, 37, 40, 42-45, 47-49, 51, 53, 71, 72, 77, 126, 137, 145, 153, 158, 165, 168-170, 172-178, 184, 185, 188, 189, 191, 193-198, 200-203, 209, 210, 212-215, 221-223, 225-227, 231, 232, 236, 239, 251-253, 255-259, 262, 263, 267, 275, 278, 294, 304, 305, 320, 366, 369-375, 377-380, 383, 387
triaxial test, 20, 21, 45, 57, 63, 64, 88, 89, 94-98, 100-104, 109, 118, 126, 127, 131, 145, 153, 165, 168, 170, 172-178, 191, 193, 195, 196, 199, 201-204, 208-211, 215, 251, 253, 256, 257, 265, 298, 309
waste rocks, 83, 88, 90, 94, 97, 99, 100, 109